Springer Series in

MATERIALS SCIENCE 130

Springer Series in

MATERIALS SCIENCE

The Springer Series in Materials Science covers the complete spectrum of materials physics, including fundamental principles, physical properties, materials theory and design. Recognizing the increasing importance of materials science in future device technologies, the book titles in this series reflect the state-of-the-art in understanding and controlling the structure and properties of all important classes of materials.

Please view available titles in *Springer Series in Materials Science* on series homepage http://www.springer.com/series/856

Antonio Miotello
Paolo M. Ossi
Editors

Laser-Surface Interactions for New Materials Production

Tailoring Structure and Properties

With 206 Figures

Springer

Editors

Professor Antonio Miotello
Università di Trento
Dipartimento di Fisica
Via Sommarive 14, 38050 Povo, Italy
E-mail: miotello@science.unitn.it

Professor Paolo M. Ossi
Politecnico di Milano
Dipartimento di Energia
Centre for NanoEngineered Materials
and Surfaces
via Ponzio 34-3, 20133 Milano, Italy
E-mail: paolo.ossi@polimi.it

Series Editors:

Professor Robert Hull
University of Virginia
Dept. of Materials Science and Engineering
Thornton Hall
Charlottesville, VA 22903-2442, USA

Professor Jürgen Parisi
Universität Oldenburg, Fachbereich Physik
Abt. Energie- und Halbleiterforschung
Carl-von-Ossietzky-Straße 9–11
26129 Oldenburg, Germany

Professor Chennupati Jagadish
Australian National University
Research School of Physics and Engineering
J4-22, Carver Building
Canberra ACT 0200, Australia

Dr. Zhiming Wang
University of Arkansas
Department of Physics
835 W. Dicknson St.
Fayetteville, AR 72701, USA

Professor R. M. Osgood, Jr.
Microelectronics Science Laboratory
Department of Electrical Engineering
Columbia University
Seeley W. Mudd Building
New York, NY 10027, USA

Professor Hans Warlimont
DSL Dresden Material-Innovation GmbH
Pirnaer Landstr. 176
01257 Dresden, Germany

Springer Series in Materials Science ISSN 0933-033X
ISBN 978-3-642-03306-3 e-ISBN 978-3-642-03307-0
DOI 10.1007/978-3-642-03307-0
Springer Heidelberg Dordrecht London New York

Library of Congress Control Number: 2009934001

Cover design: SPi Publisher Services

Printed on acid-free paper

Springer is part of Springer Science+Business Media (www.springer.com)

Preface

This book originates from lectures delivered at the First International School "*Laser-surface interactions for new materials production: tailoring structure and properties*" that was held in San Servolo Island, Venice (Italy) from 13 to 20 July, 2008 under the direction of A. Miotello and P.M. Ossi. The purpose of the School was to provide the students (mainly PhD) with a comprehensive overview of basic aspects and applications connected to the laser–matter interaction both to modify surface properties and to prepare new materials by pulsed laser deposition (PLD) at the nanometer scale. The field is relatively young and grew rapidly in the last 10 years because of the possibility of depositing virtually any material, including multi-component films, preserving the composition of the ablated target and generally avoiding post-deposition thermal treatments. In addition, the experimental setup for PLD is compatible with in situ diagnostics of both the plasma and the growing film.

The basic laser–surface interaction mechanisms, possibly in an ambient atmosphere, either chemically reactive or inert, are a challenge to scientists, while engineers are mostly interested in the characteristics of the deposited materials and the possibility of tailoring their properties through an appropriate tuning of the deposition parameters.

The School was motivated by the fact that while well established international conferences bring together many researchers every year and allow for extensive scientific exchange, the laser community was lacking a "teaching" event, specifically addressed to doctorate students and young post-docs to favour study of the deepening of the principles of laser–surface interactions, and to highlight the strong interplay between experimental and theoretical investigations of laser-induced phenomena.

Lecturers, coming from both the academy and leading research centers are actively contributing to research topics addressed during the School; we are grateful to them for the attention they gave to arranging presentations having a truly didactic, though high level, character. In addition, they maintained constructive interactions with the students throughout the School duration and prepared texts of their lectures in time for this book.

The result is an updated overview concerning laser induced phenomena on both the nanosecond and ultra-short timescale, together with pertinent diagnostics; material classes span from polymers to ceramics and metals, including piezoelectrics, ferroelectrics, biomaterials, glasses, and functional coatings. Laser direct writing, lasers in cultural heritage and MAPLE are considered and computer modelling is focussed both on atomic-level simulations and on continuum models.

Highlights of the present book reflect the guidelines of the School: they include topics that gained relevance in the scientific community in recent years, such as ultra-short laser pulses to explore electronic excitation in solids and its relaxation with phonons in highly non equilibrium conditions, surface melting, vapourisation, superheating, homogeneous and possibly heterogeneous nucleation, the synthesis of nanometer scale clusters and their assembling to prepare nanocrystalline films.

The School was hosted by Venice International University (VIU) at its quarters at S. Servolo Island, a site in the centre of the city, with a fascinating, long standing history. The site was recently restored to be used for cultural events providing a highly agreeable working ambient. The directors are grateful to the staff of VIU for the excellent organisation and hospitality.

To facilitate the exchange of scientific experiences and to benefit from the inspiring atmosphere enjoyed at S. Servolo, the number of students was limited. A total of 42 participants, most of them Ph.D. students, or young post-doc researchers, were selected from 22 Countries; although most of them originated from EU, students from Russia, USA, India, Pakistan, and Japan attended the School.

All students contributed to the activities of the School during the discussions throughout the lectures, and by bringing posters of their research activity. The posters were exhibited in the lecture hall for the entire duration of the School and were extensively discussed during three poster sessions. Students' participation in the School was facilitated by the support of the Politecnico di Milano, the University of Trento, and several industrial sponsors.

The positive evaluation of the students convinced the organising committee to plan the Second International School on "*Laser-surface interactions for new materials production*," to be held in S. Servolo Island from 11 to 18 July 2010, under the direction of C. Boulmer-Leborgne, M. Dinescu, T. Dickinson and P.M. Ossi.

Trento, Milano
October 2009

A. Miotello
P.M. Ossi

Contents

List of Contributors

Ratiba Benzerga
Université d'Orléans-CNRS, GREMI, Polytech, BP 6744, Orléans cedex2, France

Adriana Bigi
Department of Chemistry "G. Ciamician," University of Bologna, via Selmi, 2, Bologna 40126, Italy

Chantal Boulmer-Leborgne
Université d'Orléans-CNRS, GREMI, Polytech, BP 6744, Orléans cedex2, France

Nadezhda M. Bulgakova
Institute of Thermophysics SB RAS, prosp. Lavrentyev, 1, 630090 Novosibirsk, Russia, nbul@itp.nsc.ru

Anna Paola Caricato
Università del Salento, Dipartimento di Fisica, 73100 Lecce, Italy

Douglas B. Chrisey
Material Science and Engineering, Rensselaer Polytechnic Institute, 110 Eighth Street, Troy, NY 12180, USA

David T. Corr
Departments of Biomedical Engineering, Rensselaer Polytechnic Institute, 110 Eighth Street, Troy, NY 12180, USA

Maria Dinescu
National Institute for Lasers, Plasma and Radiation Physics, Bucharest, Romania, dinescum@ifin.nipne.ro

William H. Duff
Department of Materials Science & Engineering, University of Virginia, 395 McCormick Road, Charlottesville, VA 22904-4745, USA

Gyula Eres
Materials Sciences and Technology Divisions, Oak Ridge National Laboratory, Oak Ridge, TN 37831, USA

David B. Geohegan
Center for Nanophase Materials Sciences, Oak Ridge National Laboratory, Oak Ridge, TN 37831, USA
and
Materials Sciences and Technology Divisions, Oak Ridge National

Laboratory, Oak Ridge, TN 37831, USA
geohegandb@ornl.gov

Stephen J. Guy
Department of Materials Science & Engineering, University of Virginia, 395 McCormick Road, Charlottesville, VA 22904–4745, USA

Richard F. Haglund
Department of Physics and Astronomy, Vanderbilt University, 2201 West End Avenue, Nashville, TN 37240, USA

Ingolf V. Hertel
Department of Physics, Free University of Berlin, Arnimallee 14, 14195 Berlin, Germany
and
Max-Born-Institut für Nichtlineare Optik und Kurzzeitspektroskopie, Max-Born Str. 2a, 12489 Berlin, Germany

Daniel Höche
Universität Göttingen, Zweites Physikalisches Institut, Friedrich-Hund-Platz 1, 37077 Göttingen, Germany

Hui Hu
Materials Sciences and Technology Divisions, Oak Ridge National Laboratory, Oak Ridge, TN 37831, USA

Ilia Ivanov
Center for Nanophase Materials Sciences, Oak Ridge National Laboratory, Oak Ridge, TN 37831, USA
and
Materials Sciences and Technology Divisions, Oak Ridge National Laboratory, Oak Ridge, TN 37831, USA

Dmitriy S. Ivanov
Department of Materials Science & Engineering, University of Virginia, 395 McCormick Road, Charlottesville, VA 22904-4745, USA

Jeremy Jackson
Center for Nanophase Materials Sciences, Oak Ridge National Laboratory, Oak Ridge, TN 37831, USA
and
Materials Sciences and Technology Divisions, Oak Ridge National Laboratory, Oak Ridge, TN 37831, USA

Stephen L. Johnson
Department of Physics, University of Kentucky, Lexingtion, KY 40506, USA

Wolfgang Kautek
University of Vienna, Department of Physical Chemistry, Waehringer Strasse 42, A-1090 Vienna, Austria,
wolfgang.kautek@univie.ac.at

Elodie Leveugle
Department of Materials Science & Engineering, University of Virginia, 395 McCormick Road, Charlottesville, VA 22904–4745, USA

Zhibin Lin
Department of Materials Science & Engineering, University of Virginia, 395 McCormick Road, Charlottesville, VA 22904-4745, USA

Thomas Lippert
General Energy Department, Paul Scherrer Institut, CH-5232 Villigen PSI, Switzerland

Zuqin Liu
Center for Nanophase Materials Sciences, Oak Ridge National Laboratory, Oak Ridge, TN 37831, USA

Armando Luches
Università del Salento, Dipartimento di Fisica, 73100 Lecce, Italy

Isaac Mayer
Institute of Chemistry, The Hebrew University of Jerusalem, 91904 Jerusalem, Israel

Ion N. Mihailescu
National Institute for Lasers, Plasma and Radiation Physics, Box MG-54, RO-77125 Bucharest, Magurele, Romania,
`ion.mihailescu@inflpr.ro`

Karren More
Materials Sciences and Technology Divisions, Oak Ridge National Laboratory, Oak Ridge, TN 37831, USA

Paolo M. Ossi
Dipartimento di Energia, Politecnico di Milano, via Ponzio, 34–3, 20133 Milano, Italy,
`paolo.ossi@polimi.it`

Michael R. Papantonakis
Naval Research Laboratory, 4555 Overlook Avenue, SW, Washington, DC 20375, USA

Jacques Perrière
INSP, Université Pierre et Marie Curie-Paris 6, CNRS UMR 7588, Campus Boucicaut, 140 rue de Lourmel, 75015 Paris, France

Alex A. Puretzky
Center for Nanophase Materials Sciences, Oak Ridge National Laboratory, Oak Ridge, TN 37831, USA
and
Materials Sciences and Technology Divisions, Oak Ridge National Laboratory, Oak Ridge, TN 37831, USA

Juergen Reif
Brandenburgische Technische Universität, BTU Cottbus *and* Cottbus JointLab, Universitätsstrasse 1, 03046 Cottbus, Germany, `reif@tu-cottbus.de`

Carmen Ristoscu
National Institute for Lasers, Plasma and Radiation Physics, Box MG-54, RO-77125 Bucharest, Magurele, Romania

Arkadi Rosenfeld
Max-Born-Institut für Nichtlineare Optik und Kurzzeitspektroskopie, Max-Born Str. 2a, 12489 Berlin, Germany

Chris Rouleau
Center for Nanophase Materials Sciences, Oak Ridge National Laboratory, Oak Ridge, TN 37831, USA
and
Materials Sciences and Technology Divisions, Oak Ridge National Laboratory, Oak Ridge, TN 37831, USA

Peter Schaaf
TU Ilmenau, Institut für Werkstofftechnik, Werkstoffe der Elektrotechnik, Postfach 100565, 98684 Ilmenau, Germany,
`peter.schaaf@tu-ilmenau.de`

Nathan R. Schiele
Departments of Biomedical Engineering, Rensselaer Polytechnic Institute, 110 Eighth Street, Troy, NY 12180, USA

Carlos Sevilla
Department of Materials Science & Engineering, University of Virginia, 395 McCormick Road, Charlottesville, VA 22904-4745, USA

Razvan Stoian
Laboratoire Hubert Curien (UMR 5516 CNRS), Université Jean Monnet, 18 rue Benoit Lauras, 42000 Saint Etienne, France

David Styers-Barnett
Center for Nanophase Materials Sciences, Oak Ridge National Laboratory, Oak Ridge, TN 37831, USA

Koji Sugioka
RIKEN – The Institute of Physical and Chemical Research
2-1 Hirosawa,Wako, Saitama 351-01, Japan,
ksugioka@postman.riken.jp

Derek Thomas
Department of Materials Science & Engineering, University of Virginia, 395 McCormick Road, Charlottesville, VA 22904-4745, USA

Kai Xiao
Center for Nanophase Materials Sciences, Oak Ridge National Laboratory 1 Bethel Valley Road, Oak Ridge, TN 37831-6030, USA

Bin Zhao
Center for Nanophase Materials Sciences, Oak Ridge National Laboratory, Oak Ridge, TN 37831, USA

Leonid V. Zhigilei
Department of Materials Science & Engineering, University of Virginia, 395 McCormick Road, Charlottesville, VA 22904-4745, USA
lz2n@virginia.edu

1

Laser Interactions in Nanomaterials Synthesis

David B. Geohegan, Alex A. Puretzky, Chris Rouleau, Jeremy Jackson, Gyula Eres, Zuqin Liu, David Styers-Barnett, Hui Hu, Bin Zhao, Ilia Ivanov, Kai Xiao, and Karren More

Summary. Laser interactions with materials have unique advantages for exploring the rapid synthesis, processing, and in situ characterization of high-quality and novel nanoparticles, nanotubes, and nanowires. For example, laser vaporization of solids into background gases provides a wide range of processing conditions for the formation of nanomaterials by both catalyst-free and catalyst-assisted growth processes. Laser interactions with the growing nanomaterials provide remote in situ characterization of their size, structure, and composition with unprecedented temporal resolution. In this article, laser interactions involved in the synthesis of primarily carbon nanostructures are reviewed, including the catalyst-free synthesis of single-walled carbon nanohorns and quantum dots, to the catalyst-assisted growth of single- and multi-walled carbon nanotubes.

1.1 Introduction

Laser vaporization of solid targets has long been a tool for the synthesis and discovery of clusters by mass spectrometry [1], resulting in the discovery of C_{60} and higher fullerenes in 1985 [2]. Two years later, yttrium–barium–copper oxide, high-temperature superconductors were discovered, and commercial excimer lasers were found to congruently vaporize multicomponent targets to grow thin films of these materials [3], fueling a resurgence of interest in pulsed laser deposition (PLD) for materials discovery, and a need to more fully understand the laser vaporization process [4]. In 1996, while trying to develop a catalyst-assisted process for the mass production of fullerenes, laser vaporization of a multicomponent (carbon and metal catalyst) target into flowing argon gas at high temperatures ($1,100^{\circ}$C) resulted in the synthesis of single-wall carbon nanotubes (SWNTs), a major breakthrough in their production [5]. In 1998, this laser vaporization technique was generalized for the VLS-synthesis of semiconducting nanowires [6,7], further emphasizing the role of lasers in the exploration of new nanomaterials. These discoveries were highly instrumental in the development of an understanding of the synthesis of nanomaterials. In this article, we will outline some of the key processes governing the synthesis of nanomaterials by laser-driven interactions, with a special emphasis on carbon materials.

1.2 Laser Ablation and Plume Thermalization at Low Pressures

The virtues of laser ablation for the PLD of thin films primarily involve the rapid, stoichiometric removal and atomization of a solid, and the formation of an energetic beam of neutrals, ions, small molecules, and clusters [4]. The laser interaction with the solid usually forms a dense laser plasma ($T_e \sim 1$–$10\,\mathrm{eV}$), which expands and cools during a period of collisions near the target surface in which fast ions, slower neutrals, and even slower molecules and clusters emerge with a shifted, center-of-mass Maxwell-Boltzmann velocity distribution. Despite disparate masses, atoms in a multicomponent target often travel at nearly the same velocity when they emerge from this collisional "Knudsen layer," with atoms near the peak of the distribution typically moving at velocities $v \sim 1\,\mathrm{cm}\,\mu\mathrm{s}^{-1}$, corresponding to significant kinetic energies ($\sim$10–100 eV).

However, immediately following laser vaporization, oxidation and other chemical reactions can occur in the early portions of the plume expansion to form new molecules and clusters. In addition, since nanosecond or longer pulses are typically utilized, the laser may interact with the ejecta as they expand, resulting in photodissociation of clusters, photoionization of neutrals, and other processes that result in regional heating and secondary plume dynamics. An example of this is shown in Fig. 1.1, where pyrolytic graphite is ablated by ArF (193 nm) and KrF (248 nm) lasers in vacuum [8,9].

Stepwise increases in laser intensity results in the appearance of distinct regions of plasma luminescence: first, from excited primary ejecta C_3 and C_2; second, from atomic carbon resulting from photodissociation of C_2; and

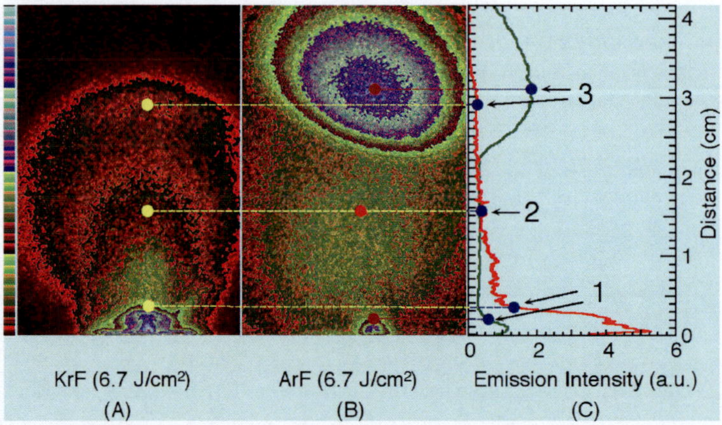

Fig. 1.1. ICCD images of visible plume emission from KrF-laser (248 nm) and ArF-laser (193 nm) ablated pyrolytic graphite in vacuum, taken $\Delta t = 1.0\ \mu\mathrm{s}$ following ablation. Three regions of plume emission are observed, corresponding to (1) C_2 and C_3, (2) C, and (3) C^+. (Reproduced with permission from [8])

third, a fast ball of C^+ ions resulting from two-photon, resonant ionization of atomic C (Reproduced with permission from [10]). The interplume dynamics, which result in the selective acceleration of the C and C^+, are observed to retard the expansion of the slower C_2 and C_3, inducing additional collisions and more clustering, and redeposition of these materials on the target surface. Thus, the choice of laser wavelength can influence the composition, kinetic energies, and trajectories of the initial ejecta from the target.

The addition of a low-pressure background gas results in collisions which slow the plume and confine it, often with the inadvertent formation of nanoparticles. Fig. 1.2a shows a sequence of images of the plume resulting from

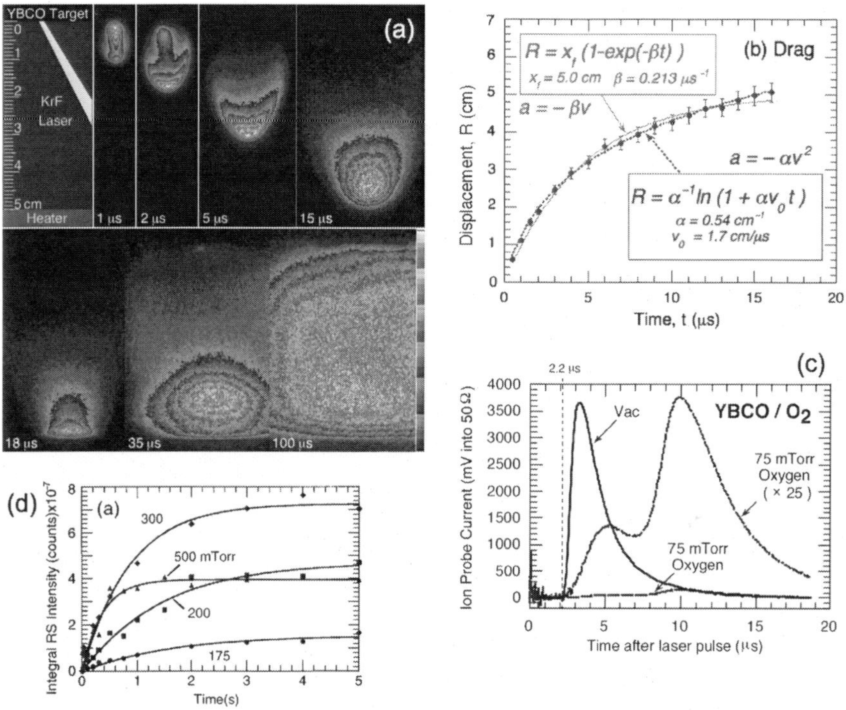

Fig. 1.2. (**a**) Side-on, false-color ICCD images of visible plume emission from YBCO ablated in 200 mTorr oxygen at the indicated times. Although initially moving at leading edge velocities of 1 cm/µs, the plume arrives at a heater surface 5 cm away at $\Delta t = 15$ µs. The plume does not entirely deposit, but rebounds to fill the region between the heater and target [10]. (**b**) The propagation of the leading edge of the plume is adequately represented by phenomenologic drag models. (**c**) However, ion probe flux measurements reveal a "splitting" of the plume at certain distances and pressures which has only been adequately explained by an elastic collision model [10]. (**d**) Integrated intensities from Rayleigh scattering images of the region between the target and heater show the time dependences of nanoparticle growth at pressures typically used for PLD. [Adapted from 8, 9, 11]

YBCO ablation into 200-mTorr oxygen. Collisions of the plume atoms and ions with the background gas lead to bright, recombination-fed fluorescence. Although this bright "shock front" progression can be adequately represented by shock and drag models [4], two components of the plume coexist for a given range of distances for a particular background pressure, as revealed by ion flux measurements as in Fig. 1.2c. This "plume splitting" has been analyzed and modeled to result from elastic collisions, which scatter and delay the plume atoms [11,12]. The two peaks roughly correspond to a fast distribution of material, exponentially decaying with distance or pressure, of original plume material which has undergone few if any collisions – and a slowed peak which has undergone one or more collisions. After all the plume atoms have undergone several collisions, they form a slowed, propagating front of material which collides with a cold heater surface in Fig. 1.2a (*lower panel*). A large fraction of the material does not stick to the heater surface, and slowly it rebounds. During the next several seconds (Fig. 1.2d), laser-induced fluorescence imaging and Rayleigh-scattering (RS) imaging (not shown) reveal that oxide clusters and nanoparticles slowly grow from this residual material for pressures above 175 mTorr under typical experimental conditions used for PLD film growth [13]. Interestingly, the imaging of Rayleigh-scattered light from a time-delayed, 308-nm laser sheet revealed that this process is highly quenched by the application of a small-temperature gradient, which flushes the nanoparticles from the region as they begin to form [13].

1.3 Synthesis of Nanoparticles by Laser Vaporization

Novel-new nanomaterials can be formed by laser vaporization into high-pressure background gases [14, 15]. The process can be modeled by an isentropic expansion of a gas [16]; however, the actual dynamics are of interest in order to control the synthesis process. Figure 1.3 shows the plume expansion following laser vaporization of Si into 10 Torr He, for the formation of brightly photoluminescent SiO_x nanoparticles. For the first 400 µs, the plasma emission can be directly imaged; however, for longer times, a second, time-delayed (308 nm) laser is used to induce luminescence from the plume. In this case, for times $>$200 µs, the photoluminescence from small clusters and nanoparticles formed in the plume is used to reveal their position and dynamics [17].

As the images show, a very bright region of photoluminescent clusters is formed behind the leading edge of the plume. These clusters were too small, however, to scatter light sufficiently for RS imaging. The nanoparticles grow and consolidate on the leading edge of the plume within 1 ms, and the swirling, forward-moving, vortex dynamics segregate the particles within a "smoke ring". The smoke ring continues forward to encounter a stationary Si wafer at room temperature however the nanoparticles do not stick, but remain there for several seconds until they agglomerate, at which point photoluminescence is quenched.

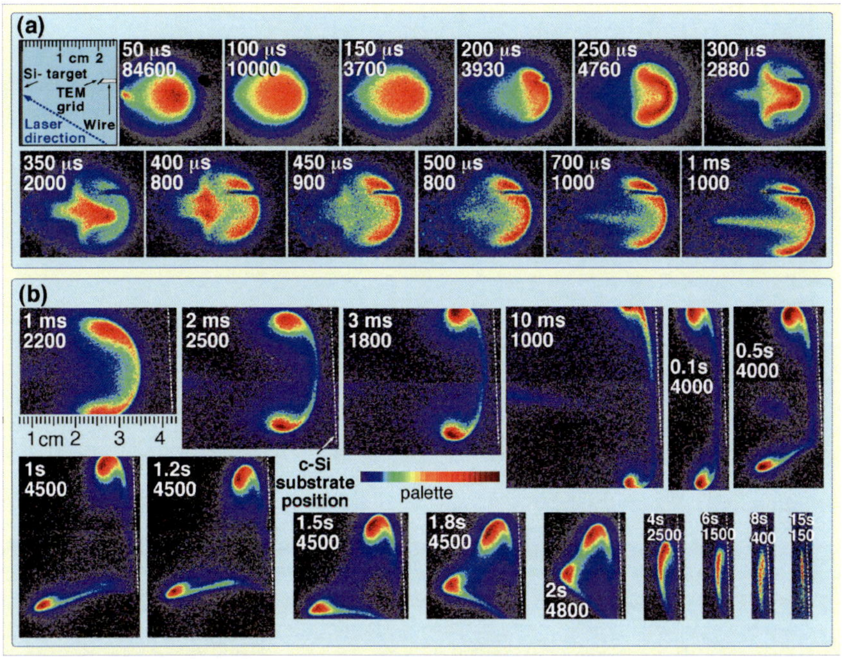

Fig. 1.3. (**a**) ICCD images of plasma luminescence ($\Delta < 400$ μs) plus photoluminescence ($\Delta t > 200$ μs) from nanoparticles produced by silicon ablation into 10 Torr He (3 μs exposures) at the indicated times and peak image intensities. (**b**) PL images utilizing a sheet of 308-nm laser light at later times show a slice through a swirling smoke ring of nanoparticles, and the nanoparticles encountering a room-temperature silicon wafer (at the *dashed line* position). The movement of the lower portion of the nanoparticle cloud is due to a very weak gas flow in the chamber caused by the gas introduction [17]

These dynamics are quite unlike the expansion of ablated Si into background Argon (not shown). The high-relative atomic mass of Ar vs. Si (40 vs. 28) induces a significant slowing of the plume compared to the Si/He case (28 vs. 4). Just 1 Torr of Ar produces a stopped and stationary cloud of nanoparticles (as revealed by RS imaging) without the turbulent motion needed to draw in oxygen required for oxidation into SiOx. Thus, without an intentional flow of Ar to introduce trace impurities of oxygen, no PL is observed. The choice of background gas can, therefore, significantly affect the propagation of the plume and its chemistry.

1.4 Self-Assembly of Carbon Fullerenes and Nanohorns

Carbon fullerenes were discovered in 1985 by the laser ablation of carbon into the high-pressure background gas within a specially constructed, windowed-pulsed nozzle source [2]. Soon after, laser vaporization of graphite targets within a hot tube furnace was used to scale the production of fullerenes to laboratory scale, which was followed by electric arc vaporization for mass production [18]. Theoretical modeling of the synthesis process has shown that high temperatures of ∼3,000 K are required to induce the curvature that is necessary for the formation of fullerenes and other curved carbon nanostructures. Synthesis temperatures of ∼1,000–2,000 K produce flat carbon chain structures and sheets. Yet, fullerenes and other larger nanostructures can be produced by laser vaporization into room-temperature background ambients.

To understand the timescales, temperatures, and dynamics that are involved in fullerene production, time-resolved imaging and spectroscopy of the laser vaporization of carbon into room temperature 300 Torr Ar gas were performed (Fig. 1.4). The images show a confined plume with a series of highly reproducible shock waves which correspond to regions of plume expansion and cooling. The initial expansion of high-density C atoms and ions is rapidly stopped (300 ns) and a backward-propagating rarefaction wave is formed. This

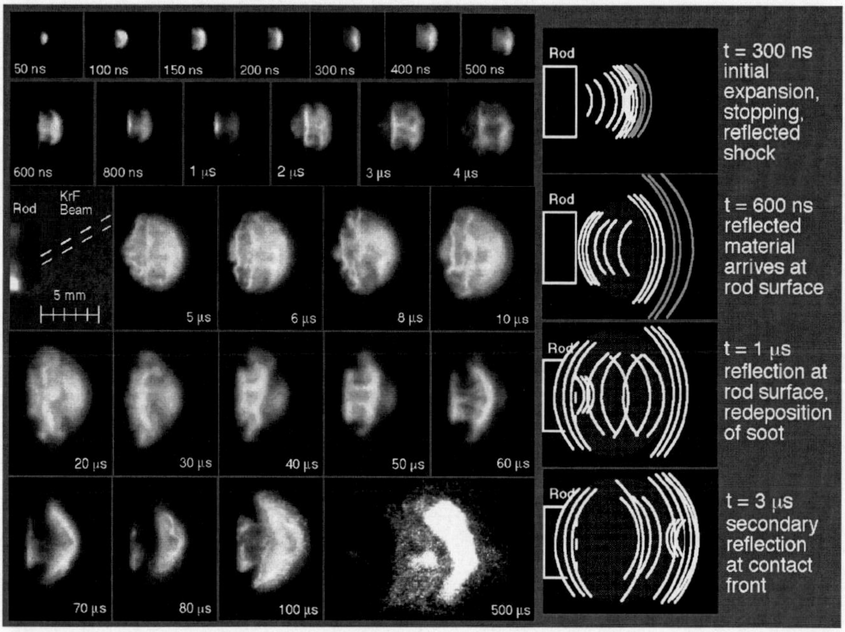

Fig. 1.4. (**a**) ICCD images of the interplume shock dynamics resulting from laser vaporization of C into 300 Torr Ar at room temperature for the formation of fullerenes. The small quantity of ablated C is quickly (300 ns) stopped, and a reflected shock drives material back toward the target. Reflected shocks continue, the plume expanding in oscillations, until a final push occurs in a mushroom cloud expansion where glowing clusters can be observed (at 500 μs) [19]

wave arrives at and reflects from the target surface from $\Delta t = 0.6$–$1.0\,\mu$s, and the plume is observed to oscillate and expand in stages as the material oscillates between the contact front with the ambient gas and the target surface. During the process, the material deposits on the target; however, no fullerenes are found there. The growth of the fullerenes occurs over extended times, during the final expansion of the plume for $t > 30\,\mu$s after ablation. During this time, the plume cools from ~3,000 K to ~1,000 K, as recorded by blackbody emission from hot clusters and particulates in the plume (as in 500 μs image in Fig. 1.4). Experimentally, the choice of background gas and pressure is found to govern the extent of plume confinement and the rate of cooling within the volume, which serves as the substrateless microreactor where nanoparticle growth takes place [19].

In 1999, much larger carbon nanostructures – single-wall carbon nanohorns (SWNHs) – were reported by a similar laser vaporization process, however at much higher laser power [20]. SWNHs are tubular shaped, single-wall carbon nanostructures (like SWNTs); however, they are produced *without catalysts.* The synthesis process was not understood; however, similar multi-walled tubular structures were formed in 1994 when "fullerene soot" from an arc reactor was annealed at high temperatures *ex situ*, indicating that in addition to completed fullerenes, incomplete carbon structures had been formed and were capable of further assembly [21]. The ablation of C targets into room temperature, and atmospheric pressure background gases of He and Ar were found to form different flower-shaped aggregates of the nanohorns, including "dahlia-like" and "bud-like" nanohorns [22].

Recently, we applied tunable laser pulses to investigate the timescales and dynamics of SWNH growth [23,24]. By varying both the energy and the pulse width of a high-power (600-W average power) laser, different ablation regimes could be explored. To explore the carbon nanostructures formed under long, *continuous* heating, and ablation, the laser pulse width was adjusted to multi-millisecond lengths, and high energies (up to 100 J per pulse) were used. To explore nanostructures formed under shorter plume lifetimes, sub-millisecond pulses and low (~1–5 J per pulse) laser energies were used. The temperature of the target surface was recorded by fast, optical pyrometry during laser irradiation, and compared to a three-dimensional, finite-element model simulation that included heating with a laser beam, heat losses due to heat conduction, target evaporation, blackbody radiation, and cooling by the surrounding buffer gas. The results are summarized in Fig. 1.5. Cumulative laser vaporization with 1 J pulses was found to require ~10 laser pulses before the surface temperature was sufficient (3,750°C) to vaporize C; however, once achieved a steady ablation rate of $\sim 6\,\mathrm{g\,h^{-1}}$ was found to be very comparable to that using high-energy individual pulses for the same ~500 W average laser power. On the other hand, individual high-energy (~100 J) pulses of 10–20 ms duration were sufficient to rapidly heat the target to 4, 200°C, and maintain vaporization in a *continuous* ablation mode. High-speed videography was used to record the heating and cooling times of the plume for

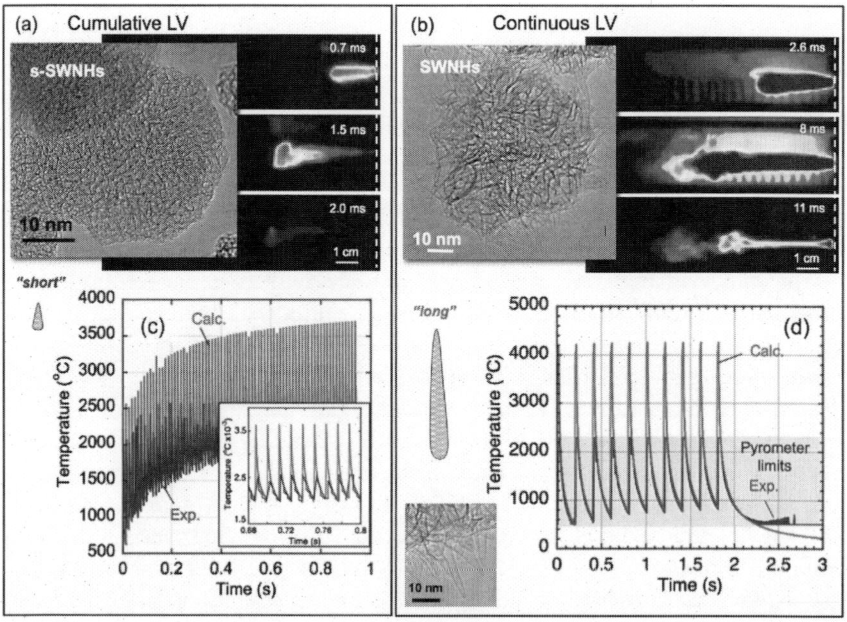

Fig. 1.5. (**a, b**) Selected frames from high-speed (50,000 fps) video images recorded in situ from within a 1,150 °C tube furnace during high-power laser vaporization of C targets using (**a**) *cumulative* ablation (from 1 ms, 9 J laser pulses, at 50 Hz) and (**b**) *continuous* ablation (10 ms, 90 J laser pulses, 5 Hz). Variation of the laser pulse widths and energies can be used to adjust the times and temperatures available for single-wall carbon nanotube and nanohorn growth. HRTEM images show representative materials collected outside the furnace following the synthesis events illustrated by the time-resolved image sequences. (**c**) and (**d**) illustrate in situ pyrometry of the target surface and calculated temperature profiles from a 3D heat transfer simulation of the target heating. Parameters are (**c**) (20 ms pulses, 100 J/pulse, at 5 Hz) and (**d**) short pulses (0.5 ms pulses, 5 J/pulse at 80 Hz). The highlighted horizontal band in (**d**) shows the pyrometer limits. After [23, 24]

comparison with the quite different, nanohorn structures obtained in the different modes. As indicated in Fig. 1.5a, b, high-resolution TEM images show a variation in both the size of the individual nanohorn subunit, as well as the size of the aggregate structures which are formed. The length of nanohorn was found to correlate well with the time spent within the high-temperature growth zone, with the length increasing at a rate of $\sim 1\,\text{nm ms}^{-1}$ of the available growth time. This rate is highly comparable to the ~ 1–$5\,\text{cm}\,\mu\text{s}^{-1}$ rates found for catalyst-assisted SWNT growth, indicating that C can self-assemble into nanostructures at rates comparable to those using catalyst assistance [24].

1.5 Catalyst-Assisted Synthesis of SWNTs

Laser vaporization of carbon targets containing ∼1–2 at % metal catalyst powders (e.g. Ni and Co), is a very effective technique to produce exclusively SWNTs at ∼1,200°C in flowing Ar [25]. As summarized in Fig. 1.6, in situ

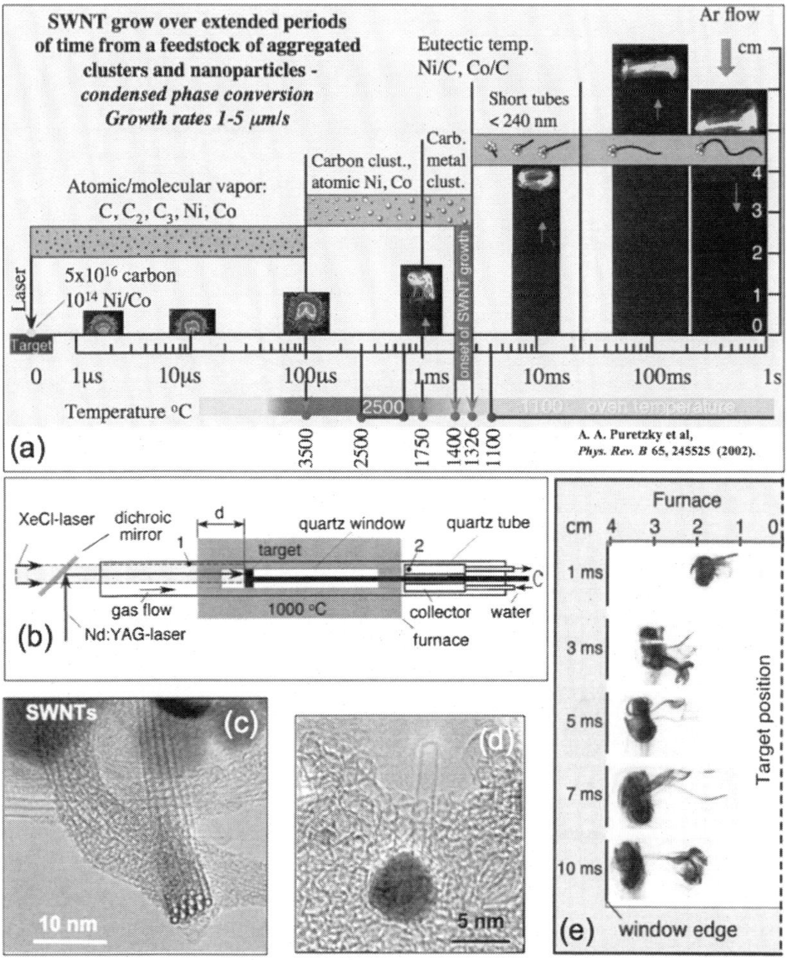

Fig. 1.6. (**a**) Summary of time-resolved imaging, spectroscopy, and temperature measurements of SWNT synthesis by laser vaporization. SWNT growth occurs at extended times from condensed carbon confined within a vortex ring at rates of 1–5 μ/s. (**b**) Schematic of the windowed laser oven used in the time-restricted growth experiments incorporating a second, time-delayed XeCl laser. (**c**) SWNT bundle typical of extended growth times (**d**) Short SWNT "seed" emanating from a 5 nm NiCo nanoparticle resulting from time-restricted growth (**e**) Rayleigh scattering images of the plume formed within the windowed portion of the furnace, just prior to exiting the furnace for rapid quenching of the growth. After [27]

imaging and spectroscopy studies of the ns-laser vaporization process revealed that (a) both carbon and metal are principally in the form of atoms and molecules (C, C_2, C_3, Ni, Co) during the first 100 µs, when the plume of ejecta are within ∼1 cm of the target, (b) that carbon forms clusters within 1 ms after laser vaporization, as the hot plasma cools, and that (c) Ni and Co form clusters later in time (1 ms $< t <$ 2 ms) after laser ablation [10, 26]. Through stop-growth experiments, where the plume was ejected from the hot oven after different growth periods (as revealed by imaging the plume via Rayleigh scattering shown in Fig. 1.5e), it was learned that only short SWNT "seeds" or nuclei had formed after 15–20 ms of growth time. By adjustment of this time, a growth rate in the range of 1–5 $\mu m\, s^{-1}$ could be inferred for SWNT growth by laser vaporization [27].

It was concluded that one of the main conditions to achieve a high yield of SWNTs was confinement of the ejected material inside the propagating laser plume, and that the main mechanism of this confinement was formation of a vortex ring. We recently showed that the confined volume could be significantly reduced if *cumulative ablation* using a sequence of pulses with a relatively low peak power (described above) was used to ablate the target, instead of individual ns-laser pulses with high-peak powers. The detailed study of this laser ablation regime revealed that preheating of the target with approximately 10 laser pulses is required to achieve stationary ablation. Weight analysis of the target and HRTEM of the products revealed that, averaged over many pulses the same ablation rates were achieved for the same input total energy between single- and multi-shot ablation, but higher conversion efficiencies of carbon to SWNTs were obtained when the ejected material was confined in a smaller volume [23]. Therefore, this cumulative regime of laser ablation is very useful for synthesis of SWNTs and other nanomaterials when long-term confinement of the ablated material is required.

1.6 Laser Diagnostics and Controlled Chemical Vapor Deposition of Carbon Nanotubes

As described in Fig. 1.7, laser-based diagnostics have also been applied recently to understand and control the growth of carbon nanotubes by chemical vapor deposition (CVD), providing some of the first direct kinetics measurements and growth rates measured in situ [28, 29].

Using the results from in situ growth rate measurements in which temperature, gas flow, and hydrocarbon concentration were varied, a kinetics model was developed to fit the measured growth rates and terminal lengths of vertically-aligned carbon nanotube arrays (VANTAs). Activation energies for the different processes were determined, and the optimal growth conditions to produce long nanotube arrays were predicted [29]. By measuring the number of walls for the nanotubes grown under different conditions, it was possible

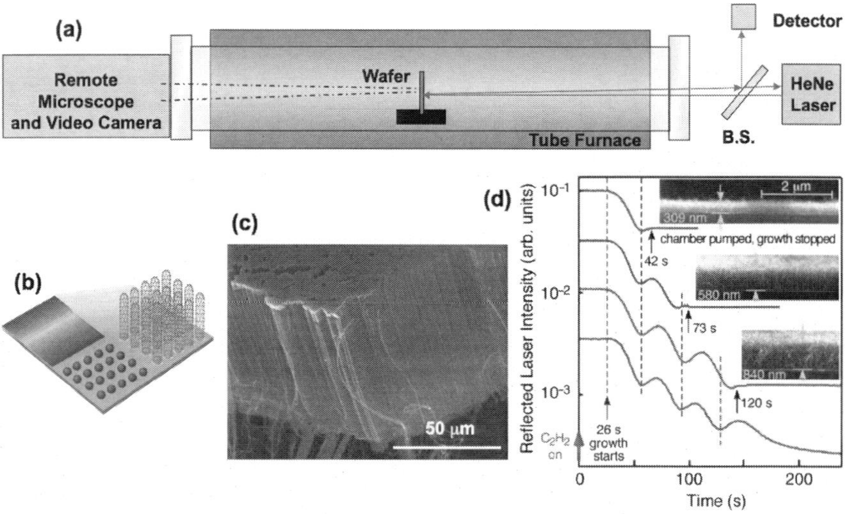

Fig. 1.7. (**a**) Schematic of apparatus used for in situ measurement of carbon nanotube growth kinetics. A CW-HeNe laser beam is reflected from a vertically-standing substrate through the end window of a tube furnace. A remote microscope and video camera may be used from the opposite window to record growth to millimeters lengths. (**b**) Schematic of chemical vapor deposition (CVD) growth of vertically-aligned nanotube arrays (VANTAs). A thin film catalyst is deposited (usually 10 nm of Al as a buffer layer on Si, then ∼1 nm of Fe as catalyst, and sometimes ∼0.2 nm of Mo as a mixed catalyst). During heating in a tube furnace to 550–950 °C in Ar/H_2 mixtures, the catalyst film roughens into nanoparticles. A mixture of hydrogen, argon and acetylene is then introduced (or another hydrocarbon such as methane, ethylene, etc.) and nanotubes nucleate and grow from the metal catalyst nanoparticles to form dense, self-aligned arrays. (**c**) SEM micrograph of a cleaved VANTA. The Si wafer is at the bottom, and the top of the array indicates the porous nature of the block of continuous nanotubes. Most VANTAs are <10 vol.% dense. (**d**) As the nanotubes begin to grow, the HeNe laser beam is reflected from both the metallized Si substrate and the top of the growing nanotube array, resulting in Fabry-Perot interference fringes measured at the detector (in addition to signal attenuation due to absorption). Each fringe corresponds to ∼300 nm of array height. The growth rate of the nanotubes can be directly measured in situ,and the length of the nanotube arrays can be controlled. After [23, 24]

to understand how the number of walls of a nanotube grown from a catalyst nanoparticle depends on the feedstock supply. The model predicts that for a particular catalyst the fastest growing nanotube is a SWNT at a given temperature and feedstock supply; however, with an oversupply of feedstock more nanotube walls are formed [30]. Typically, the number of walls found in continuously-grown VANTAs changes with time, as revealed by Raman spectroscopy in Fig. 1.8.

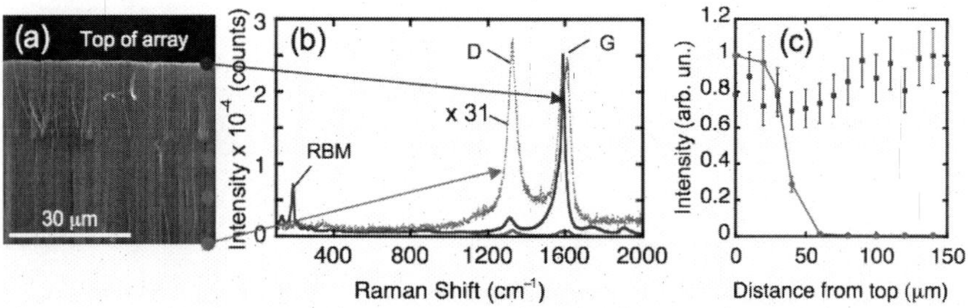

Fig. 1.8. (**a**) SEM image of the top of a VANTA array grown with different partial pressures of acetylene at 760 Torr 750° C in 2,500 sccm Ar/H_2 gas mixture. Since the nanotubes grow from catalyst anchored at the substrate, the top of the array (grown with 1 sccm C_2H_2) reflects nanotubes which grew first, and display a high SWNT fraction displaying (**b**) Raman spectra ($\lambda_{ex} = 633$ nm) with pronounced RBM modes and a high G/D Raman band ratio (*blue curve*). The number of walls in the array can be adjusted, in accordance with the growth model, by an oversupply in feedstock. Thus, the bottom part of the array (grown with 10 sccm C_2H_2) displays a lack of SWNTs and a Raman spectrum reflecting MWNTs (*red curves*, actual intensity and scaled by a factor of 31). (**c**) Raman profiling of the array (laser polarization parallel to the nanotube alignment) shows a dropoff in RBM intensity (*red circles and line*) following the change to 10 sccm feedstock supply after 15 µ of initial growth. An array grown at 1 sccm constant supply (*blue square points*) is shown for comparison [24]

Lasers, therefore, permit in situ remote characterization of nanotube growth kinetics via time-resolved reflectivity. Moreover, through Raman spectroscopy, the presence and diameter of SWNTs can be assessed through the presence of the radial breathing modes (RBMs) in micro-Raman profiling of nanotubes grown under different conditions (Fig. 1.8b). Similarly, the number of defects in the nanotubes can be assessed by a comparison of the G:D Raman band ratio intensity (Fig. 1.8c).

However, laser irradiation can also be used to alter the activity of the metal catalysts that are used for nanotube growth. Through KrF-laser processing of multilayer metal catalyst films prior to CVD, remarkable changes in subsequent VANTA growth rates, terminal heights, nanotube diameters, and wall numbers were observed [31]. Depending upon the fluence, growth was either stunted or enhanced; however, in the case of Fig. 1.9a the laser-processed regions resulted in over three times the growth rate and terminal length of the unprocessed regions, resulting in 1.4 cm-tall nanotube pillars. HRTEM analysis of the nanotubes in the tall pillars and shorter mats revealed a much narrower distribution of nanotube diameters and wall numbers in the laser-processed regions, corresponding to slimmer, faster-growing nanotubes. Despite their narrow diameter, the laser-processed regions were more densely packed; and weight measurements showed that on a *per unit substrate area* basis, the processed regions were *far* more catalytically active

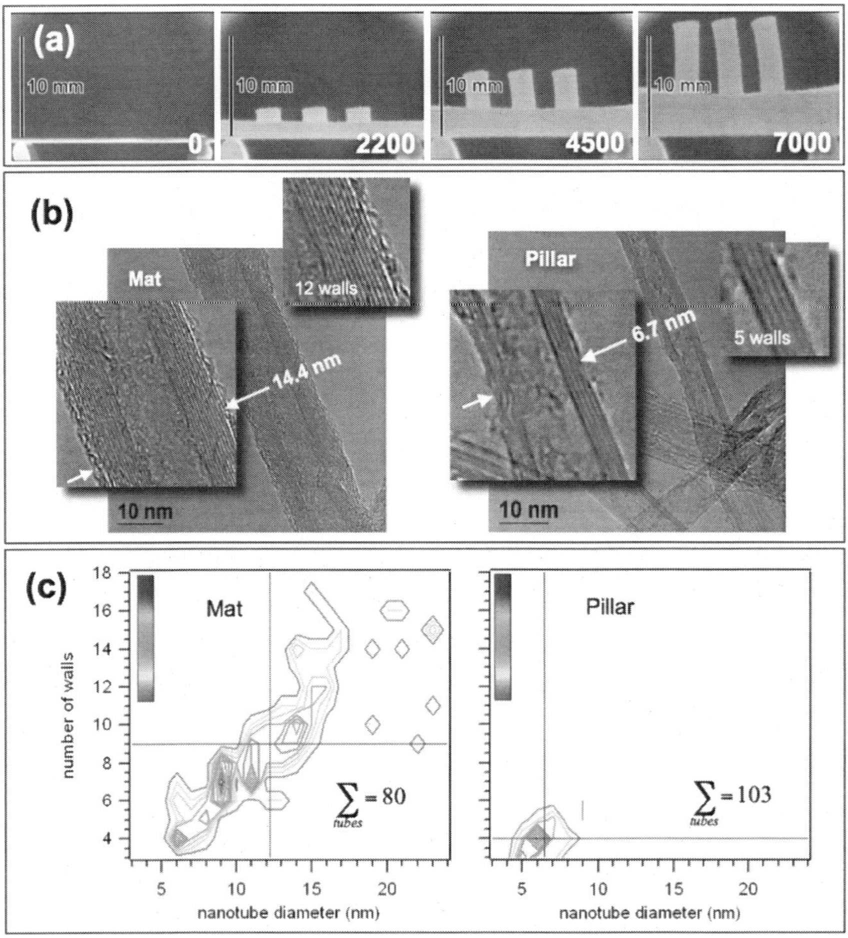

Fig. 1.9. **(a)** Time lapse images of VANTA growth on unprocessed and laser-irradiated Fe(1 nm)/Mo(0.2 nm) films on 10 nm Al-coated Si wafers, at the indicated time in seconds. **(b)**HRTEM images reveal that the taller pillars of nanotubes in the laser processed areas have ewer walls and are narrower in diameter than those in the mat (unirradiated area). **(c)** Distributions of nanotube wall number vs. nanotube diameter shows that the laser processed areas in the pillars have greatly reduced diameter distributions and smaller diameters. (Reproduced with permission from [31])

than the unprocessed area. Thus, laser processing appears highly promising to influence and control the catalytic activity of metal alloy films that are used for CVD.

Lasers can also be used to provide unique growth conditions for CVD. Recently, we utilized infrared laser pulses to provide well-defined growth periods for carbon nanotubes on Si wafers and TEM grids. As shown in Fig. 1.10,

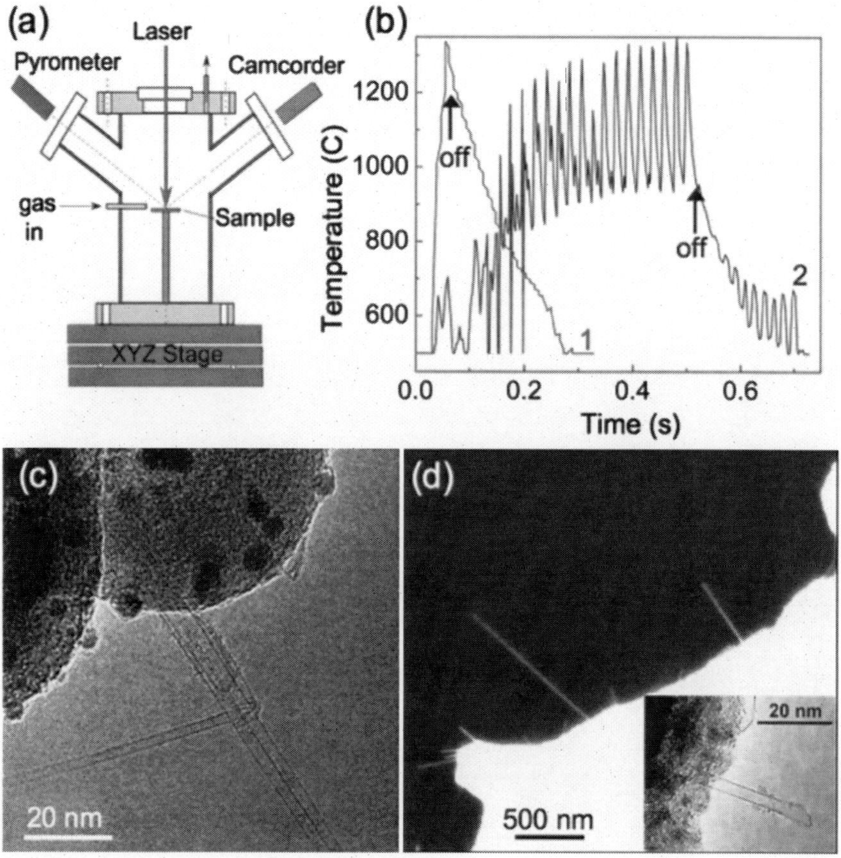

Fig. 1.10. **(a)** Schematic of PLA-CVD vacuum chamber. **(b)** Time-dependent temperature profile of a 1 cm^2 Si/SiO_2 wafer by a single 50 ms laser pulse (1), and a Mo TEM grid from 25 pulses of 5 ms width (2). *Arrows* show the time when laser irradiation is terminated. **(c)** TEM image of CNTs grown on a Mo grid coated with 1 nm Fe/Al_2O_3 by 1,500 laser pulses. **(d)** SEM of nanotubes grown using 20 pulses on an identically prepared grid as **(c)**. *Inset* shows a TEM image of the end of a nanotube free of catalyst particle [32]

laser heating of the substrates within a CVD chamber was monitored in situ by fast, optical pyrometry. The study found that exclusively SWNTs form by rapid laser heating, and at the highest recorded rates of 100 $\mu m\,sec^{-1}$ [32]. Interestingly, growth was found *not* to occur incrementally on successive laser pulses; that is, once the catalyst particle was cycled it was catalytically inactive. Nevertheless, on successive laser pulses new catalyst particles may nucleate and grow a nanotube. This feature was used to demonstrate the direct writing of SWNT field-effect transistors on prepatterned electrodes decorated with a catalyst [32].

1.7 Summary

In summary, a variety of laser interactions for nanomaterial synthesis have been described. Laser vaporization is a powerful exploratory tool in nanomaterial synthesis, providing congruent and complete vaporization of solids to permit the clean self-assembly of nanomaterials in gases, without interactions with substrates. Fullerenes, photoluminescent silicon quantum dots, and SWNHs are all examples of novel nanomaterials discovered by laser vaporization in different fluence ranges and background gas pressures.

Lasers are also used in the catalyst-assisted growth of nanomaterials, such as carbon nanotubes. Here, carbon nanotube growth by both laser vaporization and CVD were described. In both cases, laser interactions are used to remotely characterize the size, composition, and electronic structure of intermediate species and track their dynamics, as well as remotely provide some of the first-growth kinetics information of the nanotubes as they grow. During laser vaporization, laser-induced fluorescence and Rayleigh scattering utilizing time-delayed probe laser pulses were used to understand the timescales for growth. During CVD, time-resolved laser reflectivity from the growing nanotube arrays provides both length and density information of the arrays, and in situ Raman spectroscopy of SWNTs can be used to understand diameters and defects.

Lasers are emerging as instruments to modify and control growth, for example in the described laser processing of catalysts used for CVD to alter their activity, and also in the pulsed heating of metal catalysts to nucleate and grow discrete SWNTs in precise locations. A great variety of other effects remain to be described and explored; however, it is certain that lasers – with their remote delivery of energy – will be used to alter the synthesis conditions and characterize the effects for the growth of novel nanomaterials of the future.

Acknowledgments

The authors gratefully acknowledge the U.S. Dept. of Energy, Basic Energy Sciences Division of Materials Sciences and Engineering, for support of the synthesis science and the Scientific User Facilities Division for the development and support of the advanced characterization tools utilized in this work.

References

1. T.G. Dietz, M.A. Duncan, D.E. Powers, R.E. Smalley, J. Chem. Phys. **74**, 6511 (1981)
2. H.W. Kroto, J.R. Heath, S.C. O'Brien, R.F. Curl, R.E. Smalley, Nature **318**, 162 (1985)
3. A. Inam, X.D. Wu, T. Venkatesan, S.B. Ogale, C.C. Chang, D. Dijkkamp, Appl. Phys. Lett. **51**, 619 (1987)

4. D.B. Geohegan, Chap.4 in *Pulsed Laser Deposition of Thin Films*, ed. by D.B. Chrisey, G.K. Hubler (Wiley, New York, 1994)
5. A. Thess, R. Lee, P. Nikolaev, H. Dai, P. Petit, J. Robert, C. Xu, Y.H. Lee, S.G. Kim, A.G. Rinzler, D.T. Colbert, G.E. Scuseria, D. Tomanek, J.E. Fischer, R.E. Smalley, Science **273**, 483 (1996)
6. A.M. Morales, C.M. Lieber, Science **279**, 208 (1998)
7. X. Duan, C.M. Lieber, Adv. Mater. **12**, 298 (2000)
8. D.H. Lowndes, D.B. Geohegan, A.A. Puretzky, D.P. Norton, C.M. Rouleau, Science **273**, 898 (1996)
9. A.A. Puretzky, D.B. Geohegan, G.E. Jellison, M.M. McGibbon, Appl. Surf. Sci. **96–98**, 859 (1996)
10. A.A. Puretzky, D.B. Geohegan, X. Fan, S.J. Pennycook, Appl. Phys. Lett. **76**, 182 (2000)
11. R.F. Wood, K.R. Chen, J.N. Leboeuf, A.A. Puretzky, D.B. Geohegan, Phys. Rev. Lett. **79**, 1571 (1997)
12. R.F. Wood, J.N. Leboeuf, D.B. Geohegan, A.A. Puretzky, K.R. Chen, Phys. Rev. B **58**, 1533 (1998)
13. D.B. Geohegan, A.A. Puretzky, D.J. Rader, Appl. Phys. Lett. **74**, 3788 (1999)
14. H. Shinohara, Rep. Prog. Phys. **63**, 843 (2000)
15. C.N.R. Rao, G.U. Kulkarni, P.J. Thomas, Chap. 2 in *Springer Series in Materials Science 95*, ed. by C.N.R. Rao, G.U. Kulkarni, P.J. Thomas (Springer, Berlin, 2007)
16. Y.B. Zel'dowich, Y.P. Raizer, *Physics of Shock Waves and High-Temperature Hydrodynamic Phenomena* (Academic, New York, 1966)
17. D.B. Geohegan, A.A. Puretzky, G. Duscher, S.J. Pennycook, Appl. Phys. Lett. **72**, 2987 (1998); Appl. Phys. Lett. **73**, 438 (1998)
18. R.E. Smalley, Acc. Chem. Res. **25**, 98 (1992)
19. D.B. Geohegan, A.A. Puretzky, R.L. Hettich, X.-Y. Zheng, R.E. Haufler, R.N. Compton, Trans. Mat. Res. Soc. Jpn. **17**, 349 (1994)
20. S. Iijima, M. Yudasaka, R. Yamada, S. Bandow, K. Suenaga, F. Kokai, K. Takahashi, Chem. Phys. Lett. **309**, 165 (1999)
21. P.J.F. Harris, S.C. Tsang, J.B. Claridge, S.C. Sang, J.B. Claridge, M.L. Green, J. Chem. Soc. Faraday Trans. **90**, 2799 (1994)
22. D. Kasuya, M. Yudasaka, K. Takahashi, F. Kokai, S. Iijima, J. Phys. Chem. B **106**, 4947 (2002)
23. A.A. Puretzky, D.J. Styers-Barnett, C.M. Rouleau, H. Hu, B. Zhao, I.N. Ivanov, D.B. Geohegan, Appl. Phys. A. **93**(4), 849–855 (2008)
24. D. B. Geohegan, A.A. Puretzky, D. Styers-Barnett, H. Hu, B. Zhao, H. Cui, C.M. Rouleau, G. Eres, J.J. Jackson, R.F. Wood, S. Pannala, J.C. Wells, Phys. Stat. Sol. B **244**, 3944 (2007)
25. A. Thess, R. Lee, P. Nikolaev, H. Dai, P. Petit, J. Robert, C. Xu, Y.H. Lee, S.G. Kim, A.G. Rinzler, D.T. Colbert, G.E. Scuseria, D. Tomanek, J.E. Fischer, R.E. Smalley, Science **273**(5274), 483 (1996)
26. A.A. Puretzky, D.B. Geohegan, X. Fan, S.J. Pennycook, Appl. Phys. A **70**, 153 (2000)
27. A.A. Puretzky, H. Schittenhelm, X. Fan, M.J. Lance, L.F. Allard, D.B. Geohegan, Phys. Rev. B **65**, 245425/1 (2002)
28. D.B. Geohegan, A.A. Puretzky, I.N. Ivanov, S. Jesse, G. Eres, J.Y. Howe, Appl. Phys. Lett. **83**, 1851 (2003)

29. A.A. Puretzky, D.B. Geohegan, S. Jesse, I.N. Ivanov, G. Eres, Appl. Phys. A **81**, 223 (2005)
30. R.F. Wood, S. Pannala, J.C. Wells, A.A. Puretzky, D.B. Geohegan, Phys. Rev. B **75**, 235446 (2007)
31. C.M. Rouleau, G. Eres, H. Cui, H.M. Christen, A.A. Puretzky, D.B. Geohegan, Appl. Phys. A **93**(4), 1005–1009 (2008)
32. Z. Liu, D.J. Styers-Barnett, A.A. Puretzky, C.M. Rouleau, D. Yuan, I.N. Ivanov, K. Xiao, J. Liu, D.B. Geohegan, Appl. Phys. A **93**, 987 (2008)

2

Basic Physics of Femtosecond Laser Ablation

Juergen Reif

Summary. Laser ablation being the basic process for many prominent applications of lasers in present day high technology, medicine, and other fields, its basic physics is reviewed in this chapter. In order to distinguish the fundamental, laser–material interaction from any secondary effects, we concentrate on ultrashort laser pulses ($\approx$100 fs duration) at comparably low intensities, below the commonly indicated threshold for massive material removal. It is shown that – for these conditions – the principal light/matter coupling occurs via multiphoton excitation of electrons into the conduction band or the vacuum. The resulting perturbation of the target lattice results in the emission of positive particles, from atomic ions to larger clusters of more than ten atoms. With the increasing number of incident pulses, the light/material coupling is facilitated by the accumulation of transient crystal defects resulting from particle removal. On the other hand, the lattice destabilization, upon excitation and ablation, relaxes via self-organized formation of regular nanostructures at the irradiated area. The strong influence of laser polarization on the structural order is still not at all understood.

2.1 Introduction

Already in the early days of lasers, it had been observed that the concentrated light energy could affect the irradiated material considerably: Damage to optical components occurred because of evaporation and removal of material from the component surface. In fact, that was the first manifestation of laser ablation, i.e. the removal of material from a target upon laser impact. Rapidly, this effect became exploited in a more controlled way to process materials, e.g. in tool-free cutting and drilling, which are standard technologies in the car industry and related fields today.

Most of the early research and applications were performed using CO_2 lasers, and understood in terms of the classical thermodynamic processes, the laser being considered merely as a very concentrated heat source. About three decades ago, however, based on the observation of UV-laser ablation from organic polymers [1], it was suggested that the ablation process might be more complicated than had been assumed until then, involving not only rapid melting and evaporation but also electronic transitions. In the following years, the ablation phenomenon began to attract increasing interest, from the point of view of its fundamental study as well as its applications[2].

Today, laser ablation is used in a wide range of high technologies other than cutting and drilling, such as, for instance, surface processing, thin film deposition by PLD, or laser cleaning, from silicon wafers to artworks. In medicine also, many applications are based on laser ablation, e.g. in ophthalmology (amongst others, laser correction of ametropia – LASIK), dermatology (tattoo removal), surgery, etc. However, undesirable effects like the destruction of laser-irradiated biological tissue or laser damage to optical components also result from laser ablation.

This chapter is devoted to the study of the fundamental aspects of laser ablation in order to gain a better understanding of the physics underlying the phenomenon. In order to distinguish the basic processes from any secondary effects, such as laser interaction with ablation products and the like, we concentrate on the interaction of ultrashort (i.e. duration of ≈100 fs) laser pulses with solid targets. To further reduce secondary effects, the incident fluence is so low that single pulses do not result in significant, target surface modification. Usually, only after several 10,000 pulses, significant ablation craters are observed. Our sample materials will be mainly dielectrics (BaF_2, CaF_2, Al_2O_3), and silicon. We first review some basic features of the energy input, the laser–material coupling. Then we consider the follow-up processes, transiently modifying the material. Finally, we report on self-organized, nanostructure formation at the target surface as a consequence of material relaxation after ablation.

2.2 Energy Input

In principle, all decomposition or material removal from a solid target is the consequence of an energy input into the target, resulting in overcoming the solid's binding energy. In a classical process, which is slow enough to proceed in thermodynamic equilibrium, this means that the energy input ΔE is fully transferred into an increase of internal energy ΔU and thus, to an increase in temperature ΔT:

$$\Delta E = \Delta U = cm\Delta T \tag{2.1}$$

(with heat capacity c and mass m of the heated target material).

As shown in Fig. 2.1, this internal energy increase results in a classical phase transition and, occasionally, in a dissolution of the heated volume. On a microscopic scale, the temperature increase corresponds to an increase of atomic kinetic energy. In contrast to energy input by classical heating, via a global phonon bath, or by ion impact, addressing the core motion directly via a momentum transfer, energy input from laser pulses is inherently different:

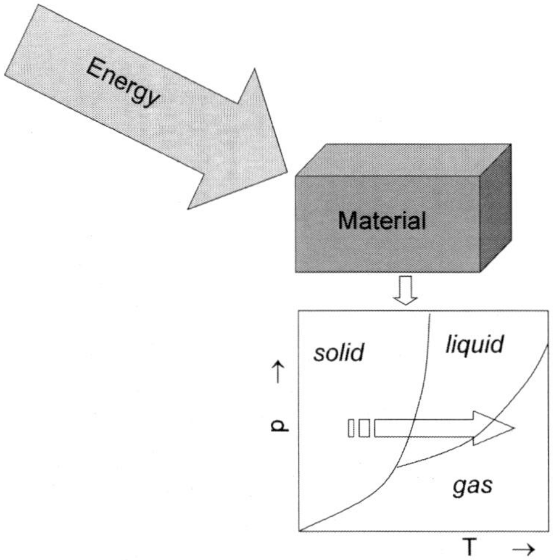

Fig. 2.1. Schematic representation of energy input into a solid material

Table 2.1. Process chart for laser ablation

Time scale	Material response	Observation
Femtoseconds	Electronic excitation	Electron emission
Picoseconds	Energy dissipation/ core motion	
	Bond breaking	Atom/ion emission
Nanoseconds		(Plasma) plume
	Surface relaxation/ reorganization	

the incident light "speaks" only to the electrons of the system, and all core motion is only a *secondary* process.[1] This allows establishing, conceptually, a history of processes starting from light absorption, leading to particle removal and, finally, to target relaxation, as is given in Table 2.1.

From these time scales, the choice of ultrashort laser pulses with duration below 150 fs for the study of fundamentals becomes justified: then, the laser light interacts only with an almost passive target. All significant target modification (e.g. transient changes in band-structure, removal of particles) occurs only *after* the laser pulse and, thus, should not affect the absorption properties. More important, the laser does not interact with ablated material. (For longer pulses, significant amounts of laser energy may be absorbed in the ablated plasma plume. This hot plasma might then, in turn, sputter the target surface.)

[1] Even a direct coupling to a vibration is, in fact, promoted via the *electronic* system, related to the cores only by electron–phonon coupling.

2.2.1 Multiphoton Excitation

In solid target, usually, the electrons will be excited across the band-gap[2] to the conduction band. For the typical situation of ultrafast Ti:Sapphire laser pulses at a photon energy of 1.55 eV, for many materials this means that the excitation occurs via multiphoton absorption,[3] with a number n of necessary photons given by

$$n = \bmod\{E_{\text{gap}}/1.55\,\text{eV}\} + 1. \qquad (2.2)$$

The resulting electron kinetic energy is, then, $0 < E_{\text{kin}} \leq 1.55\,\text{eV}$.

As the lower edge of the conduction band is close to the vacuum level, the electrons located near the target surface, at a depth less than their mean free path (cf. Fig. 2.2), can leave the target.

Indeed, the emission of electrons from the target can be observed (Fig. 2.3). As is expected for multiphoton ionization, the electron yield $Y_{\text{electrons}}$ as a function of incident intensity follows a power law:

$$Y_{\text{electrons}} \propto I^n. \qquad (2.3)$$

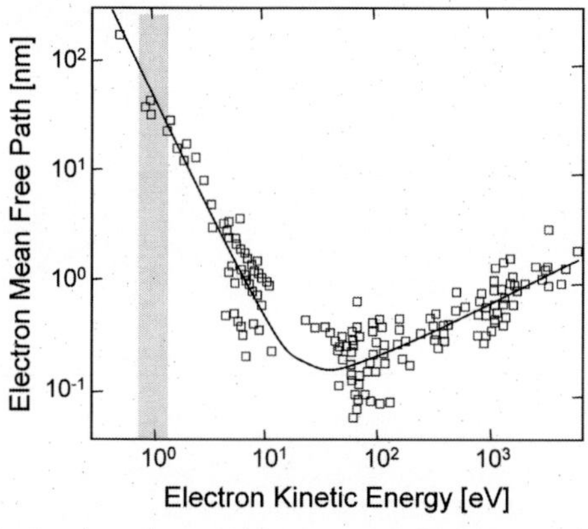

Fig. 2.2. The "Universal Curve" of electron mean free path vs. electron kinetic energy [4]. The *shaded area* indicates the region of typical kinetic energies in the presented situation

---

[2] In a metal, the light absorption may, occasionally, just increase the electron-kinetic energy/temperature inside the conduction band.

[3] Generally, in "linear optics" the absorbed energy E_{abs} is given by $E_{abs} = \sigma \cdot F_{pulse} = \sigma \cdot I \cdot \tau_{pulse}$ (F: fluence {areal density of incident laser energy}, σ interaction cross-section, I: intensity τ pulse duration). For higher intensity, however, the quantum system becomes more and more perturbed when the electrical light field approaches the atomic binding field. Perturbation theory then makes the cross-section intensity-dependent: $\sigma(I) = \sigma^{(n)} \cdot I^{n-1}$; hence, the absorbed energy will be $E_{abs} = \sigma^{(n)} \cdot I^n \cdot \tau_{pulse}$. For longer pulses, additional tunneling ionization can take place [3].

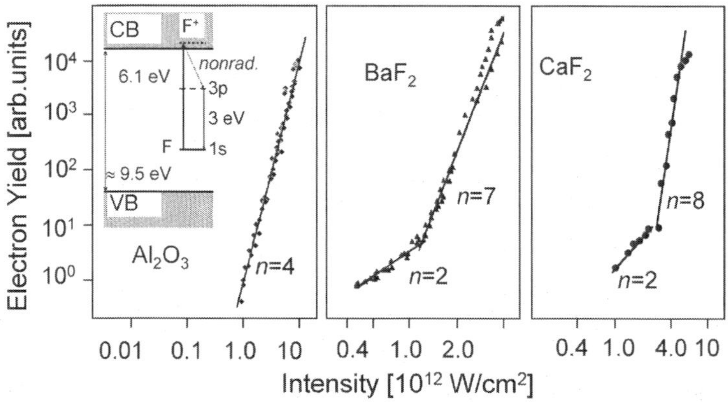

Fig. 2.3. Double logarithmic plot of electron yield vs. laser intensity at 1.55 eV for three different dielectric targets: Al_2O_3 ($E_{gap} \approx 9.5$ eV), BaF_2 ($E_{gap} \approx 10.5$ eV), and CaF_2 ($E_{gap} \approx 12$ eV). The power law (2.3) results in straight lines with the slope corresponding to the number of photons n. The insert in the Al_2O_3 panel sketches the band-structure (*VB*: valence band; *CB*: conduction band) and indicates the energy levels of a localized F-centre with a 1s ground state, a 3p excited state, and the ionised F^+-centre energy in the conduction band

A closer inspection of typical results as shown in Fig. 2.3 [5,6] reveals, however, that the nonlinearity may be, occasionally, less than expected from the band-gap. A probable reason is indicated as an insert in the Al_2O_3 panel: localized energy states within the band-gap, e.g. due to an F-centre (i.e. an anion vacancy) may act as intermediate resonances and thus reduce the order of nonlinearity. Indeed, we could observe blue radiation of the 3p–1s transition of such F-centres [7]. A similar influence of localized or transient defect states [8] should also be responsible for the reduced nonlinearity at low intensities on the fluorides shown in Fig. 2.3.

Later (Sect. 2.4), we will come back to such defects, in particular those generated by the laser interaction itself.

When the incident fluence becomes so high that the density of conduction band electrons inside the target bulk overcomes a critical density, additional coupling mechanisms like hot carrier absorption occurs [9] (via electron–photon–electron – three-body collisions), resulting in a significant increase of ablation efficiency.

2.3 Ion Emission: Ablation

2.3.1 Experimental Observation

The actual ablation, i.e. the emission of heavier target constituents, can be monitored by Time-of-Flight (ToF) mass spectrometry [5, 6]. In fact, under

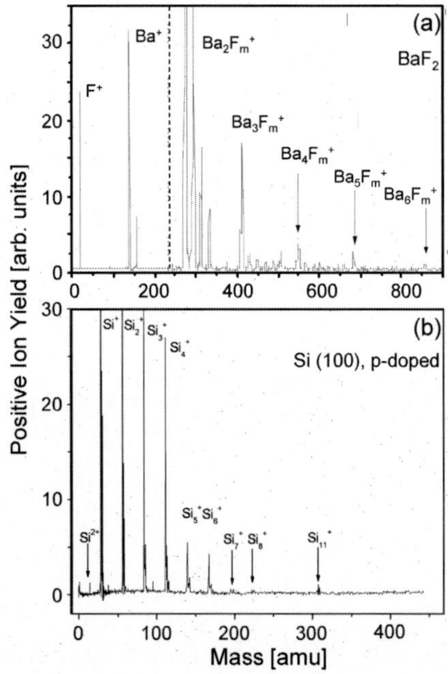

Fig. 2.4. ToF-spectra of positive ions emitted from BaF_2 [7] and single crystalline, p-doped Si(100) [11] at laser intensities around $10^{12}\,W\,cm^{-2}$ and a photon energy of 1.55 eV ($\lambda = 800\,nm$), below the threshold for a massive ablation/damage for the respective material [12]. Not only positive ions of all surface constituents are observed, but also larger clusters of up to more than ten atoms (in panel (**a**): index $m = 0, 1, 2, \ldots$)

the conditions we are dealing with, an abundant emission of positive ions is observed (Fig. 2.4). In particular, from ionic crystals not only the usual cations are emitted but also positive ions of the "anionic" species are detected. In fact, almost all ablated particles, under the considered low-laser intensities, appear to be positive ions, with neutrals and negatives occurring only at a substantially higher excitation [10].

Not only individual atomic ions are desorbed,[4] but also large cluster ions of up to more than ten atoms. This is a general effect, not depending on the target material, as can be seen in Fig. 2.4.

The positive ions exhibit an unexpectedly high kinetic energy. From experiments using a retarding voltage (Fig. 2.5) [13], as well as from experiments using the ToF-spectrometer in "drift mode," i.e. with a field-free region (without collecting field) between target and spectrometer [5,6], ion kinetic energies between $\approx$10 eV and over 100 eV are measured, depending on the target material and irradiation conditions.

[4] By "desorption" we denote an ablation rate at the individual particle level.

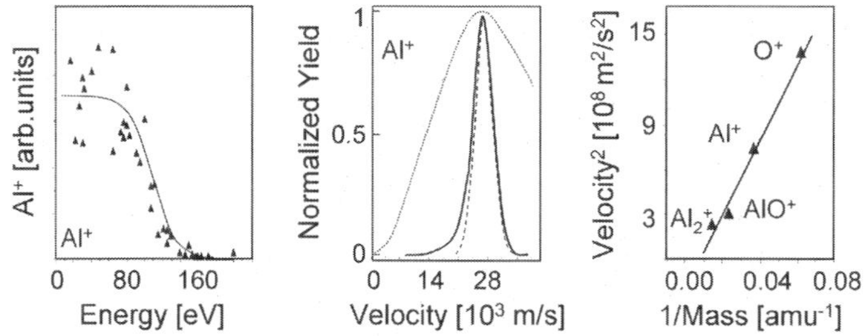

Fig. 2.5. Kinetic energy of positive ions desorbed from Al_2O_3 after irradiation with 1.55 eV laser pulses at $\approx 10^{12}$ W cm^{-2} [6, 13]. The left panel shows the Al^+-ion yield at the detector vs. the retardation energy, fitted by a sigmoid function. The central panel shows the ion velocity distribution, derived from the sigmoid of the left panel (*solid line*). For comparison, a Maxwell–Boltzmann distribution (*dotted line*) and a shifted Maxwellian (cf. (2.1)) (*dashed line*) are shown, both with the same center velocity. On the right panel, the kinetic energies of the different desorbed ion species are compared by plotting the squared velocity vs. the reciprocal mass. (From $E_{\text{kin}} = 1/2 mv^2$ the plot shows $v^2 = 2E_{\text{kin}}\, m^{-1}$ and the straight slope indicates a constant kinetic energy)

The ion velocity distribution (solid line in the central panel of Fig. 2.5), derived from the sigmoidal fit to the retardation energies in the left panel, is, obviously, not compatible with the assumption of a thermodynamic Maxwell–Boltzmann distribution (*dotted line*). Instead, it resembles very well a much "colder" distribution, centered on a high drift velocity, a situation which is typical for "seeded beams," and can be described by a so-called "shifted" Maxwellian for the distribution function $f(v)$:

$$f(v) = cv^2 \exp\left(-\frac{m\,(v-u)^2}{2kT}\right), \tag{2.4}$$

where v is the total velocity, u the center of mass or drift velocity, k Boltzmann's constant, m the particle mass, and T the temperature.

Interestingly, all desorbed species have about the same kinetic energy (*right panel* of Fig. 2.5). This indicates, at first, that all species originate from the target and are not formed by reactions in the ablation "plume," as they all travel with different velocities and thus, do not "meet" during their flight. Second, it leads to a possible understanding of the desorption mechanism, which will be described in the following subsection.

2.3.2 Desorption Mechanism – Coulomb Explosion

We have already shown in Sect. 2.2.1 that laser impact results in the emission of a considerable amount of electrons from the target. This is assumed to occur instantaneously during the laser pulse, leaving behind positive holes in the surface-near region. In dielectrics, and even in semiconductors [14, 15], the time for bulk electrons to fill these holes is at the order of about a picosecond or even longer. Thus, the surface-near region will be positively charged for a considerable time. This charged region may become electrostatically instable, so that, at a sufficiently high hole density, the surface will break apart by emitting positive particles, which will be accelerated in the residual field. This electrostatic repulsion is known by the name of Coulomb explosion [16–18].[5]

At a still higher excitation, when hot carrier absorption becomes important, a hot electron bath is generated [3], rapidly coupling to the lattice via electron–phonon collisions in a large volume, and bulk super-heating starts to set in. Then, additional, efficient desorption mechanisms like phase explosion [19] become effective.

2.4 Transient, Local Target Modification

Obviously, with interaction pulse the remanent target surface will change. Already the electron excitation and ionization may influence the absorption properties. This change can be transient, i.e. after relaxation the electronic system returns to its previous state. Additionally, however, desorption will change the local surface composition. As was indicated in Fig. 2.3, (surface) vacancies will perturb the normal band structure and produce localized states within the band-gap.

Depending on diffusion probabilities, this local disorder can last for a considerable time [20] and thus modify the absorption probabilities for subsequent pulses. As indicated in Fig. 2.3, the order of the absorption nonlinearity may be reduced, and, thus, the cross-section is increased.

2.4.1 Incubation

For multiple-pulse irradiation ("N-on-1"), this effect of (transiently) increased ablation efficiency – and thus, decreased threshold intensity – can accumulate from pulse to pulse, if the defect lifetime is sufficiently long. This "incubation" has been known for more than two decades [21, 22].

As a consequence of this incubation, the desorption yield at a fixed intensity, typically, increases continuously from pulse to pulse, and finally saturates [6] when the surface defect density reaches a steady state between the generation of new and the annihilation of previous defects.

In the first attempt to understand this incubation a merely statistical model was introduced [21], assuming the ablation threshold $I_{\rm t}$ to decrease inversely proportional to a power α of the number of pulses N: $I_{\rm t}\,(N) \propto 1/N^{\alpha}$. A more "physical" model, assuming an exponential increase in the number

[5] In contrast to Coulomb explosion in molecular physics, it is not evident whether here momentum conservation – via recoil [17] – or energy conservation – via the same repulsive field [18] – is a good parameter; obviously, the experimental data are compatible with both models within experimental error.

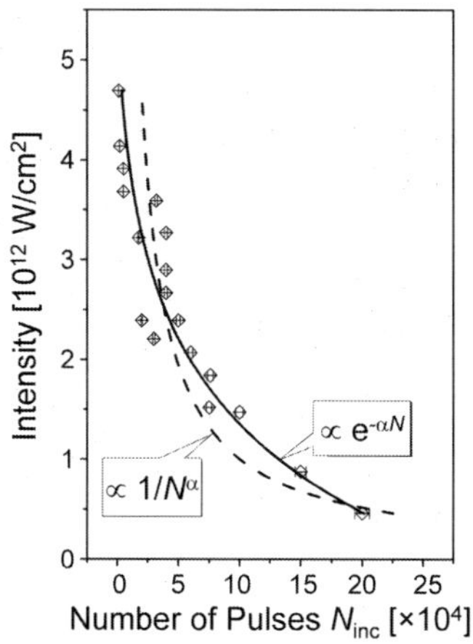

Fig. 2.6. Multi-pulse (N-on-1) desorption threshold intensity as a function of number N of subsequent pulses [23]. The comparison with the "physical" exponential model (*solid line*) and the statistical model (*dashed line*) confirms the idea of increasing defect generation as the origin of incubation

of defects (responsible for the threshold reduction), was presented in [22], postulating $I_t(N) \propto e^{-\alpha N}$. In fact, as is shown in Fig. 2.6, recent experimental results favor this exponential model [23].

Indeed, the lifetime of such structural defects, responsible for incubation, has been observed to extend to several 10 ms, as could be shown by interrupting the irradiation pulse series for controlled periods [24]. Also, the luminescence of bulk defects has been observed for more than 50 ms after excitation, thus persisting over several pulses at a repetition rate of, typically, 1 kHz. This is demonstrated in Fig. 2.7 [23].

2.4.2 Transient Dynamics

As already indicated, not only relatively long-lived defects are produced, due to the local particle removal during desorption, but also electronic excitation and ionization have an influence on the absorption properties. The lifetime of such modifications is expected, however, to be much shorter and closer related to the exciting pulse. Information at the relevant time scale can be obtained by appropriately designed pump-probe experiments, as displayed in Fig. 2.8.

The target is "prepared" by a pump pulse and then, after a defined delay time, "interrogated" by a second pulse, the probe. As shown in Fig. 2.8, both pulses may be derived from one parent pulse in a Michelson interferometer; the interdelay can be adjusted by a variation of one-arm length. The combined action of both pulses can be monitored by detecting the emitted particles.

To investigate the transient target behavior alone, care must be taken to exclude longer lived modifications like those considered in Sect. 2.4.1. In

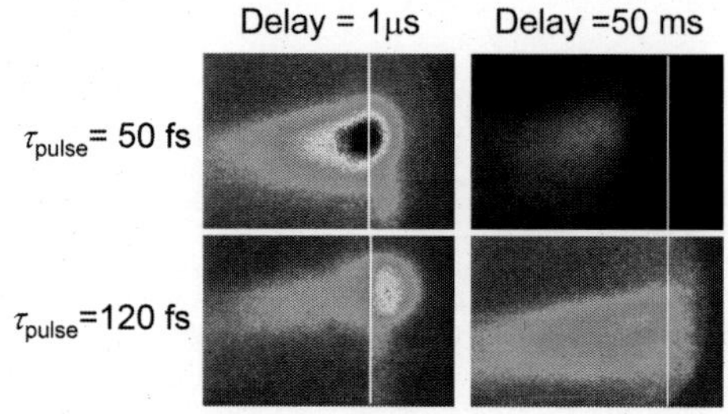

Fig. 2.7. Bulk and plasma fluorescence from CaF_2 irradiated with a single 1.55 eV pulse at a fluence of $\approx 10^{15}$ W cm^{-2} (well above the single-shot ablation threshold). The pulse duration was either 50 fs (*upper panels*) or 120 fs (*lower panels*). The fluorescence was observed – without spectral resolution – at two different delays after the excitation: 1 μs (*left panels*) and 50 ms (*right panels*). In all panels, the target surface is indicated by the vertical white line, with the bulk at the left and vacuum at the right. While at 1 μs strong plasma emission is visible (more intense for the shorter pulse), at 50 ms delay strong bulk defect luminescence survives, much more pronounced for the longer pulse

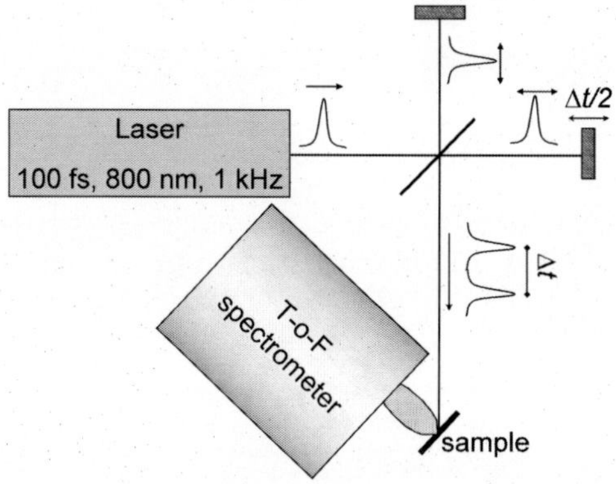

Fig. 2.8. Typical pump-probe setup to investigate the transient desorption dynamics. The laser beam is split into two partial beams of equal properties in a Michelson configuration. By varying the length of one interferometer arm, a temporal separation between both beams at the target can be adjusted. The combined effect of both pulses is monitored – in our case – by a ToF-spectrometer, capable to measure positively- and negatively charged particles (target and detection are placed in UHV at $\approx 10^{-10}$ hPa)

particular, the combined intensity from both pulses at zero delay, i.e. at full temporal overlap, must be sufficiently low [25]. If both pulses are of equal intensity, this "soft" interaction can be verified by a result which is symmetrical with respect to negative[6] and positive[7] delays.

[6] Pulse A is pump, pulse B is probe.
[7] Pulse B is pump, pulse A is probe.

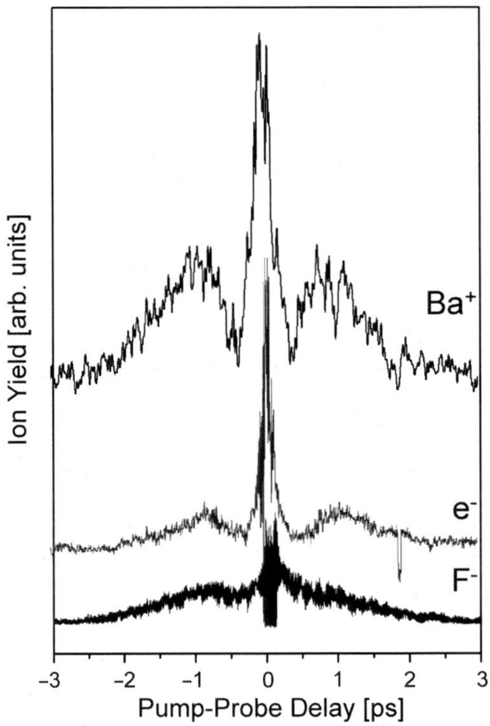

Fig. 2.9. Pump-probe experiment on particle emission from BaF_2 using pairs of 1.55 eV pulses at $0.5 \times 10^{12}\,\mathrm{W\,cm^{-2}}$. The single-pulse signal is negligibly small (cf. delay of ± 3 ps). Around zero delay the observed signal is influenced by interference effects from the Michelson setup

A typical result of such pump-probe experiments [26] is shown in Fig. 2.9 for BaF_2, monitoring positive ions $\left(Ba^+\right)$, electrons, and negative ions (as secondary products [9]). Both pulses were of the same intensity, yielding each only a very weak signal of any type of particles, as can be seen at delays of ± 3 ps, where the pulses are obviously separated far enough to act independently. At zero delay, i.e. at optimal pulse overlap, both pulses act coherently and the combined intensity[8] results in a strong excitation. Consequently, strong electron emission and, in turn, strong positive ion emission occurs, as expected. With increasing pulse separation, the emission yield decreases and reaches almost the level of an independent pulse action.

After a certain delay, however, i.e. from a certain pulse separation on, the signal increases again, reaches a maximum (for BaF_2 at a pulse separation of $\approx$700 fs) and then slowly diminishes again toward the individual pulse result for pulses separations larger than $\approx$2.5 ps. This behavior, generalized

[8] In the Michelson interferometer, this corresponds to four times the single arm intensity.

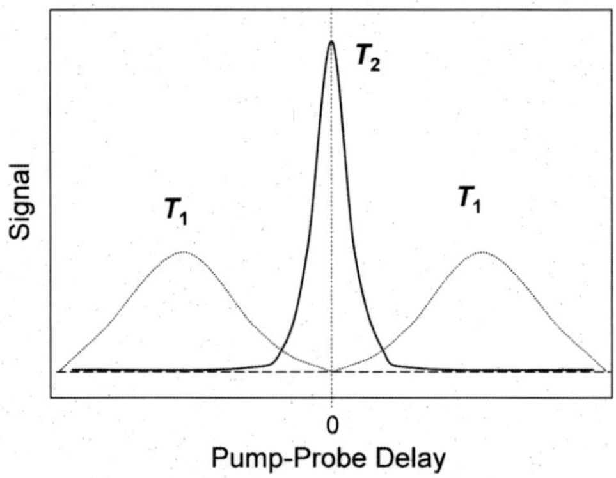

Fig. 2.10. Generalized dissipation dynamics in a target irradiated by ultrashort laser pulses. The *dashed* ground line indicates the level of single-pulse excitation, the solid line shows the coherent interaction of both pulses, labeled "T_2," and the *dotted line* is due to an increased coupling through a transient, modified absorption cross-section, labeled "T_1"

in Fig. 2.10, indicates that the energy deposition from the pump pulse evolves in the target in such a way, that the absorption probability increases to reach a maximum after ≈700 fs. Subsequently, this transient state of enhanced absorptivity dies out again. Even secondary particles, such as negative ions [9], follow this dynamics. Interestingly, the behavior is a very general one, observed for very many types of targets, even metals [27], though the explicit nature of the transient, high-absorption state may be different.

Associating the general behavior with possible relaxation processes, it appears appropriate to compare it with the excitation decay in magnetic resonance [26]. Then, the "coherence peak" should correspond to the phase relaxation with ("transverse") time T_2, and the occurrence of the transient modification should be due to energy dissipation into the target, characterized by ("longitudinal") time T_1, as indicated in Fig. 2.10.

2.5 Transient Instability and Self-Organized Structure Formation

2.5.1 Periodic "Ripples" Structures

When inspecting the bottom of an ablated area after very many pulses, the target surface morphology appears considerably changed (Fig. 2.11): a regular pattern of almost parallel modulation lines has developed, with typical feature size below the incident wavelength.

Already about 40 years ago, similar structures were observed [28, 29] after irradiation with nanosecond pulses, termed "ripples." Until the mid-1980s, a model was developed attributing the phenomenon to a modulated energy

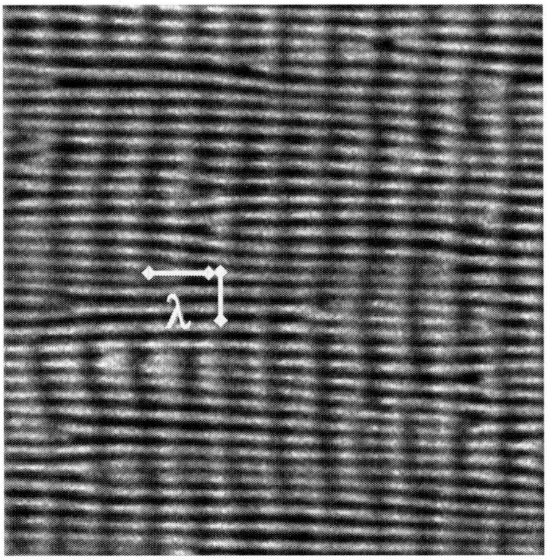

Fig. 2.11. Ripples pattern at the bottom of an ablated area on BaF_2 after 10,000 1.55 eV-pulses at $\approx 10^{12}$ W cm^{-2}. The white double arrows indicate the laser wavelength for comparison with the ripples features

input, which was ascribed to an interference of the incident wave with surface-scattered electromagnetic waves [30].

The structures observed after femtosecond illumination, however, display features which are not really compatible with that interference model. In Fig. 2.11, structures that are much finer than the wavelength, i.e. the diffraction limit, can be seen. Further, two superimposed patterns of different periodicity are observed. Finally, characteristic deviations from the regular structure, namely splitting of one line into two (bifurcations) occur.

Another, very peculiar feature is shown in Fig. 2.12. Across the ablation spot, produced by a beam of Gaussian spatial intensity distribution, the ripples periodicity changes *abruptly* between a very fine structure in the edge region of lower intensity and a coarser structure, of about double spacing, in the high-intensity central region. Again, bifurcations as well as truncations are observed which are not at all compatible with simple interference structures. Instead, the structures are very similar to those observed, typically, in the structure formation attributed to self-organization from instability, such as, for instance, large sandy areas subjected to windy erosion (Fig. 2.13). In particular, the structures are almost identical to those observed upon sputtering by an ion beam impact [6, 31, 32].

A very successful way of modeling such self-organized structure formation [33, 34] in ion sputtering and sand ripples involves the principles of nonlinear dynamics, and there have been first approaches to apply these models to the present phenomena [6, 32].

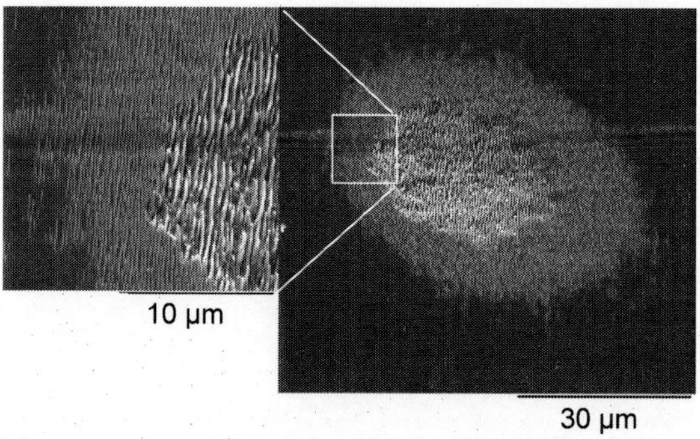

Fig. 2.12. Ripples structures in one ablation spot, irradiated by a beam of Gaussian spatial profile. The magnification in the left panel shows an abrupt transition from fine, sub-wavelength structures at the low-irradiation edge and coarser structures of about double spacing at the high-irradiation center

Fig. 2.13. Ripples formation in a sandy area subjected to windy erosion and diffusion (photo: courtesy of Gediminas Raciukaitis). Note the coexistence of coarse and fine ripples and the presence of bifurcations

2.5.2 Instability and Self-Organization

The general idea for such modeling is the assumption that the excitation and ablation create a state of extreme instability for the target:

Since – at least in dielectrics and semiconductors – the excitation involves conduction band electrons in a considerable fraction, it results in a corresponding perturbation of the target crystalline order, coordinated by the electron configuration. Moreover, the surface equilibrium order is destroyed by the emission of (individual) surface constituents. Consequently, the crystalline order is perturbed even though the target is *not* molten (i.e. the target is not in a state of thermodynamic equilibrium). Theoretical modeling indicates, in fact, a destabilization of the crystalline lattice on a several hundred to thousand femtoseconds [35,36]. Such transiently disordered or "soft" material has been, actually, observed by femtosecond, time-resolved, X-ray diffraction [37], where the short-range crystalline order disappeared shortly after excitation. The long-range order, however, persisted, excluding an equilibrium liquid state.

The corresponding instability must relax in an extremely short time, due to a steep gradient in crystalline order to the surrounding target material. This excludes "thermal" processes like crystallization or glassification. To model the relaxation [32], based on the ion-sputtering analogue [34], the surface may be assumed to be a corrugated, thin "liquid" film. In such configuration, the instability consists of a competition between corrugation increase (surface roughening due to desorption) and surface smoothing, due to diffusion. Surface corrugation and "film" thickness h can be analyzed using an approach based on the Cahn–Hilliard/Kuramoto–Sivashinsky differential equation [38], involving removed-particle speed v, sputter coefficient γ, and surface-diffusion constant D:

$$\frac{\partial h}{\partial t} = -v_0 + \gamma \nabla h + v \Delta h - D \Delta^2 h. \tag{2.5}$$

Transient solutions of this equation predict the formation of periodic structures like stripes, squares etc., increasing linearly. At a certain level of perturbation, the transient solution saturates and a nonlinear regime of structure formation sets in, showing, typically, coarsening and merging into larger, more complex structures.

Indeed, such effects are observed in experiments. This is shown, exemplary, in Fig. 2.14, which represents, similar to Fig. 2.12, *one* single ablation spot on silicon, irradiated by a Gaussian intensity distribution, increasing from left to right (the white separation lines are only inserted to guide the eye). At low-perturbation (*left*) linear, periodic ripples are observed with a period length of $\Lambda \approx 660\,\text{nm}$. At higher perturbation (*central part* of Fig. 2.14), the ripples appear merged to larger units with about double spacing, $\Lambda \approx 1{,}300\,\text{nm}$. In addition, the structures become more complex, like meandering. At a still higher perturbation (*right*), the structure is still more complex, with another period doubling to $\Lambda \approx 2{,}500\,\text{nm}$.

The microphysical details of these new surface structures are not yet clear. Some features, however, are remarkable:

- The ripples orientation is not at all correlated to the target crystal structure (cf. Sect. 2.5.3)
- Micro-Raman spectroscopy on silicon reveals that there can be a crystallographic reordering, as has been known from high-pressure modifications [6]
- The nanostructured surface exhibits peculiar electrical features [39], as shown in Fig. 2.15

Obviously, the electrical properties reflect the morphology well. The EFM phase contrast, measuring the force on the tip, at a bias of −1V follows

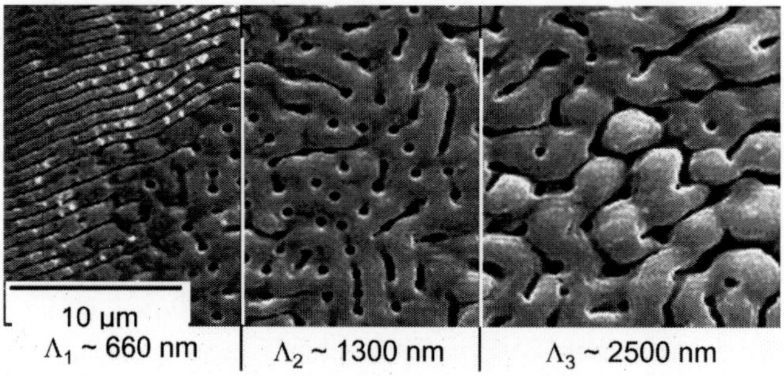

Fig. 2.14. Period doubling during self-organization of ripples-structures: *one* spot on silicon surface with three different structure sizes Λ from fine (*spot edge, left*) to coarse (*spot center, right*). The white lines are drawn to guide the eye. Note also the structure change from "simple" straight lines to more complicated, two-dimensional ordering

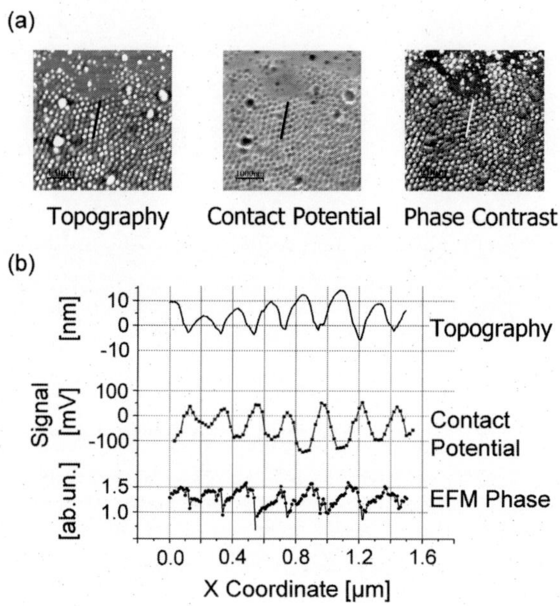

Fig. 2.15. Electrical properties of self-organized nanostructures (on p-Si surface irradiated by a circular polarized light (cf. Sect. 2.5.3), measured by electrostatic force microscopy (EFM) [39]. (**b**) line-scans as indicated in the corresponding panels of (**a**)

the morphology derivative, i.e. it is strongest at the edge of the nanospheres (cf. Sect. 2.5.3). More surprising is the high nanostructure electric charge relative to the unaffected surface, showing up in the contact potential varying between $-120\,\mathrm{mV}$ at the top of the spheres and $+50\,\mathrm{mV}$ in the valleys. A possible origin could be dopant segregation during the "soft" nonequilibrium state, but a local modification of crystalline structure (e.g. high pressure phases [6]) cannot be ruled out. More information should be gained by TEM analysis of the atomic details which is, presently, under way.

2.5.3 Polarization Dependence

Although a general picture of self-organized, ripples formation seems to be already fairly coherent, one characteristic feature of the effect is by far not yet understood. In all experiments, a dominant influence of the *laser polarization* on the shape and orientation of the nanostructures has been observed.

For linear polarization, typically, the predominant ripples are aligned perpendicular to the light electric field. Additionally, secondary structures (most often of larger spacing) can occur, aligned perpendicular to the prominent ones (i.e. parallel to the laser field). This is shown in Fig. 2.11, where the laser was polarized vertically. When rotating the polarization and keeping the target fixed, the ripples orientation follows the polarization, independent of the target crystalline structure [5, 32]. When the polarization is rotated during a long, continuing series of pulses, the ripples orientation is determined by the latest polarization direction.

An even more startling phenomenon is observed when, during repetitive irradiation, the target is moved to "write" ablation lines or areas: if the spots are slightly overlapping, the ripples structure is continued coherently over the whole line [40] resp. area [41].

To shed more light on this polarization influence, experiments were performed using circular and elliptical polarization under normal incidence.

In fact, new structures appear. As can be seen from Fig. 2.16 for irradiation with nearly circular polarization, arrays of spherical nanoparticles are generated, with a diameter of 100 nm or even less. Only a very weak linear alignment, bead-string like, is observed, due to a very slightly, non-optimal adjustment of the quarter-wave plate used for generating circular polarization.

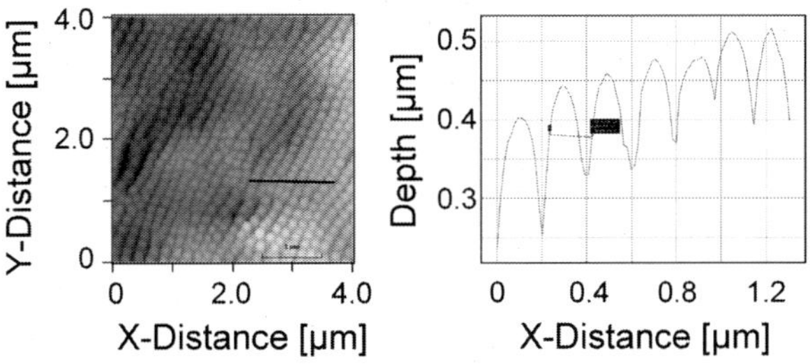

Fig. 2.16. AFM picture of self-organized nanostructures upon ablation with (almost) circular polarization: arrays of nanospheres with a diameter of ≈100 nm (right panel showing a scan along the line indicated in the full area picture, left). The apparent "alignment" of the "bead strings" is due to a very slight maladjustment of the $\lambda/4$-plate, used for generating the circular polarization, resulting in weak ellipticity (cf. Fig. 2.17)

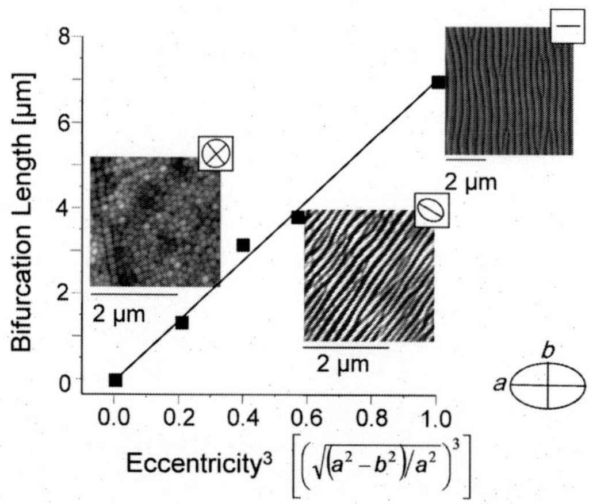

Fig. 2.17. Dependence of self-organized, long-range order on the third power of elliptical eccentricity ε of laser polarization

This loss of linear order correlates with the fact that, for circular polarization, the strong directional influence of the laser electric field on the structure formation is missing. To substantiate this observation, the directed electric field is reintroduced by polarizing the beam elliptically. Indeed, more and more oriented and ripples-like structures develop (Fig. 2.17) upon increasing the polarization eccentricity ε, resulting from the imbalance of major axis, a, and minor axis, b:

$$\varepsilon = \sqrt{a^2 - b^2/a^2} \tag{2.6}$$

Intriguingly, the structure's long-range order, measured by the average distance between two bifurcations, depends on the third power of eccentricity ε (Fig. 2.17).

The important role of the laser electric field direction is further corroborated by the result shown in Fig. 2.18, where an artificial eccentricity was introduced by changing the angle of incidence for circular (Fig. 2.18a) resp. elliptical (Fig. 2.18b) polarization. By the beam inclination, the surface spot is expanded in the plane of incidence (p-polarized fraction) though it is unchanged along the axis of beam rotation, i.e. perpendicular to the plane of incidence (s-polarized fraction). Correspondingly, the s-component of the electric field remains unchanged and the p-component is reduced by the projection. Thus, for circular polarization (at beam cross-section $E_s = E_p$), s becomes the major axis a of an artificial ellipse, with eccentricity increasing with the increasing angle of incidence. Conversely, for elliptical polarization (with minor axis b corresponding to the s-direction) the eccentricity is reduced, becoming more and more circular.

Similar to observations with linear polarization, the major axis of stronger electric field a controls the alignment of the long-range order. This is not only

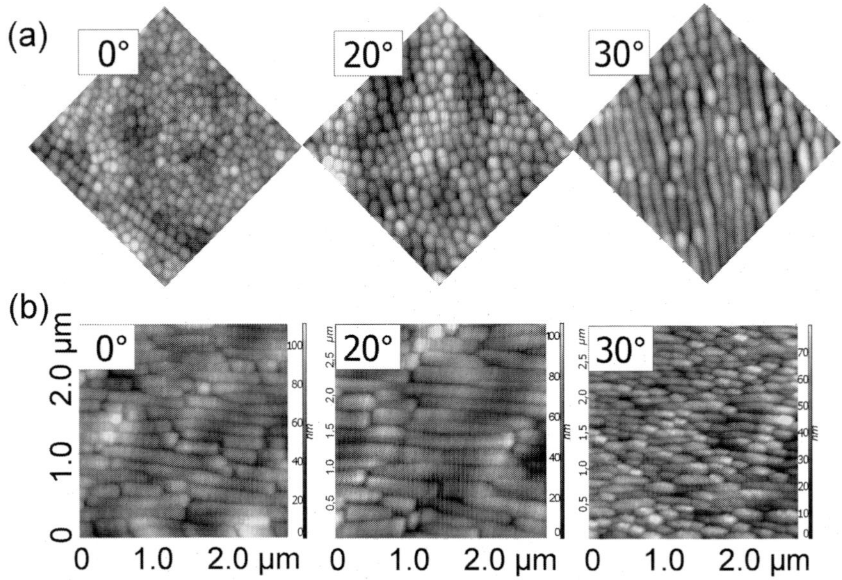

Fig. 2.18. Artificial eccentricity, introduced by oblique incidence of circularly (**a**) resp. elliptically (**b**) polarized beam. The angle of incidence, indicated in the panels, is obtained by a rotation about the horizontal, corresponding to s-polarization (all panels are $3 \times 3\,\mu m^2$ AFM pictures)

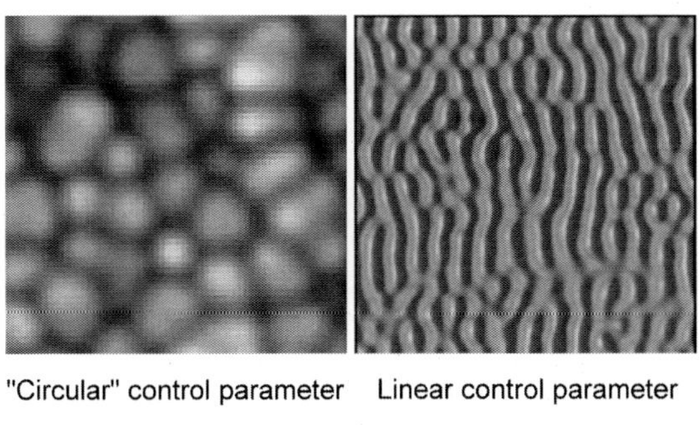

Fig. 2.19. Theoretical simulation [6, 32] of the influence of the control-parameter directionality. The simulation assumes ion beam sputtering, the control parameter being normal (*left panel*) or oblique incidence (*right panel*)

observed in Fig. 2.18, but also shows up, when the whole polarization ellipse is rotated [42] and the ripples alignment follows. Finally, the role of the electric field direction as a control parameter for ripples arrangement is confirmed by the result of first model simulations [6, 32], as shown in Fig. 2.19.

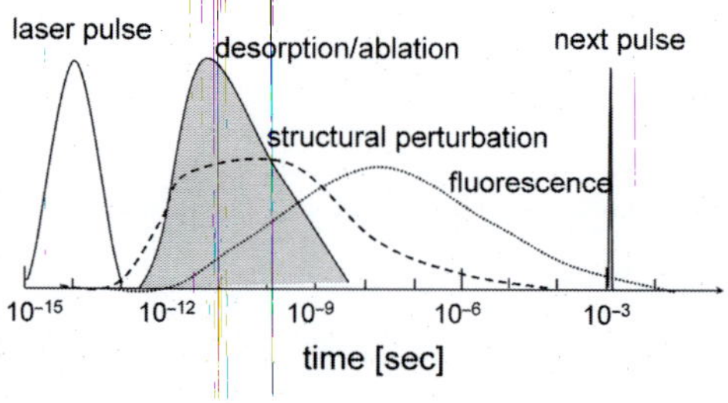

Fig. 2.20. Timing diagram of observed phenomena/processes: only after the termination of the laser pulse, desorption/ablation sets in, at the same time scale as structural perturbation takes place [26,27,35–37] (cf. also Sect. 2.4.2). Defect lifetime (cf. fluorescence in Fig. 2.7) may even extend longer than the pulse separation

2.6 Discussion

So far, empirically, the scenario for the formation of regular nanostructures upon femtosecond laser ablation appears rather rational: the short, powerful energy input first perturbs the target's electronic system, causing an instability which results in ablation/desorption. This emission further increases the instability by roughening the surface. Concomitantly, the surface diffusion tends to smooth the surface again, still further increasing the instability. All this occurring on a very fast time scale, with a tremendous order-gradient to the surrounding target material, requires a very fast relaxation, far away from equilibrium pathways. Therefore, it results in self-organized nanostructure formation. The arrangement of these ripples is controlled by the direction of the ablating laser electric field.

Closer consideration, however, reveals some severe inconsistency (as is shown in Fig. 2.20): the time scales of ongoing processes put the role of laser polarization on structure formation severely into question: when the processes of structure formation, i.e. the motion and immobilization of atoms at the surface, occur the laser pulse has terminated already for a considerably long time. So, when a structure formation takes place there is no more, electric laser field.

Consequently, the question of "polarization memory" remains still open, so far. A possible explanation may involve the excitation of surface plasmons, with a sufficient lifetime (cf. Fig. 2.7), arranging according to the incident field, and then encouraging directional diffusion during the self-organized relaxation. This picture is, however, only very speculative, so far.

Acknowledgements

The author greatly appreciates the fruitful and congenial collaboration with Florenta Costache, Olga Varlamova, Markus Ratzke, and Michael Bestehorn. Profitable discussion and interaction within the Cottbus JointLab is greatly appreciated. This work was, in part, supported by a special grant from the Land of Brandenburg.

References

1. R. Srinivasan, V. Mayne-Banton, Appl. Phys. Lett. **41**, 576 (1982); R. Srinivasan, W.J. Leigh, J. Am. Chem. Soc. **104**, 6784 (1982)
2. See, e.g. the proceedings of the *Conferences on Laser Ablation (COLA)*:
J.C. Miller, R.F. Haglund (eds), *Laser Ablation: Mechanisms and Applications.* Lecture notes in Physics, vol 389 (Springer, Heidelberg, 1991)
J.C. Miller, D.B. Geohegan (eds), *Laser Ablation: Mechanisms and Applications – II*, AIP Conference Proceedings, vol 288 (American Institute of Physics, New York, 1994)
E. Fogarassy, D. B. Geohegan, M. Stuke (eds), in *Proceedings of Symposium F: Third International Symposium on Laser Ablation of the 1995 E-MRS Spring Conference*, Appl. Surf. Sci. **96–98**, 1 (1996)
R.E. Russo, D.B. Geohegan, R.F. Haglund Jr, K. Murakami (eds), in *Proceedings of the 4th Conference on Laser Ablation,* Appl. Surf. Sci. **127–129**, 1 (1998)
J.S. Horwitz, H.U. Krebs, K. Murakami, M. Stuke (eds), *Laser Ablation V,* Appl. Phys. A **69**, Supplement 1 (1999)
K. Murakami, A. Yabe, J.S. Horwitz, C. Fotakis, J.T. Dickinson (eds), in *Proceedings of the 6th Conference on Laser Ablation,* Appl. Surf. Sci. **197–198**, 1 (2002)
C. Fotakis, H. Koinuma, D. Lowndes, M. Stuke (eds), *Laser Ablation VII*, Appl. Phys A **79**, 713 (2004)
B. Luk'yanchuk, S. Juodkazis, T. Lippert (eds), *Special Issue: Laser Ablation Fundamentals*, Appl. Phys. A **92**, 743 (2008)
3. A. Kaiser, B. Rethfeld, M. Vicanek, G. Simon, Phys. Rev. B, **61**, 11437 (2000); L.V. Keldysh, Zh. Eksp. Teor. Fiz. **47**, 1945 (1964); [Sov. Phys.-JETP **20**, 1307 (1965)]
4. A. Zangwill, *Physics at Surfaces* (Cambridge University Press, 1988)
5. J. Reif, F. Costache, O. Varlamova, S. Eckert, in *Advanced Laser Technologies 2006,* ed. by D.C. Dumitras, M. Dinescu, V.I. Konov (SPIE-International Society for Optical Engineering, Bellingham, 2007)
6. J. Reif, F. Costache, in Advances in Atomic, Molecular and Optical Physics, 53, ed. by M.O. Scully, G. Rempe (Academic Press, 2006), pp. 228–249
7. M. Henyk, D. Wolfframm, J. Reif, Appl. Surf. Sci. **168**, 263–266 (2000)
8. E. Matthias, H.B. Nielsen, J. Reif, A. Rosén, E. Westin, J. Vac. Sci. Technol. **B5**, 1415 (1987); A. Rosén, E. Westin, E. Matthias, H.B. Nielsen, J. Reif, Phys. Scripta **T23**, 184 (1988)
9. B. Rethfeld, K. Sokolowski-Tinten, D. von der Linde, S.I. Anisimov, Phys. Rev. B **65**, 092103 (2002)
10. M. Henyk, J. Reif, Appl. Surf. Sci. **208–209**, 71 (2003)
11. F. Costache, S. Kouteva-Arguirova, J. Reif, Appl. Phys. A **79**, 1429 (2004)
12. B.C. Stuart, M.D. Feit, S. Herman, A.M. Rubenchik, B.W. Shore, M.D. Perry, Phys. Rev. B **53**, 1749 (1996); J. Bonse, S. Baudach, J. Krüger, W. Kautek, M. Lenzner, Appl. Phys. A **74**, 19 (2002)

13. M. Henyk, R. Mitzner, D. Wolfframm, J. Reif, Appl. Surf. Sci. **154–155**, 249 (2000)
14. M. Terasawa, Z.A. Insepov, T. Sekioka, A.A. Valuev, T. Mitamura, Nucl. Instr. Meth. B **212**, 436 (2003); H.P. Cheng, J.D. Gillaspy, Phys. Rev. B **55**, 2628 (1997)
15. W.G. Roeterdink, L.B.F. Juurlink, O.P.H. Vaughan, J. Dura Diez, M. Bonn, A.W. Kleyn, Appl. Phys. Lett. **82**, 4190 (2003)
16. J. Reif, Opt. Eng. **28**, 1122 (1989)
17. J. Reif, M. Henyk, D. Wolfframm, SPIE Proceeding Series 3933, 26 (2000)
18. R. Stoian, D. Ashkenasi, A. Rosenfeld, E.E.B. Campbell, Phys. Rev. B **62**, 13167 (2000); R. Stoian, A. Rosenfeld, D. Ashkenasi, I.V. Hertel, N.M. Bulgakova, E.E.B. Campbell, Phys. Rev. Lett. **88**, 097603 (2002)
19. R. Kelly, A. Miotello, Appl. Surf. Sci. **96**, 205 (1995)
20. N. Itoh, Adv. Phys. **31**, 491 (1982); N. Itoh, T. Nakayama, Phys. Lett. **92A**, 471 (1982); K. Tanimura, Phys. Rev. B **63**, 184303 (2001)
21. Y. Jee, K. Becker, R.M. Walser, J. Opt. Soc. Am. B **5**, 648 (1988)
22. S. Petzoldt, A.P. Elg, J. Reif, E. Matthias, SPIE Proceeding Series 1438, 180 (1989)
23. F. Costache, S. Eckert, J. Reif, Appl. Phys. A **92**, 897 (2008)
24. M. Henyk, D. Wolfframm, J. Reif, Nucl. Instr. Meth. B **166–167**, 716 (2000)
25. M. Henyk, F. Costache, J. Reif, Appl. Surf. Sci. **186**, 381 (2002)
26. J. Reif, F. Costache, S. Eckert, J. Phys.: Conf. Ser. **59**, 1 (2007)
27. V. Schmidt, W. Husinsky, G. Betz, Phys. Rev. Lett. **85**, 3516 (2000)
28. M. Birnbaum, J. Appl. Phys. **36**, 3688 (1965)
29. M. Sigrist, G. Kaech, F.K. Kneubühl, Appl. Phys. **2**, 45 (1973)
30. J.E. Sipe, J.F. Young, J.S. Preston, H.M. van Driel, Phys. Rev. B **27**, 1141 (1983); J.F. Young, J.S. Preston, H.M. van Driel, J.E. Sipe, Phys. Rev. B **27**, 1155 (1983); J.F. Young, J.E. Sipe, H.M. van Driel, Phys. Rev. B **30**, 2002 (1984)
31. J. Erlebacher, M.J. Aziz, E. Chason, M.B. Sinclair, J.A. Floro, Phys. Rev. Lett. **82**, 2330 (1999)
32. J. Reif, F. Costache, M. Bestehorn, Chap. 9 in *Recent Advances in Laser Processing of Materials,* ed. by J. Perrière, E. Millon, E. Fogarassy (E-MRS/Elsevier, Oxford, 2006)
33. T. Aste, U. Valbusa, New J. Phys. **7**, 122 (2005)
34. P. Sigmund, Phys. Rev. **184**, 383 (1969); P. Sigmund, J. Mater. Sci. **8**, 1545 (1973)
35. H.O. Jeschke, M.E. Garcia, M. Lenzner, J. Bonse, J. Krüger, W. Kautek, Appl. Surf. Sci. **197–198**, 839 (2002)
36. P. Lorazo, L.J. Lewis, M. Meunier, Phys. Rev. Lett. **91**, 225502 (2003)
37. A. Lindenberg, S. Engemann, K. Gaffney, K. Sokolowski-Tinten, J. Larsson, P. Hillyard, D. Reis, D. Fritz, J. Arthur, R. Akre, M. George, A. Deb, P. Bucksbaum, J. Hajdu, D. Meyer, M. Nicoul, C. Blome, Th. Tschentscher, A. Cavalieri, R. Falcone, S. Lee, R. Pahl, J. Rudati, P. Fuoss, A. Nelson, P. Krejcik, D. Siddons, P. Lorazo, J. Hastings, Phys. Rev. Lett. **100**, 135502 (2008)
38. Y. Kuramoto, *Chemical Oscillations, Waves, and Turbulence* (Springer, Berlin, 1984); J.W. Cahn, J.E. Hilliard, J. Chem. Phys. **28**, 258 (1958)
39. M. Ratzke, O. Varlamova, F. Costache, J. Reif, Mat. Sci. Eng. B **134**, 114 (2006)

40. R. Wagner, J. Gottmann, A. Horn, E.W. Kreutz, Appl. Surf. Sci. **252**, 8576 (2006)
41. N. Sanner, N. Huot, E. Audouard, C. Larat, J.-P. Huignard, Opt. Lasers Eng. **45**, 737 (2007)
42. O. Varlamova, F. Costache, M. Ratzke, J. Reif, Appl. Surf. Sci. **253**, 7932 (2007)

3

Atomic/Molecular-Level Simulations of Laser–Materials Interactions

Leonid V. Zhigilei, Zhibin Lin, Dmitriy S. Ivanov, Elodie Leveugle, William H. Duff, Derek Thomas, Carlos Sevilla, and Stephen J. Guy

Summary. Molecular/atomic-level computer modeling of laser–materials interactions is playing an increasingly important role in the investigation of complex and highly nonequilibrium processes involved in short-pulse laser processing and surface modification. This chapter provides an overview of recent progress in the development of computational methods for simulation of laser interactions with organic materials and metals. The capabilities, advantages, and limitations of the molecular dynamics simulation technique are discussed and illustrated by representative examples. The results obtained in the investigations of the laser-induced generation and accumulation of crystal defects, mechanisms of laser melting, photomechanical effects and spallation, as well as phase explosion and massive material removal from the target (ablation) are outlined and related to the irradiation conditions and properties of the target material. The implications of the computational predictions for practical applications, as well as for the theoretical description of the laser-induced processes are discussed.

3.1 Introduction

Short-pulse lasers are used in a diverse range of applications, from advanced materials processing, cutting, drilling, and surface micro- and nano-structuring [1,2] to pulsed-laser deposition of thin films and coatings [3], laser surgery [4,5], and artwork restoration [6,7], and to the exploration of the conditions for inertial confinement fusion, with the world's most energetic laser system being built at the National Ignition Facility at Lawrence Livermore National Laboratory [8]. At the fundamental science level, short-pulse laser irradiation has the ability to bring material into a highly nonequilibrium state and provides a unique opportunity to probe the material behavior under extreme conditions. In particular, optical pump-probe experiments have been used to investigate transient changes in the electronic structure of the irradiated surface with high (often subpicosecond) temporal resolution [9–13], whereas recent advances in time-resolved X-ray and electron diffraction techniques [14–22] provide an opportunity to directly probe the ultrafast atomic dynamics in laser-induced structural transformations. Further optimization of experimental parameters in current applications, the emergence of new techniques, and interpretation of the results of probing the transient atomic dynamics in materials and at surfaces can be facilitated by computational modeling of laser–materials interactions.

One of the main challenges in the computational description of short-pulse laser interactions with materials is presented by the complex multiscale character of the cascade of interrelated processes triggered by the laser excitation. These processes, schematically illustrated in Fig. 3.1, include laser excitation of optically active states in the target material (electronic or vibrational, linear or nonlinear/multiphoton), relaxation/thermalization of the absorbed laser energy (electron–phonon energy transfer, intramolecular and intermolecular vibrational equilibration, nonthermal atomic dynamics), active structural and phase transformations occurring in the high-temperature/high-pressure region of the laser energy deposition (melting/denaturation/charring, generation of crystal defects, fracture/spallation, explosive boiling, and surface vaporization), as well as long-term evolution of multicomponent ablation plume (evaporation/condensation of clusters, chemical and ionization reactions). Computer modeling of this diverse range of processes is challenging and requires a combination of different models/techniques.

The description of the effect of the electronic excitation on the material properties is typically performed by means of computationally expensive electronic structure calculations, e.g., [23–36]. Simulations based on electronic structure calculations provide information on the changes in the interatomic bonding and the ultrafast atomic dynamics induced by the electronic excitation. The size of the systems used in electronic structure calculations, however, is typically limited to several hundreds of atoms and does not allow for a realistic representation of the transition from the electronic excitation to the *collective atomic dynamics* responsible for the structural transformations in the irradiated material.

Continuum-level simulations, on the other hand, are often used to study the laser heating, melting, evaporation, and ablation on realistic, experimental time and length scales. The most straightforward and computationally efficient continuum approach is based on the solution of a set of partial differential equations describing the laser energy deposition and evolution of temperature in the irradiated target. Various descriptions of melting, resolidification, surface vaporization, and ablation can be incorporated into such models, albeit at a rather simplified level. In particular, laser melting and resolidification are often described with a phase-change model based on an assumption of local equilibrium at the solid–liquid interface (heat-flow limited, interface kinetics formulated within the framework of the Stephan problem), e.g., [37–39], or using a kinetic equation relating the interface velocity to the interface temperature, e.g., [40–44]. The latter nonequilibrium kinetic description has been shown to be necessary for subnanosecond pulses, when

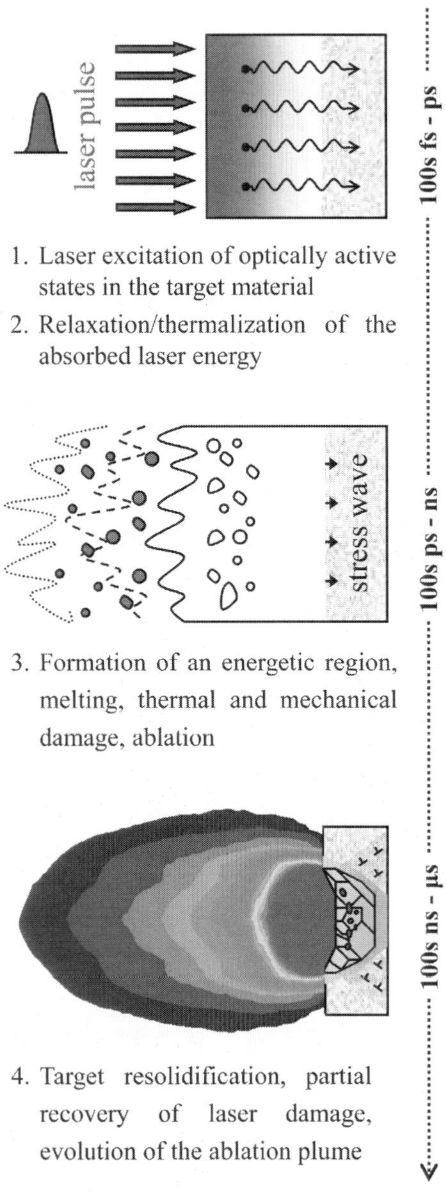

Fig. 3.1. Schematic representation of processes involved in short-pulse laser interactions with materials

a fast thermal energy flow to/from the liquid–solid interface creates conditions for significant overheating/undercooling of the interface [43, 44]. The material removal from the target can be incorporated into continuum models in the form of surface or volumetric vaporization models, e.g., [45–49], whereas the expansion of the vaporized plume is commonly described by solving gas dynamics equations, e.g., [45–49] or using the Direct Simulation Monte Carlo (DSMC) technique, e.g., [50–55]. Hydrodynamic computational models based on multiphase equations-of-state have also been used for simulation of laser melting, spallation, and ablation [56–62]. The empirical equations-of-state provide a powerful framework for the description of the evolution of the thermodynamic parameters of the material (temperature, pressure, density, internal energy), as well as the conditions for the laser-induced phase transitions.

The common strength of the continuum models is in their computational efficiency and the ability to simulate laser-induced processes at experimental time and length scales. The highly nonequilibrium nature of the processes induced in the target material by the fast laser energy deposition, however, is challenging some of the basic assumptions of the continuum descriptions that are commonly designed based on the equilibrium material behavior and properties. Although incorporation of kinetic models and metastable states, e.g., [43, 44, 60, 61], is possible within the continuum approach, the predictive power of the models is limited by the necessity to make a priori assumptions on the mechanisms and kinetics of all the processes that may take place during the simulation. The investigation of the generation of crystal defects and microscopic mechanisms of laser melting, nucleation and growth of voids in photomechanical spallation, the characteristics of the explosive volume ablation and the parameters of the ejected multicomponent and multiphase ablation plume is difficult if not impossible with continuum models.

In a situation where the continuum modeling of laser–materials interactions is hindered by the complexity and the highly nonequilibrium nature of the phenomenon, the classical molecular dynamics (MD) computer simulation technique has emerged as a promising alternative approach, which is capable of providing atomic-level insights into the laser-induced processes. A quickly expanding range of applications of MD simulations includes investigations of laser-induced thermoelastic deformation, melting, and resolidification [63–74]; photomechanical damage and spallation [65, 68, 70, 75–82]; as well as laser ablation of various material systems [70, 78, 79, 83–115].

In the remaining part of this chapter, the capabilities and limitations of MD simulations of laser–materials interactions are discussed and illustrated by the results obtained in several recent computational studies. The basic ideas of the classical MD method and some of the recent developments of computational methodology that enable simulations of laser interactions with molecular systems and metals are presented next, in Sect. 3.2. Some of the results obtained in MD simulations of laser-induced generation of crystal defects, melting, photomechanical spallation, and ablation are discussed in Sect. 3.3. Finally, in Sect. 3.4, some of the promising directions for future computational exploration are discussed.

3.2 Molecular Dynamics Method for Simulation of Laser–Materials Interactions

In this section, we start from a brief introduction to the classical MD method and highlight the advantages and limitations of the MD technique with respect to addressing research questions relevant to laser ablation and laser material-processing applications. The MD models developed for simulation of laser interaction with molecular systems and metals are presented next, followed by a discussion of the boundary conditions that can provide a realistic description of the energy dissipation from the absorption region to the bulk of the target.

3.2.1 Molecular Dynamics Method

Molecular dynamics (MD) is a computer simulation technique that allows one to predict the time evolution of a system of interacting particles (atoms, molecules, granules). A detailed discussion of this method and the areas of its applicability can be found in several books devoted to atomistic simulation techniques, e.g., [116, 117]. Briefly, MD allows one to follow the evolution of a system of N particles in time by solving a set of classical equations of motion for all particles in the system,

$$m_i \frac{\mathrm{d}^2 \mathbf{r}_i}{\mathrm{d}t^2} = \mathbf{F}_i, i = 1, 2, \ldots, \mathrm{N}, \tag{3.1}$$

where m_i and $\mathbf{r}_i$ are the mass and position of a particle i and $\mathbf{F}_i$ is the force acting on this particle due to the interaction with other particles in the system. The force acting on the ith particle at a given time can be obtained from the interparticle interaction potential $U(\mathbf{r}_1, \mathbf{r}_2, \mathbf{r}_3, \ldots, \mathbf{r}_N)$ that, in general, is a function of the positions of all the particles:

$$\mathbf{F}_i = -\boldsymbol{\nabla}_i U \left(\mathbf{r}_1, \mathbf{r}_2, \mathbf{r}_3, \ldots, \mathbf{r}_N\right). \tag{3.2}$$

Once the initial conditions (initial positions and velocities of all particles in the system) and the interaction potential are defined, the equations of motion, (3.1), can be solved numerically. The result of the solution is the trajectories (positions and velocities) of all the particles as a function of time, $\mathbf{r}_i(t), \mathbf{v}_i(t)$, which is the only direct output of an MD simulation. From the trajectories of all particles in the system, one can easily calculate the spatial and time evolution of structural and thermodynamic parameters of the system. For example, the atomic-level analysis of the development of the defect structures or phase transformations can be performed and related to the changes in temperature and pressure in the system.

The main strength of the MD method is that the only input in the model is the function describing the interparticle interaction, $U(\mathbf{r}_1, \mathbf{r}_2, \mathbf{r}_3, \ldots, \mathbf{r}_N)$, and no assumptions are made about the character of the processes under study. This is an important advantage that makes the MD method to be capable of discovering new physical phenomena or processes in the course of a "computer experiment." Moreover, unlike real experiments, the analysis of fast nonequilibrium processes in MD simulations can be performed with unlimited atomic-level resolution, providing complete information on the phenomena of interest. The predictive power of the MD method, however, comes at a price of a high computational cost, which imposes severe limitations on the time and length scales accessible for the simulation. The record length-scale MD simulations of systems containing more than 10^{11} atoms (micron-size cubic samples) have been performed with the use of thousands of processors on one of the world's largest supercomputers [118], whereas long time-scale (up to hundreds of microseconds) simulations of protein folding have been performed through distributed computing [119].

The limitations on the time- and length scales that are accessible for MD simulations present a serious challenge for the modeling of laser-induced processes that typically involve a collective motion of a large number of atoms or molecules in the surface region of the irradiated target. Moreover, since the electrons and quantum effects are not explicitly included in the classical MD, the optical properties of the irradiated material cannot be obtained in the course of the simulation, but have to be assumed in advance and provided as input to the model. Thus, the design of novel approaches aimed at extending the time- and/or length-scales of MD simulations and incorporating a description of the laser excitation into the MD model is required for an adequate modeling of laser–materials interactions. Two examples of computational models developed for MD simulations of laser interactions with molecular systems and metals are discussed next, in Sects. 3.2.2 and 3.2.3.

3.2.2 Coarse-Grained MD Model for Simulation of Laser Interactions with Molecular Systems

In an atomic-level MD model, a typical small molecule or a monomer unit can include tens of atoms, and a time-step of the integration of the equations of motion of 0.1 fs or smaller must be used to follow high-frequency atomic vibrations. In order to overcome the limitations of the atomistic MD model and to address collective processes responsible for laser-induced material modification or ablation, a coarse-grained "breathing sphere" MD model has been developed [98, 101].

The breathing sphere model assumes that each molecule can be represented by a single particle, Fig. 3.2a. The parameters of interparticle interaction are chosen to approximately reproduce the physical properties of a molecular target. The equilibrium distance in the interparticle potential is defined as the distance between the edges of the spherical particles rather than their centers, Fig. 3.2c. This choice of equilibrium distance is based on the physical

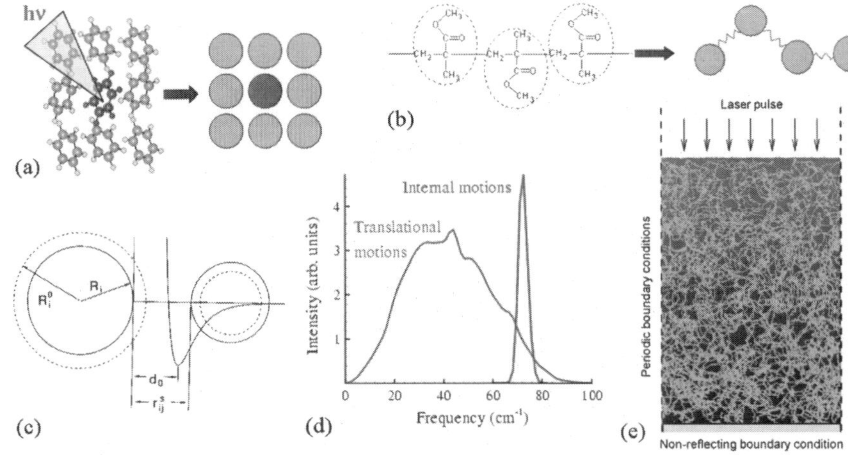

Fig. 3.2. Schematic representation of the approximations used in the design of coarse-grained breathing sphere (**a**) and bead-and-spring (**b**) MD models. Potential of intermolecular interaction in the breathing sphere model is shown in (**c**), where R_i^0 and R_i are the equilibrium and instantaneous radii of the particle i; d_0 and r_{ij}^{s} are the equilibrium and instantaneous distances between the edges of the spherical particles. The radii of the breathing spheres, R_{i}, are dynamic variables for which equations of motion are solved during the simulation. Vibrational spectrum of an organic solid represented by the breathing sphere model and the vibrational peak corresponding to the internal breathing mode are shown in (**d**). Schematic sketch of a simulation setup for modeling of laser ablation of a 3 wt.% polymer solution is shown in (**e**). The polymer chains are shown in light grey color and are superimposed on top of the image of matrix molecules shown in the background. Figures shown in (**c**) and (**d**) are from [98] and the image in (**e**) is from [113]

concept that the sublimation or cohesive energy of an organic solid is governed primarily by the interactions among atoms on the outside of the molecule. This representation of intermolecular interactions allows an easy means of simulating multicomponent molecular systems [98, 102, 105, 106].

In order to simulate molecular excitation by photon absorption and vibrational relaxation of the excited molecules, an additional internal degree of freedom is attributed to each molecule. The internal degree of freedom, or breathing mode, is implemented by allowing the particles to change their sizes. In the case of UV laser irradiation, the breathing mode can be considered as the recipient of the energy released by an internal conversion from electronically excited states. The parameters of a potential function attributed to the internal motion control the characteristic frequency of the breathing mode, Fig. 3.2d; and, as a result, define the rate of the conversion of the internal energy of the molecules excited by the laser to the translational and internal motions of the surrounding molecules. The rate of the vibrational relaxation of excited molecules is an input parameter in the model and can be either estimated from pump-probe experiments [120, 121] or obtained in atomistic [122] or *ab initio* [123] MD simulations.

The laser irradiation is simulated by vibrational excitation of molecules that are randomly chosen during the laser pulse duration within the penetration depth appropriate for a given wavelength. Vibrational excitation is modeled by depositing a quantum of energy equal to the photon energy into the kinetic energy of internal motion of a given molecule. An alternative result of the photon absorption, photofragmentation of the excited molecule into fragments that can subsequently participate in chemical reactions, can also be reproduced within the model [78, 106, 111]. A description of the processes leading to the ionization of molecules in laser ablation has recently been incorporated into the breathing sphere model and the mechanisms responsible for the ion formation in matrix-assisted laser desorption ionization (MALDI) mass spectrometry technique have been explored [112].

In order to enable simulations of laser interaction with polymer solutions [107, 113, 114, 124–126], the breathing sphere model has recently been combined with the bead-and-spring model, commonly used in polymer modeling [127]. In the bead-and-spring model, schematically illustrated in Fig. 3.2b, the "beads" representing the functional groups of a polymer molecule (monomers) are connected by anharmonic springs with strengths appropriate for chemical bonding. An example of the computational setup used in simulations of laser ablation of polymer solutions [113, 124, 125] is given in Fig. 3.2e.

Since both the breathing sphere model and the bead-and-spring model adopt a coarse-grained representation of molecules, in which each molecule or monomer unit is represented by a single particle, the system size can be sufficiently large to reproduce the collective dynamics in a molecular system leading to laser ablation or damage. Moreover, since explicit atomic vibrations are not followed, the time-step in the numerical integration of the equations of motion can be much longer and the dynamics in the irradiated sample can be followed for as long as nanoseconds. The limitations of the breathing sphere model are related to the approximation of all the internal degrees of freedom of a molecule by one internal mode. The rates of intermolecular energy transfer cannot be studied within the model, but have to be specified through the input parameters, as discussed earlier. The accuracy in quantitative description of the thermodynamic and transport properties of the materials represented at the coarse-grained level is limited, and the model is appropriate for investigation of general, rather than material-specific, characteristics of the laser-induced processes. A smaller number of degrees of freedom in the model system should also be taken into account when performing a quantitative comparison with experimental data, e.g., of the threshold fluence for the ablation onset [79, 103].

3.2.3 Combined Continuum-Atomistic Model for Simulation of Laser Interactions with Metals

In metals, laser light is absorbed by the conduction band electrons. The deposited energy quickly, within femtoseconds, is equilibrated among the electrons and, more slowly, is transferred to the lattice vibrations. The later process is controlled by the strength of the electron–phonon coupling and can take from fractions of a picosecond to several tens of picoseconds. Finally, a thermal equilibrium is established between the electrons and phonons, and the conventional heat conduction equation can be used to describe the heat flow into the bulk of the irradiated target. The classical MD technique does not include an explicit representation of electrons and, therefore, cannot be used, in its conventional formulation, for simulation of the laser light interaction with the target material, the relaxation/thermalization of the absorbed laser energy, and the fast electron heat conduction to the bulk of the irradiated target.

To enable atomic-level simulations of processes involving electronic excitations of metal targets by short-pulse laser irradiation (or energetic ion bombardment), several computational approaches have been proposed [64, 65, 89, 108, 128, 129]. In particular, the model described in [65] combines classical MD method with a continuum description of the laser excitation and subsequent relaxation of the conduction band electrons, based on the so-called two-temperature model (TTM) [130]. In the original TTM, the time evolution of the lattice and electron temperatures, T_l and T_e, is described by two coupled nonlinear differential equations. In the combined TTM–MD method, schematically illustrated in Fig. 3.3, MD partially substitutes the TTM equation for the lattice temperature. The diffusion equation for the electron temperature is solved by a finite difference method simultaneously with MD integration of the equations of motion of atoms. The electron temperature enters a coupling term that is added to the MD equations of motion to account for the energy exchange between the electrons and the lattice. The MD method is used only in the very surface region of the target, where active processes of laser melting, resolidification, and/or ablation take place, whereas the diffusion equation for the electron temperature is solved in a much wider region affected by the thermal conduction. A special pressure-transmitting boundary condition applied at the bottom of the MD part of the computational region, as well as the periodic boundary conditions imposed in the directions parallel to the surface, is briefly discussed later, in Sect. 3.2.4. In the part of the computational cell beyond the MD region (left part in Fig. 3.3), the energy exchange between the electrons and the lattice is described by the conventional TTM.

The hybrid continuum-atomistic model, briefly described above, combines the advantages of TTM and MD methods. TTM provides an adequate description of the laser energy deposition into the electronic system, the energy exchange between the electrons and phonons, and the fast electron heat

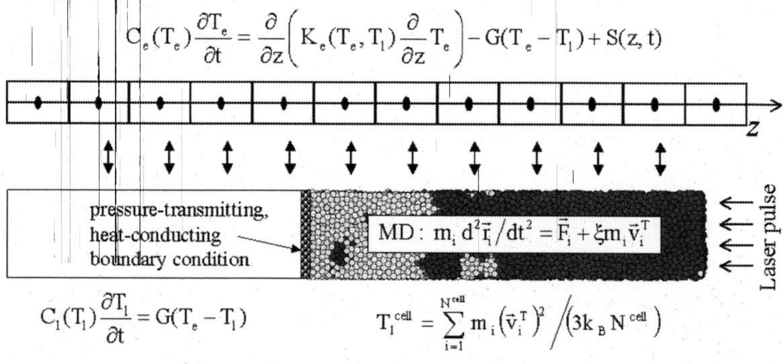

Fig. 3.3. Schematic representation of the combined continuum-atomistic model for simulation of laser interaction with a metal target. The evolution of electron temperature, T_e, is described by a nonlinear differential equation, whereas the atomic motions are described by the MD method with an additional term, $\xi m_i \mathbf{v}_i^T$, added to the ordinary MD equations of motion to account for the electron–phonon coupling. Spatial discretization in the continuum model (typically ~1 nm) and size of the atomistic region are not drawn to scale. The cells in the finite difference discretization are related to the corresponding volumes of the MD system and the local lattice temperature, T_l^{cell}, is defined for each cell from the average kinetic energy of thermal motion of atoms. Thermal velocity of an atom is defined as $\mathbf{v}_i^T = \mathbf{v}_i - \mathbf{v}^c$, where $\mathbf{v}_i$ is the actual velocity of an atom i and $\mathbf{v}^c$ is the velocity of the center of mass of a cell to which the atom i belongs. A Gaussian temporal profile, $S(z,t)$, is used to describe the laser excitation of the conduction band electrons. The expansion, density variations and, at higher fluences, disintegration of the irradiated target predicted in the MD part of the model are accounted for in the continuum part of the model. A complete description of the combined TTM–MD model is given in [65]

conduction in metals, whereas the MD method is appropriate for simulation of rapid nonequilibrium phase transformations, damage, and ablation.

The results of the recent investigation of the electron temperature dependence of the electron–phonon coupling factor G, the electron heat capacity C_e, and the heat conductivity K_e (thermophysical material properties included in the TTM equation for the electron temperature, see Fig. 3.3) suggest that the effect of the thermal excitation from the electron states below the Fermi level should be accounted for in a model aimed at a quantitative description of the laser-induced processes in metals [131–134]. Indeed, a computational analysis based on the first-principles electronic structure calculations of the electron density of states reveals that these thermophysical materials properties are very sensitive to details of the electronic structure of the material and can exhibit large deviations (up to an order of magnitude) from the commonly used approximations of a linear temperature dependence of the electron heat capacity and a constant electron–phonon coupling. A number of practically important characteristics of the laser–material interactions, such as the threshold fluences for the onset of melting and ablation, the strength of the laser-induced stress wave, the emission of electrons from the irradiated surface, and the depth of the melting and/or heat-affected zone, can all be significantly altered by the transient changes of the thermophysical properties occurring during the time of electron–phonon equilibration. It has been shown, in particular, that incorporation of the new electron temperature dependences of the thermophysical properties [133, 134] into TTM or TTM–MD models results in an improved agreement between the computational predictions and experimental observations [20, 131, 132, 135].

In the examples considered in Sect. 3.3 of this chapter, the interatomic interactions in the MD part of the TTM–MD model are described by the embedded atom method (EAM) potential [136, 137] that provides a computationally simple but rather realistic description of bonding in metallic systems. In particular, the functional form and parameters of the EAM potential for Ni, Al, Cu, and Au are given in [138], whereas a recently developed potential for Cr is described in [74].

3.2.4 Boundary Conditions: Pressure Waves and Heat Conduction

The severe limitations on the length scales in MD method make it impossible to directly simulate processes occurring within the whole laser spot. For a laser spot of 10–100 μm in diameter and an ablation depth of 10–100 nm, one can estimate that the number of molecules/atoms ejected from an irradiated target in a single laser shot is in the range from tens of billions to trillions. These numbers are much beyond the limits of the MD simulation technique (see Sect. 3.2.1). In this situation, the MD computational cell is typically assumed to represent a local volume within the laser spot and the material response to local laser energy deposition is investigated, as schematically shown in Fig. 3.4. The periodic boundary conditions in the lateral directions, parallel to the surface of the target, are used in this case to reproduce the interaction of molecules or atoms in the MD computational cell with the surrounding material. This approach is appropriate for a situation in which the laser spot diameter is much larger than the depth of the laser energy deposition, so that any effects related to the lateral variations of the irradiation and thermal conditions can be neglected and the simulated part of the system remains laterally confined by the surrounding material during the time of the simulation. The information on the material ejection from the whole laser spot can then be obtained by integrating over the results of a series of MD simulations performed for a range of "local fluences," Fig. 3.4.

In the direction normal to the surface of the irradiated target, a free boundary condition, allowing for a natural expansion of the irradiated target and the ejection of atoms, molecules, and clusters in laser ablation, is the natural choice for the irradiated (top) surface. More complex boundary conditions, however, accounting for the thermal conduction and pressure wave propagation from the absorption region deeper into the bulk of the target, have to be used at the bottom of the MD computational cell.

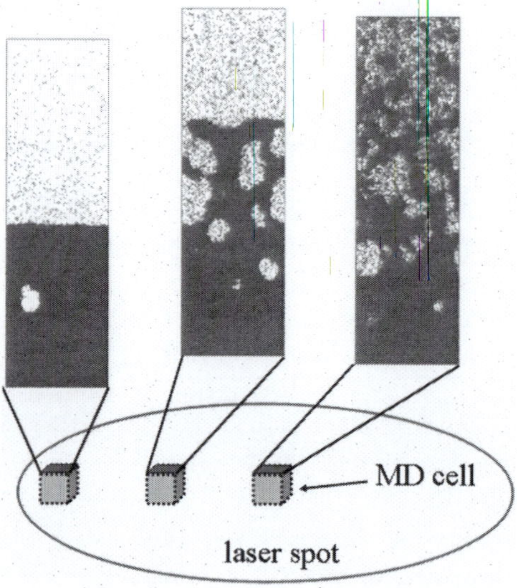

Fig. 3.4. Schematic illustration of the local areas represented in MD simulations of laser ablation at different locations within a laser spot. Snapshots used in this figure are from simulations of laser interactions with a molecular target [78]

In order to evaluate the necessity for the introduction of the heat-conductive and pressure-transmitting boundary conditions, one can consider characteristic times of the heat transfer and pressure wave propagation across a typical size of the MD computational cell, $L_{\mathrm{MD}} \approx 100\,\mathrm{nm}$. The timescale of the heat conduction across the computational cell can be evaluated as $\tau_{\mathrm{th}} \approx \left(L_{\mathrm{MD}}^2\right)/\left(2D_{\mathrm{T}}\right)$, where D_{T} is the thermal diffusivity of the target material. For molecular systems $D_{\mathrm{T}} \approx 10^{-7}\,\mathrm{m^2\,s^{-1}}$, and the timescale of the heat conduction is $\tau_{\mathrm{th}} \approx 50\,\mathrm{ns}$, much longer than the time-scale of a typical MD simulation. In metals, however, the heat conduction, dominated by the electron heat transport, is much larger, e.g., $D_{\mathrm{T}} \approx 10^{-4}\,\mathrm{m^2\,s^{-1}}$ for gold. This yields $\tau_{\mathrm{th}} \approx 50\,\mathrm{ps}$, a time that is shorter than the time needed for an adequate simulation of laser-induced structural transformations. Therefore, we can conclude that, while the effect of heat transfer through the bottom of a sufficiently large computational cell can be neglected in simulations of molecular systems, the boundary conditions in simulations of laser interactions with metals must account for the heat conduction. The combined TTM–MD model, discussed in Sect. 3.2.3, provides a natural description of the electron heat conduction from the surface region of the target, represented with atomic-level resolution, to the deeper part of the target, represented at the continuum level, Fig. 3.3. Indeed, a seamless transition in the temperature field from the atomistic to the continuum regions can be seen in Fig. 3.9 (Sect. 3.3.3), illustrating the evolution of temperature in a simulation performed for a bulk Ni target irradiated with a 1 ps laser pulse.

The pressure waves, generated as a result of the relaxation of laser-induced thermoelastic stresses and, above the threshold for the ablation onset, recoil pressure from the ejected material, present an additional challenge for simulations of short-pulse laser–materials interactions. In order to simulate a propagation of the laser-induced pressure wave into the bulk of the sample, the size of the MD computational cell should be increased linearly with the time of the simulation. For times longer than a hundred of picoseconds, the size of the model required to follow the wave propagation becomes computationally prohibitive. If large computational cells are not used, however, artificial border effects can interfere with the simulation results, as both rigid and free boundary conditions lead to the complete reflection of the pressure wave [78, 82]. The free boundary condition at the bottom of the computational cell is appropriate for simulations of laser interaction with free-standing films [65–68, 70–72, 75, 76, 82, 91], whereas the rigid boundary condition can be related to experiments performed for a thin absorbing layer deposited on a hard substrate [139]. In most cases, however, we are interested in much larger systems for which the effect of the pressure wave reflection has to be avoided. To enable the simulations of laser interactions with bulk systems, special pressure-transmitting boundary condition based on an analytical evaluation of the forces acting on atoms/molecules in the boundary region from the outer "infinite medium" has been developed [140, 141]. The energy that is carried away by the stress wave though the pressure-transmitting boundary condition can be monitored, allowing for a control over the energy conservation in the model [69]. The nonreflecting boundary conditions have been successfully used in simulations of laser melting, ablation, and damage for different target materials in which both planar, e.g., [69, 78–82, 89, 92, 93, 103–113] and spherical [142] pressure waves are generated. An illustration of the nonreflective propagation of the pressure wave from the atomistic to the continuum parts of the combined TTM–MD model can be seen in Fig. 3.9 (Sect. 3.3.3).

3.3 Simulations of Laser-Induced Structural and Phase Transformations

The MD method allows one to perform a detailed analysis of the laser-induced processes in which thermodynamic parameters of the system can be correlated with microscopic dynamics at the atomic level. In this section, the ability of the MD method to provide insights into the mechanisms of laser–materials interactions is demonstrated by a representative set of recent computational results obtained in simulations of laser-induced generation of crystal defects, melting, photomechanical spallation, and ablation.

3.3.1 Generation of Crystal Defects

The understanding of the mechanisms and driving forces responsible for laser-induced generation of crystal defects is important for the advancement of laser processing applications aimed at controlled modification of surface microstructure. MD simulations are capable of providing detailed atomic-level information on the elementary processes responsible for the generation and evolution of defect configurations in irradiated targets. To illustrate this capability and highlight the sensitivity of the laser-induced defect structures to the type of the crystal structure of the target, the results of simulations performed for two metals with different crystal structures, body-centered cubic (bcc) Cr and face-centered cubic (fcc) Ni, are discussed in this section.

The fast structural changes in a Cr target irradiated with a 200-fs laser pulse have been analyzed in [74] based on the results of TTM–MD simulations. The snapshots of atomic configurations taken at different times of a simulation performed at an absorbed fluence of $638\,\mathrm{J\,m^{-2}}$ (just above the threshold for surface melting) are shown in Fig. 3.5a. Only atoms that belong to the liquid phase or are located in the vicinity of crystal defects, are shown in the snapshots, with all the atoms that have local atomic surroundings (and corresponding values of the potential energy) similar to the ones in the original bcc structure blanked. During the first 100 ps after the laser pulse, the irradiated target experiences transient melting and epitaxial resolidification of a thin (up to 3 nm) surface layer, which shows up in Fig. 3.5a as a layer of red atoms at 50 ps and reduces to a plane composed of atoms located at the surface of the recrystallized target by the time of 100 ps.

Another transient effect apparent from the snapshots shown in Fig. 3.5a is the appearance, expansion (up to 30 ps), retraction, and disappearance (by 115 ps) of a complex pattern of atomic planes with elevated energy. Detailed analysis of the atomic configurations reveals that these planes correspond to the intrinsic stacking faults generated as a result of multiple internal shifts along $\{110\}$ crystallographic planes by displacement vectors $a/8 < 110 >$ (where a is the lattice parameter). The generation of the stacking faults is activated by the rapid uniaxial expansion of the crystal in the direction normal to the irradiated surface. Calculations of the generalized stacking fault energy suggest, in agreement with earlier studies [143], that the intrinsic stacking faults are unstable in an unstrained bcc crystal but can be stabilized by a uniaxial expansion of the crystal. Indeed, the appearance of the stacking faults correlates with the lattice expansion associated with the initial relaxation of the laser-induced stresses. All stacking faults disappear by ~115 ps, shortly after the laser-induced tensile stress wave leaves the surface region of the target [74].

The disappearance of the stacking faults makes the presence of a large number of vacancies clearly visible in the surface region of the target, e.g., snapshot shown for 450 ps in Fig. 3.5a. With the visualization method used in Fig. 3.5a, where only atoms with elevated potential energy are shown, each

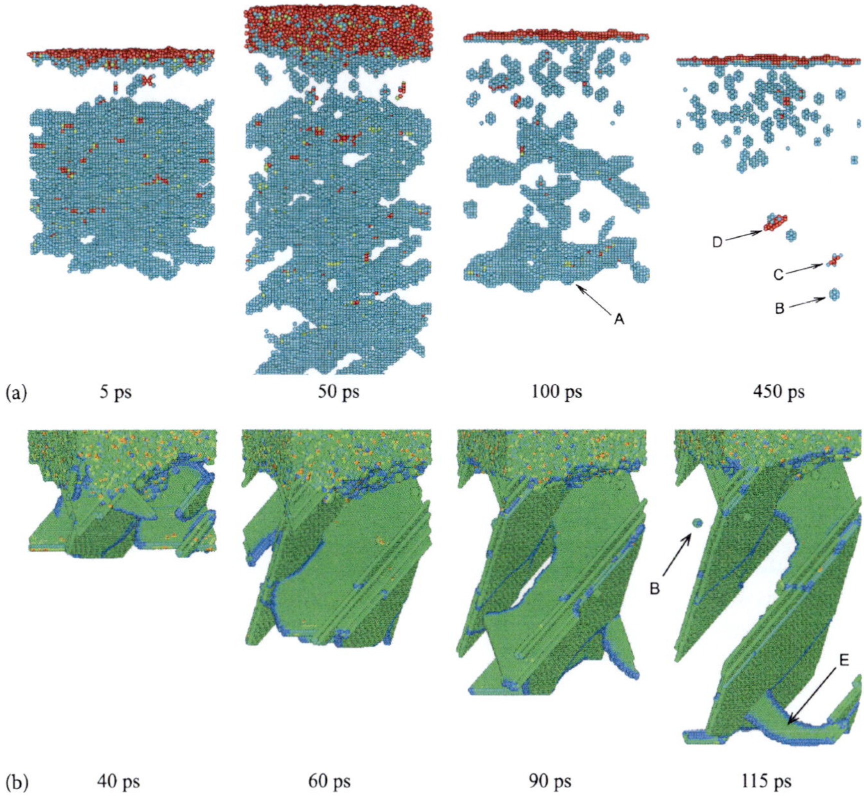

Fig. 3.5. Snapshots of the surface regions of atomic configurations obtained in TTM–MD simulations of bcc Cr (**a**) and fcc Ni (**b**) targets irradiated with a short pulse laser. The absorbed laser fluences and pulse durations are 638 J m^{-2} and 200 fs for Cr, and 645 J m^{-2} and 1 ps for Ni targets. The snapshots are shown down to the depth of 20 nm below the level of the initial surface in (**a**) and for a region located between 30 and 60 nm below the level of the initial surface in (**b**). The atoms are colored according to their potential energies in (**a**) and the centrosymmetry parameter in (**b**), with atoms that belong to local configurations corresponding to the original bcc (**a**) or fcc (**b**) structure blanked to expose crystal defects. Typical defect configurations marked in the snapshots are "A" – stacking fault with a displacement vector of $a/8$<110>, "B" – a vacancy, "C" – an interstitial in a <110>-dumbbell configuration, "D" – a four <111>-crowdion interstitial cluster, and "E" – a dislocation with a Burgers vector of $a/2$<110>, dissociated into two $a/6$<112> Shockley partial dislocations connected by a stacking fault ribbon. The snapshots shown in (**a**) are from [74]

vacancy appears as a cluster of 14 atoms that includes the eight nearest neighbors and six second-nearest neighbors of the missing atom. The number of vacancies observed in the top 5 nm surface region of the target at 450 ps corresponds to a very high vacancy concentration, more than 10^{-3} vacancies per lattice site. The thermally activated generation of vacancy-interstitial pairs during the laser-induced temperature spike serves as the initial source of the point defects. Due to the high mobility of self-interstitials, they quickly escape to the melting front or the free surface of the target, leaving behind a large number of vacancies (only one individual interstitial and one cluster of four interstitials arranged in a mobile <111>-crowdion configuration can be identified in a snapshot shown for 450 ps in Fig. 3.5a). A significant number of vacancies are also produced at the advancing solid–liquid interface during the fast resolidification process.

The strong temperature gradient created in the surface region of the target by the short-pulse laser irradiation, and the associated ultrafast cooling rates exceeding $5 \times 10^{12}\,\mathrm{K\,s^{-1}}$ at the time of resolidification, provide the conditions for stabilization of the highly nonequilibrium vacancy concentration. Indeed, an analysis of the long-term evolution of the vacancy configuration, performed in [74], suggests that the average vacancy diffusion length during tens of nanoseconds after the end of the TTM–MD simulation is very small, on the order of an interatomic distance. The configuration of mostly individual vacancies observed at the end of the TTM–MD simulation is, therefore, unlikely to undergo any significant changes during the remaining part of the cooling process.

The processes responsible for the generation of crystal defects in fcc Ni target exhibit both similarities and differences with the ones discussed above for bcc Cr target. The formation of vacancy-interstitial pairs followed by the fast escape of the interstitials is observed in both Ni and Cr targets and proceeds in a qualitatively similar manner. An important difference between the simulations performed for the two materials is a massive generation of partial dislocations observed for Ni targets, e.g., Fig. 3.5b. This observation can be related to the existence of stable low-energy stacking faults and 12 close-packed {111} <1$\bar{1}$0> slip systems with small resistance to the motion of dislocations (low Peierls stress) in fcc crystals. Unlike the transient appearance of the unstable stacking faults in Cr, the stacking faults left behind by the partial dislocations propagating from the melting front in the Ni target are stable and have relatively low energy ($110\,\mathrm{mJ\,m^{-2}}$ is predicted by the EAM Ni potential, in a reasonable agreement with the experimental value of $125\,\mathrm{mJ\,m^{-2}}$ [144]). Interactions between the dislocations propagating along the different slip planes result in the formation of immobile dislocation segments (the so-called stair-rod dislocations) that, together with the fast cooling of the surface region of the target, stabilize the dislocation configuration generated during the initial spike of temperature and thermoelastic stresses.

The supersaturation of the surface region of an irradiated target with vacancies, observed for both Ni and Cr targets, may result in the formation of nanovoids and degradation of the mechanical properties of the surface region of the target in the multipulse irradiation regime. The generation of crystal defects may be, thus, related to the incubation effect, when the laser fluence threshold for ablation/damage decreases significantly with increasing number of laser pulses applied to the same area, e.g., [145–149]. The high density of vacancies generated in the surface region should also play an important role in the redistribution of impurities or mixing/alloying in multicomponent or composite targets. The generation of dislocations and, in particular, dislocation reactions leading to the formation of immobile dislocation configurations should result in hardening of the surface region of the target.

3.3.2 Mechanisms and Kinetics of Laser Melting

Most of the methods of laser surface modification involve melting and subsequent resolidification of a surface region. It has been well established that melting starts at surfaces and internal crystal defects under minor superheating conditions or even below the equilibrium melting temperature [150, 151]. After heterogeneous nucleation of the liquid phase, the liquid–solid interface propagates into the bulk of the solid, precluding any significant superheating and making observation of an alternative mode of melting, homogeneous nucleation in the bulk of a superheated crystal, difficult. The extremely high heating rates achievable with short-pulse laser irradiation, however, create the conditions for competition between the heterogeneous and homogeneous melting mechanisms and provide unique opportunities for the investigation of the kinetic limits of achievable superheating. Moreover, the emerging time-resolved electron and X-ray diffraction experimental techniques are capable of probing the transient atomic dynamics in laser melting with subpicosecond resolution [14–22]. The complexity of the fast nonequilibrium phase transformation, however, hinders the direct translation of the diffraction profiles to the transient atomic structures.

MD simulations are well suited for investigation of the ultrafast laser melting phenomenon and are capable of providing detailed atomic-level information needed for a reliable interpretation of experimental observations. In particular, the kinetics and mechanisms of laser melting have been investigated in a series of TTM–MD simulations performed for Ni, Au, and Al thin films and bulk targets irradiated by short, from 200 fs to 150 ps, laser pulses [65–72, 109, 132]. The relative contributions of the homogeneous and heterogeneous melting mechanisms have been analyzed and related to the irradiation conditions. Except for the fluences close to the threshold for surface melting, the heterogeneous melting (melting front propagation from the surface) is found to make very limited contribution to the overall melting process, with homogeneous nucleation of multiple liquid regions being the dominant melting mechanism [65, 66, 71]. This observation has been supported by the results of recent large-scale TTM–MD simulations aimed at establishing the maximum velocity of the melting front propagation in metals [152]. A surprising result from this study is that the maximum velocity of the melting front just below the limit of the crystal stability against homogeneous melting is below 3% of the speed of sound, more than an order of magnitude lower than commonly assumed in interpretation of the results of laser melting experiments, e.g., [10, 14, 153]. The relatively low maximum velocity of the melting front, revealed in the simulations, has direct implications for interpretation of the experimental data on the kinetics of melting. For example, for thin 20-nm Au films used in recent time-resolved electron diffraction experiments [20,154], the melting time shorter than 70 ps would clearly point to the major contribution of the homogeneous nucleation to the melting process [71, 132].

A schematic map of the melting mechanisms shown in Fig. 3.6 can provide guidance in the analysis of the relative contributions of different processes to laser melting. The heterogeneous melting starts from the free surface(s) of the target as soon as the temperature exceeds the equilibrium melting temperature, T_{m}. The equilibrium melting temperature is changing with pressure according to the Clapeyron equation (increases with increasing pressure for metals having positive volume change on melting). As discussed earlier, the melting front propagation is relatively slow and the surface region of the irradiated target can be easily overheated significantly above the equilibrium melting temperature, up to the limit of superheating shown by the dashed line in Fig. 3.6. The temperature of the maximum superheating, T_{s}, is defined as a temperature at which melting starts within tens of picoseconds in a simulation performed for a perfect crystal with three-dimensional periodic boundary conditions (no external surfaces) under conditions of constant hydrostatic pressure. The values of the maximum superheating, $(T_{\mathrm{s}}\text{–}T_{\mathrm{m}})/T_{\mathrm{m}}$, predicted in MD simulations for different close-packed metals vary from 0.19 to 0.30 [155] and are somewhat smaller, below 0.15, for bcc metals [74, 156]. In the case of EAM Ni used in Fig. 3.6, the maximum superheating gradually increases from 0.21 to 0.25 as pressure increases from –5 GPa to 10 GPa.

In the area of the pressure–temperature field above the limit of superheating (red area in Fig. 3.6), rapid nucleation and growth of liquid regions inside the superheated crystal are responsible for the melting process. Note that the homogeneous melting observed above the maximum superheating does not follow the classical picture of a homogeneous phase transition – the nucleation and growth of well-defined spherical liquid regions. Rather, the melting in this regime proceeds as a collapse of the lattice superheated above the limit of its stability and takes place within just several picoseconds (several periods of atomic vibrations). Actually, the "classical" homogeneous melting has never been observed in laser melting simulations performed so far and the image showing two compact liquid regions in Fig. 3.6 is taken from a simulation of a slow heating of a crystal under well-controlled temperature and pressure conditions. Indeed, one can expect that the fast evolution of the temperature and pressure induced by short-pulse laser irradiation would readily overshoot the narrow region close to the limit of superheating (shown by green color in Fig. 3.6) where the "classical" homogeneous melting may be expected. Moreover, the temperature of the onset of homogeneous melting (the limit of superheating) can be significantly reduced by anisotropic lattice distortions associated with the relaxation of the laser-induced thermoelastic stresses [66]. Above the limit of superheating, the melting happens so fast that there is no

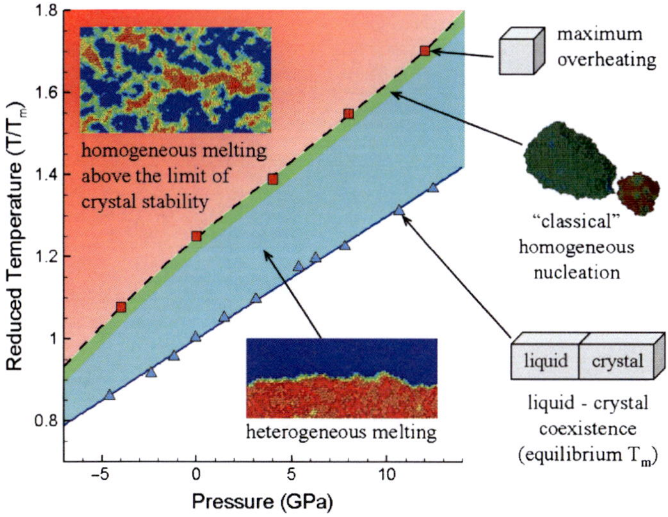

Fig. 3.6. Pressure/temperature conditions for equilibrium and nonequilibrium melting observed in simulations of laser interactions with metal targets. Blue triangles correspond to the conditions of equilibrium melting obtained in liquid-crystal coexistence simulations. Red squares connected by the black dashed line correspond to the maximum overheating of a crystal observed in simulations performed with three-dimensional periodic boundary conditions and constant hydrostatic pressure. The areas of the pressure–temperature field corresponding to the ultrafast homogeneous melting above the limit of superheating, classical homogeneous melting by nucleation and growth of individual liquid regions, and heterogeneous melting by the melting front propagation from the surface are shown by red, green, and blue colors, respectively. The data points are calculated for the EAM Ni material

time for the system to minimize the interfacial energy for the rapidly evolving liquid regions.

A typical picture of the homogeneous melting above the limit of superheating is shown in Fig. 3.7, where the snapshots from a simulation of laser melting of a 20 nm Au film are shown along with the corresponding structure functions. The fluence used in this simulation is ~75% above the fluence needed for the complete melting of a 20 nm Au film [71]. The small thickness of the film and the fast electron energy transport in Au [65,71,132] result in the even distribution of the electron temperature established shortly after the laser excitation. The electron–phonon energy transfer then leads to the increase of the lattice temperature. The lattice temperature exceeds the equilibrium melting temperature by more than 40% by the time of 6 ps, triggering a spontaneous homogeneous nucleation of a large number of small liquid regions throughout the film and a rapid collapse of the crystalline structure within the subsequent 3–4 ps ($T_s \approx 1.25T_m$ for the EAM Au). The visual analysis of the snapshots taken during the melting process shows that by ~6 ps the growth of liquid regions starts at two free surfaces of the film, where the kinetic energy barrier

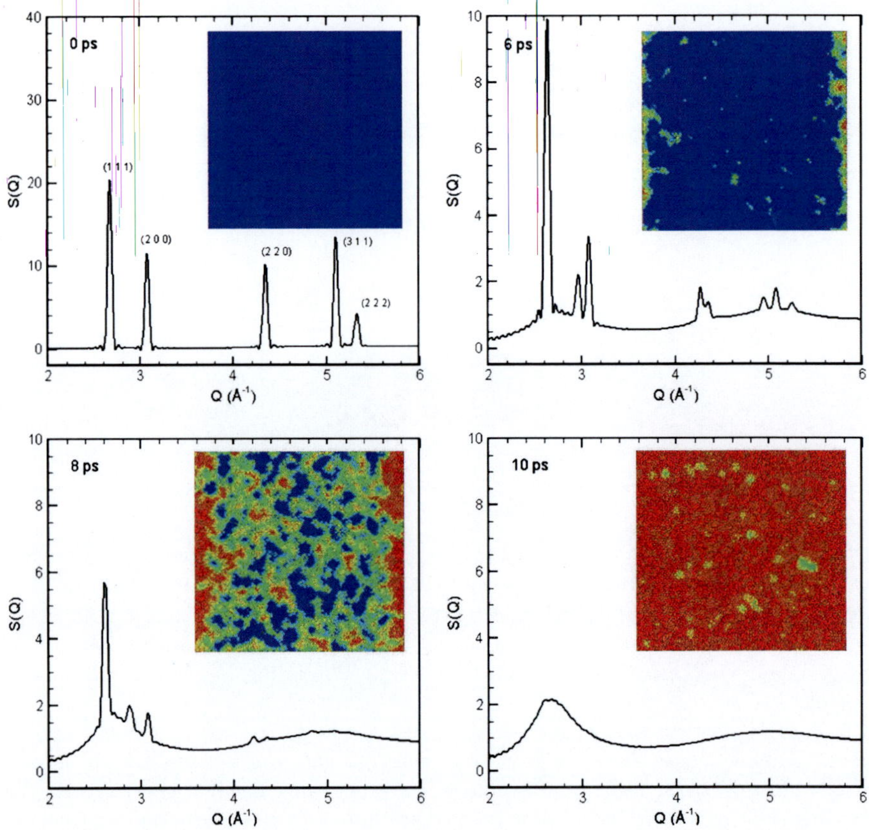

Fig. 3.7. Structure functions calculated for atomic configurations generated in a TTM–MD simulation of laser melting of a 20-nm Au film, irradiated with a 200-fs laser pulse at an absorbed fluence of $92.5\,\mathrm{J\,m^{-2}}$. The corresponding snapshots of atomic configurations are shown as insets in the plots, with the laser pulse directed from the right to the left sides of the snapshots. Atoms in the snapshots are colored according to the local order parameter [65] – blue atoms have local crystalline surroundings and red atoms belong to the liquid phase. Zero time corresponds to a perfect fcc crystal at 300 K just before the laser irradiation. The effect of the thermal excitation of *d*-band electrons on the parameters of the TTM equation for the electron temperature [133] is included in this simulation. The snapshots are from Ref. [132]

is absent for the nucleation of the liquid phase. However, due to the fast rate of the lattice heating, the propagation of the melting fronts from the free surfaces of the film does not make any significant contribution to the overall melting process.

The calculation of the diffraction profiles and density correlation functions [71,72] provides a direct connection between the results of MD simulations and time-resolved diffraction experiments. The increasing amplitude of thermal atomic vibrations (Debye–Waller factor), as well as shifts and splittings of the diffraction peaks due to the thermoelastic deformation of the film prior to melting, is found to be responsible for the initial decrease of the intensity of the diffraction peaks (from 0 to 6 ps in Fig. 3.7). The onset of the melting process at ~6 ps leads to the complete disappearance of the crystalline diffraction peaks by the time of 10 ps, Fig. 3.7.

The simulation illustrated in Fig. 3.7 is performed for the irradiation conditions similar to the ones used in recent time-resolved electron diffraction measurements performed for 20 nm Au films [20, 154]. The disappearance of the diffraction peaks corresponding to the crystal structure is found to take place between 7 ps and 10 ps after the laser pulse. This experimental observation is in an excellent agreement with the simulation results illustrated in Fig. 3.7. Note that this agreement has only been achieved by accounting for the effect of the thermal excitation of *d* band electrons on the electron temperature dependence of the electron heat capacity and electron–phonon coupling [131–134] in the TTM–MD model. Earlier simulations, performed with the commonly used approximations of the constant electron–phonon coupling factor and the linear temperature dependence of the electron heat capacity, predict a much longer, ~16 ps, delay time for the onset of melting [71, 132]. This observation supports the importance of accounting for the effects related to the thermal excitation of lower band electrons [133] for realistic modeling of laser-induced processes.

3.3.3 Photomechanical Spallation

The fast energy deposition in short-pulse laser processing application not only results in a sharp temperature rise in the surface region of the target but, unavoidably, generates strong thermoelastic stresses that can play an important role in defining the characteristics of laser melting, generation of crystal defects, and material ejection. The maximum values of the laser-induced stresses and the contribution of the so-called photomechanical effects to the material modification and damage are related to the condition of stress confinement [5, 78, 82, 157–160]. In systems with relatively slow heat conduction and fast thermalization of the deposited laser energy, the condition for the stress confinement is mainly defined by the laser penetration depth, L_p, and the laser pulse duration, τ_p. It can be written as $\tau_p \leq \tau_s \sim L_p/C_s$, where C_s is the speed of sound in the target material. In metals, the strength of the electron–phonon coupling and much faster electron heat conduction are additional factors that affect the maximum thermoelastic stresses that can be created in the target. The characteristic time of the energy transfer from the excited hot electrons to the lattice, $\tau_{\mathrm{e-ph}}$, and the diffusive/ballistic penetration depth of the excited electrons before the electron–phonon equilibration, L_c, define the condition for the stress confinement, $max\{\tau_p, \tau_{e-ph}\} \leq \tau_s \sim L_c/C_s$ [82].

The interaction of the laser-induced compressive stresses with the free surface of the irradiated sample can result in the generation of tensile stresses sufficiently high to cause mechanical fracture of a brittle material or promote cavitation and fragmentation in a metastable liquid. By analogy with the term "spallation," commonly used to describe the dynamic fracture that results from the reflection of a shock wave from a back surface of a sample [161–163], the material ejection (or partial separation of a surface layer) due to the laser-induced stresses is often called front-surface laser spallation. Although "cavitation" may be a more appropriate term when the photomechanical processes take place in the melted part of the target, in this chapter we use the term "front-surface laser spallation" for both solid and liquid/melted targets, as soon as the transient thermoelastic stresses play the dominant role in causing ablation/damage of the target. The processes of photomechanical front- and back-surface spallation are schematically illustrated in Fig. 3.8. Short-pulse laser irradiation occurring under conditions of stress confinement results in the generation of high compressive stresses in the surface region of the target, Fig. 3.8a. The interaction of the initial compressive stresses with the free surface of the target results in the development of a tensile component of the pressure wave that propagates deeper into the bulk of the target. The tensile stresses are increasing with depth and can overcome the dynamic strength of the target material, leading to the mechanical separation and ejection of a front layer of the target, Fig. 3.8b. At later times, the layer ejected from the front surface can disintegrate into clusters/droplets, whereas the pressure wave can reach the back surface of the target and cause back-surface spallation, Fig. 3.8c.

As an example, the evolution of temperature and pressure in the surface region of an irradiated target leading to the spallation is shown in Fig. 3.9 for a TTM–MD simulation of a bulk Ni target irradiated by a 1 ps laser pulse [82, 109]. The rapid heating of the lattice due to the energy transfer from the excited electrons results in the build up of high compressive stresses in the surface region of the target. The relaxation of the compressive stresses leads to the generation of an unloading tensile wave that propagates from the surface of the target and increases its strength with depth. At a certain depth under the surface the tensile stresses exceed the dynamic strength of the melted metal, leading to the separation (spallation) of ∼25-nm-thick liquid layer from the target. The ability of the liquid to withstand the dynamic loading decreases with increasing temperature, shifting the depth of the laser-induced void nucleation and spallation closer to the surface and away from the depth at which the maximum tensile stresses are reached [68, 70, 82, 109].

The microscopic mechanisms of front-surface laser spallation have been investigated in a number of MD simulations performed for molecular systems [78–82, 164], metal targets [65, 68, 70, 76, 82, 109], and "generic" systems described by Lennard–Jones interatomic potential [75,77]. Nucleation, growth, and coalescence of voids have been identified as the main processes responsible for laser spallation. A visual picture of the spallation process is provided

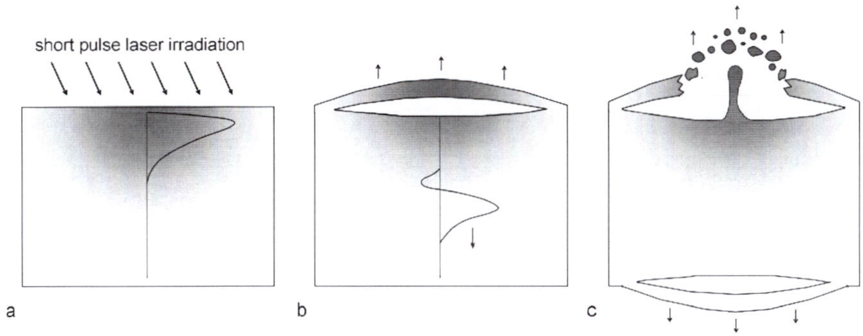

Fig. 3.8. Schematic representation of the processes involved in laser-induced front- and back-surface spallation: (**a**) generation of high compressive stresses in the surface region of the irradiated target; (**b**) propagation of the pressure wave deeper into the target, development of the tensile component of the pressure wave, separation and ejection of a front layer of the target (front-surface laser spallation) at a depth where the tensile component of the wave exceeds the dynamics strength of the (typically melted) material; (**c**) interaction of the pressure wave with the back surface of the target leading to the back-surface spallation, disintegration of the layer ejected from the front surface into clusters/droplets

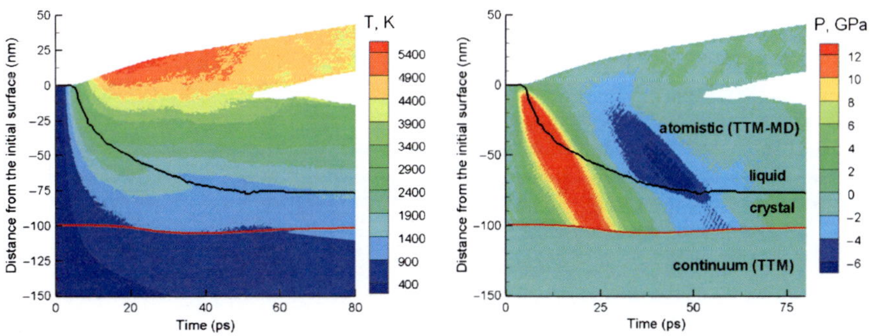

Fig. 3.9. Temperature and pressure contour plots in a simulation of a bulk Ni target irradiated with a 1 ps laser pulse at an absorbed fluence of $1935\,\mathrm{J\,m^{-2}}$. Laser pulse is directed along the Y-axis, from the top of the contour plots. Black line separates the melted region from the crystalline bulk of the target. Red line separates the atomistic and continuum parts of the combined TTM–MD model. Areas where the density of the material is less than 10% of the initial density before the irradiation are not shown in the plots. The data are from [82, 109]

in the left part of Fig. 3.10, where the evolution of voids (empty space) is shown for a simulation performed for a 100 nm Ni film irradiated by a 1 ps laser pulse at an absorbed fluence of $1623\,\mathrm{J\,m^{-2}}$. An active growth of voids starts at ∼32–35 ps, the time corresponding to the concentration of the tensile stresses associated with the interaction of the unloading stress wave, propagating from the irradiated surface, and the second tensile wave, generated

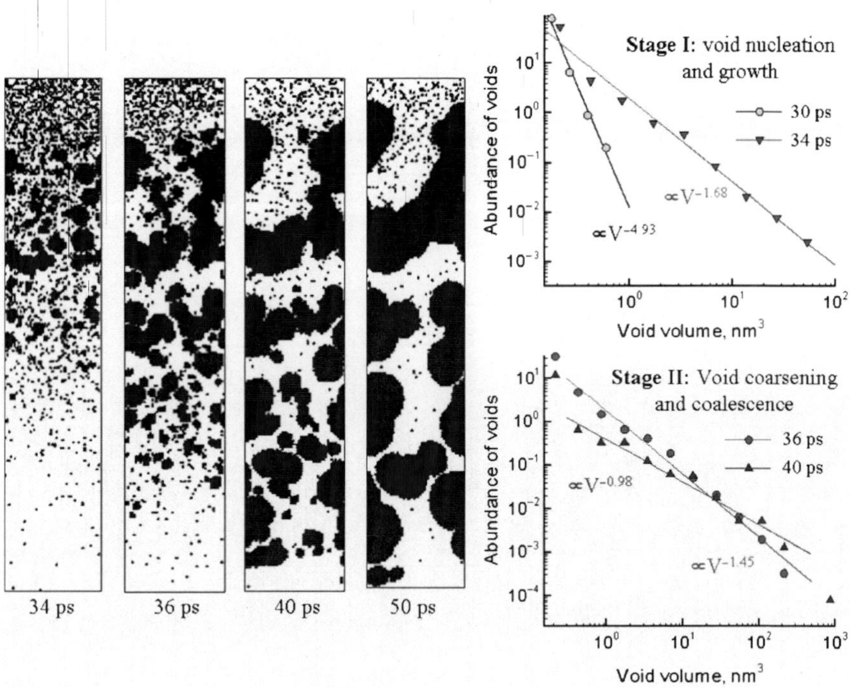

Fig. 3.10. Visual picture of the evolution of voids (empty space) in a sub-surface region of a 100 nm Ni film irradiated with a 1 ps laser pulse at an absorbed fluence of $1623\,\mathrm{J\,m^{-2}}$ and corresponding void abundance distributions. The laser pulse is directed from the top of the figure and the region shown in the snapshots is located ∼20 nm below the surface. The lines in the distributions are power law fits of the data points with the exponents indicated in the figures. The data are from [68]

upon the reflection of the original compressive wave from the back surface of the free-standing film [68, 82]. The area affected by the photomechanical damage quickly expands, and the size of the voids increases with time.

Quantitative information on the evolution of voids in the simulation discussed above is presented in the form of the void volume distributions in the right part of Fig. 3.10. All distributions can be relatively well described by a power law $\mathrm{N(V)} \sim \mathrm{V}^{-\tau}$, with an exponent $-\tau$ gradually increasing with time. Two distinct stages can be identified in the evolution of the void volume distributions. The initial stage of the *void nucleation and growth* is characterized by the increase in both the number of voids and the range of void sizes, as can be seen from the distributions shown for 30 and 34 ps after the laser pulse. The second stage of the evolution of the photomechanical damage corresponds to the *void coarsening and coalescence*, when the number of large voids increases at the expense of quickly decreasing population of small voids, e.g., compare the distributions for 36 and 40 ps. The second stage of the void evolution leads to the eventual percolation of the empty volume and ejection of large liquid layer (or droplets) from the irradiated side of the film.

The two stages in the evolution of void volume distribution, discussed above for photomechanical spallation of a metal film [68], have also been observed in simulations of laser spallation of molecular targets [82]. Moreover, the time dependences of the power law exponent predicted in the simulations performed for these two drastically different materials, amorphous molecular systems [82] and crystalline metal targets [68, 82], are in an excellent quantitative agreement with each other. The power law dependences have also been reported for the void volume distributions observed in MD simulations of shock-induced, back-surface spallation of metal targets [165]. The critical power law exponent predicted for void distribution in the MD simulations of shock-induced, back-surface spallation, $\tau \sim 2.2$, is close to the ones that separate the two regimes of void evolution observed in the simulations of laser-induced, front-surface spallation of the molecular and metal targets [68, 82]. These observations suggest that the spallation mechanisms identified in [68,82] and briefly described in this section may reflect general characteristics of the dynamic fracture at high deformation rates.

3.3.4 Phase Explosion and Laser Ablation

At a sufficiently high laser fluence, the surface region of the irradiated target can be overheated above the limit of its thermodynamic stability, leading to an explosive decomposition of the overheated material into a mixture of vapor and liquid droplets. This process, commonly called "phase explosion" or "explosive boiling," results in the ejection (ablation) of a multicomponent plume consisting of individual atoms/molecules, small clusters, and larger liquid droplets.

The mechanisms of laser ablation have been extensively investigated in MD simulations addressing various aspects of the ablation process [70, 78, 79, 83–115]. One of the findings of the simulations is the existence of a well-defined threshold fluence for the transition from surface evaporation (desorption regime) to the collective material ejection (ablation regime) [70, 79, 100, 104, 109]. The threshold behavior in laser ablation can be related to the sharp transition from a metastable superheated liquid to a two-phase mixture of liquid and vapor (explosive boiling) at a temperature of approximately 90% of the critical temperature, as predicted based on the classical nucleation theory [166–169] and confirmed in simulations [170]. Experimental observations of the existence of a threshold fluence for the onset of the droplet ejection, as well as a steep increase of the ablation rate at the threshold, have also been interpreted as evidence of the transition from normal vaporization to the phase explosion [169, 171–173].

The active processes occurring in the vicinity of the irradiated surface during the first hundreds of picoseconds after the laser irradiation are illustrated in Fig. 3.11, where snapshots from a coarse-grained MD simulation of laser

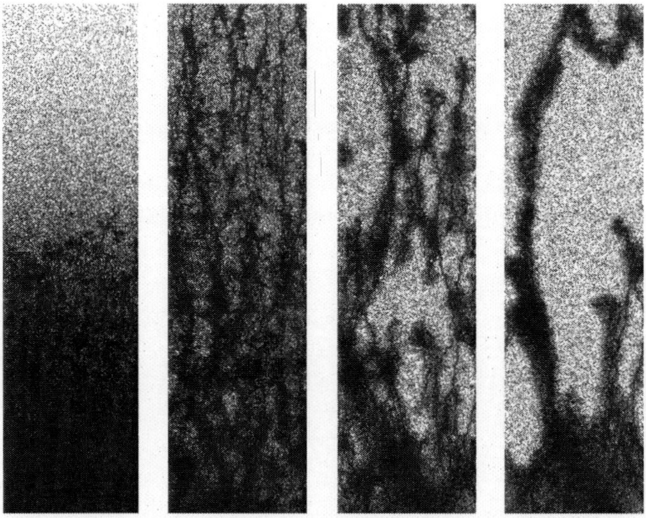

Fig. 3.11. Snapshots from a coarse-grained MD simulation of laser ablation of a polymer solution with polymer concentration of 6 wt.% [113]. The model is parameterized to represent PMMA in toluene and the simulation is performed at an absorbed laser fluence of $80\,\mathrm{J\,m^{-2}}$, pulse duration of 50 ps, and optical penetration depth of 50 nm. Matrix molecules and units of polymer chains are shown by black and blue dots, respectively

ablation of a frozen polymer solution with polymer concentration of 6 wt.% are shown. The simulation is performed with a laser pulse duration of 50 ps, optical penetration depth of 50 nm, and an absorbed laser fluence of $80\,\mathrm{J\,m^{-2}}$, about twice the ablation threshold for this model system [113]. In the first snapshot, shown for 100 ps, 50 ps after the end of the laser pulse, we see a homogeneous expansion of a significant part of the surface region. The homogeneous expansion is followed by the appearance of density fluctuations and gradual decomposition of the expanding plume into gas-phase molecules and liquid-phase regions. The decomposition of the expanding plume leads to the formation of a foamy transient structure of interconnected liquid regions, as shown in the snapshot at 200 ps. The foamy transient structure subsequently decomposes into separate liquid regions and vapor-phase molecules, forming a multicomponent ablation plume that expands away from the target.

While in the simulations performed for one-component molecular targets the liquid regions emerging from the explosive decomposition of the overheated region quickly develop into well-defined spherical liquid droplets [110,174], the entanglement of polymer chains in laser ablation of polymer solutions facilitates the formation of intricate, elongated viscous structures that extend far above the ablating surface, e.g., snapshot for 600 ps in Fig. 3.11. The elongated liquid structures that eventually separate from the target can be stabilized by evaporative cooling in the expanding plume and can reach the substrate in matrix-assisted pulsed laser evaporation (MAPLE) film deposition technique [175–177], contributing to the roughness of the deposited films [178–182] (see Chap. 9 of this book for a detailed discussion of MAPLE). Indeed, the ejection of the extended liquid structures observed in the simulations [113], can be related to "nanofiber" or "necklace" polymer surface features observed in SEM images of PMMA films deposited in MAPLE [124–126, 182], as well as in films fabricated by ablation of a polymer target involving a partial thermal decomposition of the target material into volatile species [183]. Moreover, the effect of dynamic molecular redistribution in the ejected matrix-polymer droplets, leading to the generation of transient "molecular balloons" in which polymer-rich surface layers enclose the volatile matrix material, has been identified in the simulations [114, 126, 184] as the mechanism responsible for the formation of characteristic wrinkled polymer structures observed experimentally in films deposited by MAPLE [114, 126, 182].

Regardless of the specific characteristics of the phase explosion affected by the properties of the target material and irradiation conditions, an important general conclusion that can be drawn from the results of MD simulations performed for different target materials, from metals to multicomponent molecular systems, is that particles/droplets and small atomic/molecular clusters are unavoidable products of the processes responsible for the material ejection in the ablation regime, e.g., [70, 87–89, 109, 110, 113, 125]. The energy density deposited by the laser pulse is decreasing with depth under the irradiated surface, leading to the strong dependence of the character of material decomposition from the depth of origin of the ejected material. Even when the laser fluence is sufficiently high to induce a complete vaporization of the surface layer of the target, the decrease of the energy density with depth results in the increase in the fraction of the liquid phase that emerges from the explosive phase decomposition [110, 185]. Since it is the amount of the released vapor phase that provides the driving force for the material decomposition and plume expansion, the decomposition process becomes less vigorous with depth, resulting in lower ejection velocities of droplets/clusters produced at higher depth in the target.

The difference in the characteristics of the phase explosion occurring in different parts of the target results in the effect of spatial segregation of clusters/droplets of different sizes in the plume. In particular, a detailed analysis of the dynamics of the plume formation in simulations performed for molecular targets with both long (no stress confinement) [110] and short (stress confinement) [185] laser pulses and fluences about twice the threshold for the ablation onset, reveals that only small clusters and monomers are ejected at the front of the expanding plume, medium-sized clusters are localized in the middle of the expanding plume, whereas the larger liquid droplets formed later during the plume development tend to be slower and are closer to the original surface. The cluster segregation effect, predicted in the simulations, can be related to the recent results of plume imaging experiments [186–190], where splitting of the plume into a fast component with optical emission characteristic for neutral atoms and a slow component with blackbody-like emission attributed to the presence of hot clusters [191], is observed. Similarly, and consistently with the results of the simulations discussed in [110, 185], a layered structure of the plume (vaporized layer followed by small particles and larger droplets) observed in nanosecond laser ablation of water and soft tissue [192], is attributed to the succession of phase transitions occurring at different depths in the irradiated target [192, 193]. More examples of experimental observations suggesting the spatial segregation of clusters/droplets of different sizes in the plume can be found in Chap. 6 of this book.

Despite being ejected from deeper under the surface, where the energy density deposited by the laser pulse is smaller, the larger clusters in the plume are found to have substantially higher internal temperatures when compared with the smaller clusters [110, 185]. The lower temperature of the smaller clusters can be attributed to a more vigorous phase explosion (a larger fraction of the vapor-phase molecules is released due to a higher degree of overheating) and a fast expansion of the upper part of the plume that provides a more efficient cooling when compared with a slower cooling of the larger clusters dominated by evaporation.

Depending on the irradiation conditions, as well as the thermodynamic, mechanical, and electronic properties of the target material, the thermal phase explosion may be intertwined with other processes, such as the generation of the thermoelastic stresses in the regime of stress confinement (see Sect. 3.3.3), photochemical reactions in organic systems, or optical breakdown plasma generation in dielectrics. In particular, it has been observed in MD simulations of molecular systems [78, 79] and metals [109] that larger and more numerous clusters with higher ejection velocities are produced by the explosive phase decomposition in the regime of stress confinement when compared with simulations performed at the same laser fluences, but with longer pulses, in the regime of thermal confinement. Moreover, the transient tensile stresses generated in the regime of stress confinement can bring the system deeper into the metastable region and induce nucleation and growth of vapor bubbles at fluences at which no homogeneous boiling takes place without the assistance of thermoelastic stresses [5, 193, 194], thus shifting the threshold fluence for the ablation onset to lower values [78, 79, 109].

3.4 Concluding Remarks

MD simulation technique has successfully been adopted for simulation of laser–materials interactions. Recent developments of the coarse-grained models for molecular systems and a combined continuum-atomistic TTM–MD model for metals have provided computationally efficient means for incorporation of a description of the laser energy coupling and equilibration into the classical MD method. The design of special heat-conductive, pressure-transmitting boundary conditions eliminates the need to model parts of the system where no structural transformations take place, further improving the efficiency of MD simulations of laser–materials interactions.

The examples of application of the MD simulation technique, briefly reviewed in this chapter, demonstrate the ability of atomic/molecular-level simulations to provide insights into the complex nonequilibrium processes responsible for material modification or removal in laser-processing applications. MD simulations of laser melting, generation of crystal defects, spallation, and ablation have already made contributions to the interpretation of experimental results and the advancement of theoretical understanding of laser-induced processes. With further innovative development of computational methodology and the fast growth of the available computing resources, one can expect that MD modeling will continue to play an increasingly important role in the investigation of laser interactions with materials.

One of the challenging directions of future work is the development of multiscale models for simulation of the processes occurring at the length-scale of the entire laser spot. For investigation of the long-term expansion of the ablation plume, in particular, a combination of MD with the DSMC method [195] has been demonstrated to be a promising approach capable of following the evolution of the parameters of the ablation plume on the scales, characteristic for experimental conditions, up to hundreds of microseconds and millimeters [50–55]. In the combined MD–DSMC model [78,185,187,196–199], MD is used for simulation of the initial stage of the ablation process (first nanoseconds) and provides the initial conditions for DSMC simulation of the processes occurring during the long-term expansion of the ejected plume. First applications of the combined MD–DSMC model for simulation of laser interactions with molecular systems have demonstrated the ability of the model to reveal interrelations between the processes occurring at different time- and length-scales and responsible for the evolution of the characteristics of the ablation plume [187,197–199]. In particular, the initial generation of clusters in the phase explosion, predicted in MD simulations, is found to provide cluster precursors for condensation during the long-term plume expansion, thus eliminating the three-body collision bottleneck in the cluster growth process (see Chap. 5). The presence of clusters makes a strong impact on the following collisional condensation and evaporation processes, affecting the cluster composition of the plume, as well as the overall dynamics of the plume expansion [187,197–199].

In addition to using MD model for direct simulation of laser–materials interactions, the detailed information on laser-induced structural and phase transformations, revealed in MD simulations, can help in the development of continuum-level hydrodynamic models. As briefly discussed in the introduction, the adaptation of the hydrodynamic computational models based on multiphase equations-of-state [56–62] for simulations of laser–materials interactions involve a number of assumptions on the kinetics of phase transformations, evolution of photomechanical damage under the action of laser-induced tensile stresses, characteristics of the ablation plume generated as a result of the explosive decomposition of the overheated surface region in laser ablation, etc. The results of MD simulations on the kinetics and mechanisms of melting, spallation, and ablation, e.g., [66,68,82,109,110,152,170,200], can be used to provide the necessary information for the design of a reliable description of the fast nonequilibrium processes within a continuum model [60, 61, 96].

Further expansion of the domain of applicability of the MD-type of simulations into the area of laser interactions with complex multicomponent systems, such as nanocomposite materials or biological tissue, may involve the design of novel mesoscopic models, possibly based on the dynamic elements different from spherical particles, e.g., [201, 202]. Finally, an incorporation of the information on the transient changes in the interatomic bonding and thermophysical properties of the target material in the electronically excited state (Stages 1 and 2 in Fig. 3.1), revealed in electronic structure calculations and theoretical analysis, e.g., [23–36, 131–133], into large-scale atomistic simulations is needed for investigation of the implications of the initial ultrafast atomic dynamics for the final outcome of short-pulse laser irradiation.

Acknowledgements

Financial support of this work is provided by the National Science Foundation (USA) through grants CTS-0348503, DMII-0422632, CMMI-0800786, and DMR-0907247. The authors would like to thank Barbara J. Garrison of Penn State University (USA), Aaron T. Sellinger and James M. Fitz-Gerald of the University of Virginia (USA), Antonio Miotello of the University of Trento (Italy), Nadezhda Bulgakova of the Institute of Thermophysics SB RAS (Russia), Alfred Vogel of the Institute of Biomedical Optics in Lübeck (Germany), Roland Hergenröder of the Institute for Analytical Sciences in Dortmund (Germany), and Tatiana Itina and Jörg Hermann of the CNRS Laboratory of Lasers, Plasmas, and Photonic Processing in Marseille (France), for insightful and stimulating discussions.

References

1. D. Bäuerle, *Laser Processing and Chemistry* (Springer, Berlin, 2000)
2. R.R. Gattass, E. Mazur, Nat. Photon. **2**, 219–225 (2008)
3. D.B. Chrisey, G.K. Hubler (eds), *Pulsed Laser Deposition of Thin Films* (Wiley, New York, 1994)
4. M.H. Niemz, *Laser-Tissue Interactions: Fundamentals and Applications* (Springer, Berlin, 1996)
5. A. Vogel, V. Venugopalan, Chem. Rev. **103**, 577–644 (2003)
6. V. Zafiropulos, in *Laser Cleaning*, ed. by B. Luk'yanchuk (World Scientific, Singapore, 2002), p. 343
7. A. Nevin, P. Pouli, S. Georgiou, C. Fotakis, Nat. Mater. **6**, 320–322 (2007)

8. E.I. Moses, R.E. Bonanno, C.A. Haynam, R.L. Kauffman, B.J. MacGowan, R.W. Patterson Jr, R.H. Sawicki, B.M. Van Wonterghem, Eur. Phys. J. D **44**, 215–218 (2007)
9. L. Huang, J.P. Callan, E.N. Glezer, E. Mazur, Phys. Rev. Lett. **80**, 185–188 (1998)
10. M.B. Agranat, S.I. Ashitkov, V.E. Fortov, A.V. Kirillin, A.V. Kostanovskii, S.I. Anisimov, P.S. Kondratenko, Appl. Phys. A 69, 637–640 (1999)
11. C. Guo, G. Rodriguez, A. Lobad, A.J. Taylor, Phys. Rev. Lett. **84**, 4493–4496 (2000)
12. O.P. Uteza, E.G. Gamaly, A.V. Rode, M. Samoc, B. Luther-Davies, Phys. Rev. B **70**, 054108 (2004)
13. J. Wang, C. Guo, Phys. Rev. B **75**, 184304 (2007)
14. A. Rousse, G. Rischel, S. Fourneaux, I. Uschmann, S. Sebban, G. Grillon, P.h. Balcou, E. Förster, J.P. Geindre, P. Audebert, J.C. Gauthier, D. Hulin, Nature **410**, 65–68 (2001)
15. A.M. Lindenberg, J. Larsson, K. Sokolowski-Tinten, K.J. Gaffney, C. Blome, O. Synnergren, J. Sheppard, C. Caleman, A.G. MacPhee, D. Weinstein, D.P. Lowney, T.K. Allison, T. Matthews, R.W. Falcone, A.L. Cavalieri, D.M. Fritz, S.H. Lee, P.H. Bucksbaum, D.A. Reis, J. Rudati, P.H. Fuoss, C.C. Kao, D.P. Siddons, R. Pahl, J. Als-Nielsen, S. Duesterer, R. Ischebeck, H. Schlarb, H. Schulte-Schrepping, T.H. Tschentscher, J. Schneider, D. von der Linde, O. Hignette, F. Sette, H.N. Chapman, R.W. Lee, T.N. Hansen, S. Techert, J.S. Wark, M. Bergh, G. Huldt, D. van der Spoel, N. Timneanu, J. Hajdu, R.A. Akre, E. Bong, P. Krejcik, J. Arthur, S. Brennan, K. Luening, J.B. Hastings, Science **308**, 392–395 (2005)
16. K. Sokolowski-Tinten, C. Blome, J. Blums, A. Cavalleri, C. Dietrich, A. Tarasevich, I. Uschmann, E. Förster, M. Kammler, M. Horn-von-Hoegen, D. von der Linde, Nature **422**, 287–289 (2003)
17. W.E. King, G.H. Campbell, A. Frank, B. Reed, J.F. Schmerge, B.J. Siwick, B.C. Stuart, P.M. Weber, J. Appl. Phys. **97**, 111101 (2005)
18. S. Williamson, G. Mourou, J.C.M. Li, Phys. Rev. Lett. **52**, 2364–2367 (1984)
19. B.J. Siwick, J.R. Dwyer, R.E. Jordan, R.J.D. Miller, Science **302**, 1382–1385 (2003)
20. J.R. Dwyer, R.E. Jordan, C.T. Hebeisen, M. Harb, R. Ernstorfer, T. Dartigalongue, R.J.D. Miller, J. Mod. Optics **54**, 905–922 (2007)
21. M. Harbst, T.N. Hansen, C. Caleman, W.K. Fullagar, P. Jönsson, P. Sondhauss, O. Synnergren, J. Larsson, Appl. Phys. A **81**, 893–900 (2005)
22. A.M. Lindenberg, S. Engemann, K.J. Gaffney, K. Sokolowski-Tinten, J. Larsson, P.B. Hillyard, D.A. Reis, D.M. Fritz, J. Arthur, R.A. Akre, M.J. George, A. Deb, P.H. Bucksbaum, J. Hajdu, D.A. Meyer, M. Nicoul, C. Blome, T.H. Tschentscher, A.L. Cavalieri, R.W. Falcone, S.H. Lee, R. Pahl, J. Rudati, P.H. Fuoss, A.J. Nelson, P. Krejcik, D.P. Siddons, P. Lorazo, J.B. Hastings, Phys. Rev. Lett. **100**, 135502 (2008)
23. S. Mazevet, J. Clérouin, V. Recoules, P.M. Anglade, G. Zerah, Phys. Rev. Lett. **95**, 085002 (2005)
24. V. Recoules, J. Clérouin, G. Zérah, P.M. Anglade, S. Mazevet, Phys. Rev. Lett. **96**, 055503 (2006)
25. F. Bottin, G. Zérah, Phys. Rev. B **75**, 174114 (2007)
26. D. Dundas, J. Phys. B: At. Mol. Opt. Phys. **37**, 2883–2901 (2004)

27. H.O. Jeschke, M.E. Garcia, Chap. 7 in *Nonlinear Optics, Quantum Optics, and Ultrafast Phenomena with X-Rays,* ed. by B.W. Adams (Springer, New York, 2003), p. 175
28. H.O. Jeschke, M.E. Garcia, M. Lenzner, J. Bonse, J. Krüger, W. Kautek, Appl. Surf. Sci. **197/198**, 839–844 (2002)
29. P.L. Silvestrelli, A. Alavi, M. Parrinello, D. Frenkel, Phys. Rev. Lett. **77**, 3149–3152 (1996)
30. P.L. Silvestrelli, Phys. Rev. B **60**, 16382–16388 (1999)
31. P.L. Silvestrelli, A. Alavi, M. Parrinello, D. Frenkel, Phys. Rev. B **56**, 3806–3812 (1997)
32. L.X. Benedict, C.D. Spataru, S.G. Louie, Phys. Rev. B **66**, 085116 (2002)
33. P.B. Hillyard, D.A. Reis, K.J. Gaffney, Phys. Rev. B **77**, 195213 (2008)
34. R.E. Allen, T. Dumitrica, B. Torralva, Chap. 7 in *Ultrafast Physical Processes in Semiconductors,* ed. by K.T. Tsen (Academic, New York, 2000), p. 315
35. B. Torralva, T.A. Niehaus, M. Elstner, S. Suhai, T.h. Frauenheim, R.E. Allen, Phys. Rev. B **64**, 153105 (2001)
36. T. Dumitrica, A. Burzo, Y. Dou, R.E. Allen, Phys. Stat. Solidi B **241**, 2331–2342 (2004)
37. V.A. Gnatyuk, T. Aoki, O.S. Gorodnychenko, Y. Hatanaka, Appl. Phys. Lett. **83**, 3704–3706 (2003)
38. K.N. Vonatsos, D.I. Pantelis, Appl. Phys. A **80**, 885–889 (2005)
39. V.N. Tokarev, A.F.H. Kaplan, J. Appl. Phys. **86**, 2836–2846 (1999)
40. R.Černý R. Šáršik, I. Lukeš, V. Cháb, Phys. Rev. B **44**, 4097–4102 (1991)
41. X. Xu, G. Chen, K.H. Song, Int. J. Heat Mass Transf. **42**, 1371–1382 (1999)
42. J.C. Kapat, Z. Wei, A. Kumar, Appl. Surf. Sci. **127–129**, 212–217 (1998)
43. G.X. Wang, E.F. Matthys, Trans. ASME **118**, 944–951 (1996)
44. I.H. Chowdhury, X. Xu, Num. Heat Transf. A **44**, 219–232 (2003)
45. R.K. Singh, J. Narayan, Phys. Rev. B **41**, 8843–8858 (1990)
46. A. Peterlongo, A. Miotello, R. Kelly, Phys. Rev. E **50**, 4716–4727 (1994)
47. J.R. Ho, C.P. Grigoropoulos, J.A.C. Humphrey, J. Appl. Phys. **78**, 4696–4709 (1995)
48. A. Bogaerts, Z. Chen, R. Gijbels, A. Vertes, Spectrochim. Acta. B **58**, 1867–1893 (2003)
49. Z. Chen, A. Vertes, Phys. Rev. E **77**, 036316 (2008)
50. I. NoorBatcha, R.R. Lucchese, Y. Zeiri, J. Chem. Phys. **86**, 5816–5824 (1987)
51. H.M. Urbassek, D. Sibold, Phys. Rev. Lett. **70**, 1886–1889 (1993)
52. O. Ellegaard, J. Schou, H.M. Urbassek, Appl. Phys. A **69**, S577-S582 (1999)
53. T.E. Itina, W. Marine, M. Autric, J. Appl. Phys. **82**, 3536–3542 (1997)
54. T.E. Itina, W. Marine, M. Autric, J. Appl. Phys. **85**, 7905–7908 (1999)
55. A.A. Morozov, Phys. Fluids **19**, 087101 (2007)
56. K. Eidmann, J. Meyer-ter-Vehn, T. Schlegel, S. Huller, Phys. Rev. E **62**, 1202–1214 (2000)
57. Y. Kondoh, T. Yabe, J. Maehara, T. Nakamura, Y. Ogata, Phys. Rev. E **68**, 066408 (2003)
58. F. Vidal, T.W. Johnston, J.C. Kieffer, F. Martin, Phys. Rev. B **70**, 184125 (2004)
59. J.P. Colombier, P. Combis, F. Bonneau, R. Le Harzic, E. Audouard, Phys. Rev. B **71**, 165406 (2005)
60. M.E. Povarnitsyn, T.E. Itina, M. Sentis, K.V. Khishchenko, P.R. Levashov, Phys. Rev. B **75**, 235414 (2007)

61. A.N. Volkov, L.V. Zhigilei, J. Phys. Conf. Ser. **59**, 640–645 (2007)
62. A.N. Volkov, C. Sevilla, L.V. Zhigilei, Appl. Surf. Sci. **253**, 6394–6399 (2007)
63. C.F. Richardson, P. Clancy, Mol. Simul. **7**, 335–355 (1991)
64. H. Hakkinen, U. Landman, Phys. Rev. Lett. **71**, 1023–1026 (1993)
65. D.S. Ivanov, L.V. Zhigilei, Phys. Rev. B **68**, 064114 (2003)
66. D.S. Ivanov, L.V. Zhigilei, Phys. Rev. Lett. **91**, 105701 (2003)
67. D.S. Ivanov, L.V. Zhigilei, Appl. Phys. A **79**, 977–981 (2004)
68. L.V. Zhigilei, D.S. Ivanov, E. Leveugle, B. Sadigh, E.M. Bringa, in *High-Power Laser Ablation V*, ed. by C.R. Phipps, Proc. SPIE **5448**, 505–519 (2004)
69. L.V. Zhigilei, D.S. Ivanov, Appl. Surf. Sci. **248**, 433–439 (2005)
70. L.V. Zhigilei, Z. Lin, D.S. Ivanov, in *Proceedings of the 2006 ASME International Mechanical Engineering Congress and Exposition (IMECE2006)*, ASME paper IMECE2006–16305 (2006)
71. Z. Lin, L.V. Zhigilei, Phys. Rev. B **73**, 184113 (2006)
72. Z. Lin, L.V. Zhigilei, J. Phys. Conf. Ser. **59**, 11–15 (2007)
73. W.H. Duff, L.V. Zhigilei, J. Phys. Conf. Ser. **59**, 413–417 (2007)
74. Z. Lin, R.A. Johnson, L.V. Zhigilei, Phys. Rev. B **77**, 214108 (2008)
75. S.I. Anisimov, V.V. Zhakhovskii, N.A. Inogamov, K. Nishihara, A.M. Oparin, Y.V. Petrov, Pis'ma Zh. Eksp. Teor. Fiz. **77**, 731–736 (2003) [JETP Lett. **77**, 606–610 (2003)]
76. A.K. Upadhyay, H.M. Urbassek, J. Phys. D **38**, 2933–2941 (2005)
77. H.Y. Lai, P.H. Huang, Appl. Surf. Sci. **254**, 3067–3073 (2008)
78. L.V. Zhigilei, E. Leveugle, B.J. Garrison, Y.G. Yingling, M.I. Zeifman, Chem. Rev. **103**, 321–348 (2003)
79. L.V. Zhigilei, B.J. Garrison, J. Appl. Phys. **88**, 1281–1298 (2000)
80. A.G. Zhidkov, L.V. Zhigilei, A. Sasaki, T. Tajima, Appl. Phys. A **73**, 741–747 (2001)
81. E. Leveugle, L.V. Zhigilei, Appl. Phys. A **79**, 753–756 (2004)
82. E. Leveugle, D.S. Ivanov, L.V. Zhigilei, Appl. Phys. A **79**, 1643–1655 (2004)
83. E. Ohmura, I. Fukumoto, Int. J. Japan Soc. Prec. Eng. **30**, 128–133 (1996)
84. E. Ohmura, I. Fukumoto, I. Miyamoto, Int. J. Jpn. Soc. Precis. Eng. **32**, 248–253 (1998)
85. R.F.W. Herrmann, J. Gerlach, E.E.B. Campbell, Appl. Phys. A **66**, 35–42 (1998)
86. X. Wu, M. Sadeghi, A. Vertes, J. Phys. Chem. B **102**, 4770–4778 (1998)
87. S. Kristyan, A. Bensura, A. Vertes, Theor. Chem. Acc. **107**, 319–325 (2002)
88. X. Xu, C. Cheng, I.H. Chowdhury, ASME Trans. J. Heat Transf. **126**, 727–734 (2004)
89. C. Cheng, X. Xu, Phys. Rev. B **72**, 165415 (2005)
90. X.W. Wang, X.F. Xu, J. Heat Transf. **124**, 265–274 (2002)
91. X.W. Wang, J. Phys. D **38**, 1805–1823 (2005)
92. P. Lorazo, L.J. Lewis, M. Meunier, Phys. Rev. B **73**, 134108 (2006)
93. P. Lorazo, L.J. Lewis, M. Meunier, Phys. Rev. Lett. **91**, 225502 (2003)
94. N.N. Nedialkov, P.A. Atanasov, Appl. Surf. Sci. **252**, 4411–4415 (2006)
95. N.N. Nedialkov, P.A. Atanasov, S.E. Imamova, A. Ruf, P. Berger, F. Dausinger, Appl. Phys. A **79**, 1121–1125 (2004)
96. S.I. Anisimov, V.V. Zhakhovskii, N.A. Inogamov, K. Nishihara, Y.V. Petrov, V.A. Khokhlov, J. Exp. Theor. Phys. **103**, 183–197 (2006)

97. M.B. Agranat, S.I. Anisimov, S.I. Ashitkov, V.V. Zhakhovskii, N.A. Inogamov, K. Nishihara, Y.V. Petrov, V.E. Fortov, V.A. Khokhlov, Appl. Surf. Sci. **253**, 6276–6282 (2007)
98. L.V. Zhigilei, P.B.S. Kodali, B.J. Garrison, J. Phys. Chem. B **101**, 2028–2037 (1997)
99. L.V. Zhigilei, B.J. Garrison, Appl. Phys. Lett. **71**, 551–553 (1997)
100. L.V. Zhigilei, P.B.S. Kodali, B.J. Garrison, Chem. Phys. Lett. **276**, 269–273 (1997)
101. L.V. Zhigilei, P.B.S. Kodali, B.J. Garrison, J. Phys. Chem. B **102**, 2845–2853 (1998)
102. L.V. Zhigilei, B.J. Garrison, Rapid Commun. Mass Spectrom. **12**, 1273–1277 (1998)
103. L.V. Zhigilei, B.J. Garrison, Appl. Phys. Lett. **74**, 1341–1343 (1999)
104. L.V. Zhigilei, B.J. Garrison, Appl. Phys. A **69**, S75–S80 (1999)
105. Y.G. Yingling, L.V. Zhigilei, B.J. Garrison, A. Koubenakis, J. Labrakis, S. Georgiou, Appl. Phys. Lett. **78**, 1631–1633 (2001)
106. Y.G. Yingling, L.V. Zhigilei, B.J. Garrison, J. Photochem. Photobiol. A **145**, 173–181 (2001)
107. T.E. Itina, L.V. Zhigilei, B.J. Garrison, J. Phys. Chem. B **106**, 303–310 (2002)
108. C. Schäfer, H.M. Urbassek, L.V. Zhigilei, Phys. Rev. B **66**, 115404 (2002)
109. L.V. Zhigilei, Z. Lin, D.S. Ivanov, J. Phys. Chem. C **113**, 11892–11906 (2009)
110. L.V. Zhigilei, Appl. Phys. A **76**, 339–350 (2003)
111. Y.G. Yingling, B.J. Garrison, J. Phys. Chem. B **109**, 16482–16489 (2005)
112. R. Knochenmuss, L.V. Zhigilei, J. Phys. Chem. B **109**, 22947–22957 (2005)
113. E. Leveugle, L.V. Zhigilei, J. Appl. Phys. **102**, 074914 (2007)
114. E. Leveugle, A. Sellinger, J.M. Fitz-Gerald, L.V. Zhigilei, Phys. Rev. Lett. **98**, 216101 (2007)
115. M. Prasad, P. Conforti, B.J. Garrison, J. Appl. Phys. **101**, 103113 (2007)
116. M.P. Allen, D.J. Tildesley, *Computer Simulation of Liquids* (Clarendon, Oxford, 1990, 1987)
117. D. Frenkel, B. Smit, *Understanding Molecular Simulation: From Algorithms to Applications* (Academic, San Diego, 1996)
118. K. Kadau, T.C. Germann, P.S. Lomdahl, Int. J. Mod. Phys. C **17**, 1755–1761 (2006)
119. http://folding.stanford.edu/
120. J.C. Deàk, L.K. Iwaki, S.T. Rhea, D.D. Dlott, J. Raman Spectrosc. **31**, 263–274 (2000)
121. S. Woutersen, H.J. Bakker, Nature **402**, 507–509 (1999)
122. H. Kim, D.D. Dlott, Y. Won, J. Chem. Phys. **102**, 5480–5485 (1995)
123. R.E. Wyatt, C. Iung, C. Leforestier, Acc. Chem. Res. **28**, 423–429 (1995)
124. E. Leveugle, L.V. Zhigilei, A. Sellinger, J.M. Fitz-Gerald, J. Phys. Conf. Ser. **59**, 126–131 (2007)
125. E. Leveugle, L.V. Zhigilei, A. Sellinger, J.M. Fitz-Gerald, Appl. Surf. Sci. **253**, 6456–6460 (2007)
126. A. Sellinger, E. Leveugle, J.M. Fitz-Gerald, L.V. Zhigilei, Appl. Phys. A **92**, 821 (2008)
127. E.A. Colbourn (ed.) *Computer Simulation of Polymers* (Longman Scientific and Technical, Harlow, 1994)
128. F. Gao, D.J. Bacon, P.E.J. Flewitt, T.A. Lewis, Modelling Simul. Mater. Sci. Eng. **6**, 543–556 (1998)

129. D.M. Duffy, A.M. Rutherford, J. Phys: Condens. Matter **19**, 016207 (2007)
130. S.I. Anisimov, B.L. Kapeliovich, T.L. Perel'man, Sov. Phys. JETP **39**, 375–377 (1974)
131. Z. Lin, L.V. Zhigilei, Appl. Surf. Sci. **253**, 6295–6300 (2007)
132. Z. Lin, L.V. Zhigilei, in *High-Power Laser Ablation VI*, ed. by C.R. Phipps, Proc. SPIE **6261**, 62610U (2006)
133. Z. Lin, L.V. Zhigilei, V. Celli, Phys. Rev. B **77**, 075133 (2008)
134. http://www.faculty.virginia.edu/CompMat/electron-phonon-coupling/
135. N. Rotenberg, A.D. Bristow, M. Pfeiffer, M. Betz, H.M. van Driel, Phys. Rev. B **75**, 155426 (2007)
136. M.S. Daw, S.M. Foiles, M.I. Baskes, Mater. Sci. Rep. **9**, 251–310 (1993)
137. S.M. Foiles, MRS Bull. **21**, 24–28 (1996)
138. X.W. Zhou, H.N.G. Wadley, R.A. Johnson, D.J. Larson, N. Tabat, A. Cerezo, A.K. Petford-Long, G.D.W. Smith, P.H. Clifton, R.L. Martens, T.F. Kelly, Acta Mater. **49**, 4005–4015 (2001)
139. D.S. Ivanov, B. Rethfeld, G.M. O'Connor, T.h.J. Glynn, A.N. Volkov, L.V. Zhigilei, Appl. Phys. A **92**, 791–796 (2008)
140. L.V. Zhigilei, B.J. Garrison, Mat. Res. Soc. Symp. Proc. **538**, 491–496 (1999)
141. C. Schafer, H.M. Urbassek, L.V. Zhigilei, B.J. Garrison, Comp. Mater. Sci. **24**, 421–429 (2002)
142. L.V. Zhigilei, B.J. Garrison, in *Laser-Tissue Interaction IX*, ed. by S.L. Jacques, Proc. SPIE **3254**, 135–143 (1998)
143. V. Vitek, Philos. Mag. **21**, 1275–1278 (1970)
144. J.P. Hirth, J. Lothe, *Theory of Dislocations*, 2nd edn. (Wiley, New York, 1982)
145. D. Ashkenasi, M. Lorenz, R. Stoian, A. Rosenfeld, Appl. Surf. Sci. **150**, 101–106 (1999)
146. P.T. Mannion, J. Magee, E. Coyne, G.M. O'Connor, T.J. Glynn, Appl. Surf. Sci. **233**, 275–287 (2004)
147. S.E. Kirkwood, A.C. van Popta, Y.Y. Tsui, R. Fedosejevs, Appl. Phys. A **81**, 729–735 (2005)
148. J. Krüger, D. Dufft, R. Koter, A. Hertwig, Appl. Surf. Sci. **253**, 7815–7819 (2007)
149. G. Raciukaitis, M. Brikas, P. Gecys, M. Gedvilas, Proc. SPIE **7005**, 70052L (2008)
150. J.G. Dash, Rev. Mod. Phys. **71**, 1737–1743 (1999)
151. P.R. Couchman, W.A. Jesser, Nature **269**, 481–483 (1977)
152. D.S. Ivanov, L.V. Zhigilei, Phys. Rev. Lett. **98**, 195701 (2007)
153. B. Rethfeld, K. Sokolowski-Tinten, D. von der Linde, S.I. Anisimov, Phys. Rev. B **65**, 092103 (2002)
154. J.R. Dwyer, C.T. Hebeisen, R. Ernstorfer, M. Harb, V. Deyirmenjian, R.E. Jordan, R.J.D. Miller, Phil. Trans. R. Soc. A **364**, 741–778 (2006)
155. S.N. Luo, T.J. Ahrens, T. Çağın, A. Strachan, W.A. Goddard, D.C. Swift, Phys. Rev. B **68**, 134206 (2003)
156. V. Sorkin, E. Polturak, J. Adler, Phys. Rev. B **68**, 174102 (2003)
157. G. Paltauf, P.E. Dyer, Chem. Rev. **103**, 487–518 (2003)
158. I. Itzkan, D. Albagli, M.L. Dark, L.T. Perelman, C. von Rosenberg, M.S. Feld, Proc. Natl. Acad. Sci. U S A **92**, 1960–1964 (1995)
159. A.A. Oraevsky, S.L. Jacques, F.K. Tittel, J. Appl. Phys. **78**, 1281–1290 (1995)
160. R. Cramer, R.F. Haglund Jr., F. Hillenkamp, Int. J. Mass Spectrom. Ion Proc. **169/170**, 51–67 (1997)

161. H. Tamura, T. Kohama, K. Kondo, M. Yoshida, J. Appl. Phys. **89**, 3520–3522 (2001)
162. V.E. Fortov, V.V. Kostin, S. Eliezer, J. Appl. Phys. **70**, 4524–4531 (1991)
163. G.I. Kanel', V.E. Fortov, S.V. Razorenov, Physics-Uspekhi **50**, 771–791 (2007)
164. Animated sequences of snapshots from a molecular dynamics simulation of laser spallation of a molecular target can be found at http://www.faculty.virginia.edu/CompMat/spallation/animations/
165. A. Strachan, T. Cagin, W.A. Goddard III, Phys. Rev. B **63**, 060103 (2001)
166. M.M. Martynyuk, Sov. Phys. Tech. Phys. **21**, 430–433 (1976)
167. A. Miotello, R. Kelly, Appl. Phys. A **69**, S67–S73 (1999)
168. R. Kelly, A. Miotello, J. Appl. Phys. **87**, 3177–3179 (2000)
169. N.M. Bulgakova, A.V. Bulgakov, Appl. Phys. A **73**, 199–208 (2001)
170. B.J. Garrison, T.E. Itina, L.V. Zhigilei, Phys. Rev. E **68**, 041501 (2003)
171. K.H. Song, X. Xu, Appl. Surf. Sci. **127–129**, 111–116 (1998)
172. J.H. Yoo, S.H. Jeong, X.L. Mao, R. Greif, R.E. Russo, Appl. Phys. Lett. **76**, 783–785 (2000)
173. C. Porneala, D.A. Willis, Appl. Phys. Lett. **89**, 211121 (2006)
174. Animated sequences of snapshots from a molecular dynamics simulation of laser ablation of a molecular target can be found at http://www.faculty.virginia.edu/CompMat/ablation/animations/
175. A. Piqué, R.A. McGill, D.B. Chrisey, J. Callahan, T.E. Mlsna, Mater. Res. Soc. Symp. Proc. **526**, 375–383 (1998)
176. A. Piqué, R.A. McGill, D.B. Chrisey, D. Leonhardt, T.E. Mslna, B.J. Spargo, J.H. Callahan, R.W. Vachet, R. Chung, M.A. Bucaro, Thin Solid Films **355/356**, 536–541 (1999)
177. D.B. Chrisey, A. Piqué, R.A. McGill, J.S. Horwitz, B.R. Ringeisen, D.M. Bubb, P.K. Wu, Chem. Rev. **103**, 553–576 (2003)
178. R. Fryček, M. Jelínek, T. Kocourek, P. Fitl, M. Vrřata, V. Myslík, M. Vrbová, Thin Solid Films **495**, 308–311 (2006)
179. D.M. Bubb, P.K. Wu, J.S. Horwitz, J.H. Callahan, M. Galicia, A. Vertes, R.A. McGill, E.J. Houser, B.R. Ringeisen, D.B. Chrisey, J. Appl. Phys. **91**, 2055–2058 (2002)
180. K. Rodrigo, P. Czuba, B. Toftmann, J. Schou, R. Pedrys, Appl. Surf. Sci. **252**, 4824–4828 (2006)
181. A. Gutierrez-Llorente, R. Perez-Casero, B. Pajot, J. Roussel, R.M. Defourneau, D. Defourneau, J.L. Fave, E. Millon, J. Perriere, Appl. Phys. A **77**, 785–788 (2003)
182. A.T. Sellinger, E.M. Leveugle, K. Gogick, L.V. Zhigilei, J.M. Fitz-Gerald, J. Vac. Sci. Technol. A **24**, 1618–1622 (2006)
183. J. Blazevska-Gilev et al., Polym. Degrad. Stab. **91**, 213–220 (2006)
184. Animated sequences of snapshots from a molecular dynamics simulation of the generation of "molecular balloons" in laser-induced explosive boiling of polymer solutions can be found at http://www.faculty.virginia.edu/CompMat/molecular-balloons/
185. L.V. Zhigilei, in *Advances in Materials Theory and Modeling-Bridging Over Multiple-Length and Time Scales*, ed. by V.V. Bulatov, F. Cleri, L. Colombo, L.J. Lewis, N. Mousseau, Mater. Res. Soc. Symp. Proc. **677** AA2.1.1 (2001)
186. S. Noël, J. Hermann, T. Itina, Appl. Surf. Sci. **253**, 6310–6315 (2007)
187. T.E. Itina, K. Gouriet, L.V. Zhigilei, S. Noël, J. Hermann, M. Sentis, Appl. Surf. Sci. **253**, 7656–7661 (2007)

188. S. Amoruso, R. Bruzzese, C. Pagano, X. Wang, Appl. Phys. A **89**, 1017–1024 (2007)
189. O. Albert, S. Roger, Y. Glinec, J.C. Loulergue, J. Etchepare, C. Boulmer-Leborgne, J. Perriere, E. Millon, Appl. Phys. A **76**, 319–323 (2003)
190. N. Jegenyes, J. Etchepare, B. Reynier, D. Scuderi, A. Dos-Santos, Z. Tóth, Appl. Phys. A **91**, 385–392 (2008)
191. D. Scuderi, R. Benzerga, O. Albert, B. Reynier, J. Etchepare, Appl. Surf. Sci. **252**, 4360–4363 (2006)
192. I. Apitz, A. Vogel, Appl. Phys. A **81**, 329–338 (2005)
193. A. Vogel, I. Apitz, V. Venugopalan, in *Oscillations, Waves and Interactions*, ed. by T. Kurz, U. Parlitz, U. Kaatze (Universitätsverlag Göttingen, Göttingen, 2007), p. 217
194. D. Kim, M. Ye, C.P. Grigoropoulos, Appl. Phys. A **67**, 169–181 (1998)
195. G.A. Bird, *Molecular Gas Dynamics and the Direct Simulation of Gas Flows* (Clarendon, Oxford, 1994)
196. L.V. Zhigilei, A.M. Dongare, Comput. Model. Eng. Sci **3**, 539–555 (2002)
197. M.I. Zeifman, B.J. Garrison, L.V. Zhigilei, J. Appl. Phys. **92**, 2181–2193 (2002)
198. M.I. Zeifman, B.J. Garrison, L.V. Zhigilei, Appl. Surf. Sci. **197/198**, 27–34 (2002)
199. T.E. Itina, L.V. Zhigilei, J. Phys. Conf. Ser **59**, 44–49 (2007)
200. T.E. Itina, Chem. Phys. Lett. **452**, 129–132 (2008)
201. L.V. Zhigilei, C. Wei, D. Srivastava, Phys. Rev. B **71**, 165417 (2005)
202. A.N. Volkov, K.R. Simov, L.V. Zhigilei, in *Proceedings of ASME International Mechanical Engineering Congress and Exposition*, ASME paper IMECE2008–68021 (2008)

4

Continuum Models of Ultrashort Pulsed Laser Ablation

Nadezhda M. Bulgakova, Razvan Stoian, Arkadi Rosenfeld, and Ingolf V. Hertel

Summary. The aim of this chapter is to provide a basic introduction to the principles that lay the foundation for established approaches that treating matter as a continuum model, in order to describe and comprehend the aspects of laser–matter interactions. The chapter considers relevant processes induced in solids under laser irradiation in a frame of continuum models successfully applied to quantify the laser heating and subsequent ablation processes. We intend to focus on a critical assessment of these strategies with a clear perception of their advantages and limitations. The drift-diffusion approach of laser-induced material charging is considered as an example. The time and length scales of its application in describing laser-induced modifications for different classes of materials are analyzed and further improvements are also discussed. In the final part of the chapter, we give a short overview of laser–solid interaction phenomena, which could be further treated with continuum models, and present a number of examples too.

4.1 Introduction

Femtosecond laser pulses provide unique possibilities for high-precision material processing. Due to a rapid energy delivery, heat-affected zones in the irradiated targets are strongly localized with a minimal residual damage that can allow the generation of well-defined microstructures with high quality and reproducibility [1–3]. Understanding of the underlying physics and interrelations of the processes taking place in laser-irradiated materials can facilitate an optimization of experimental parameters in current applications and development of contemporary, pulsed laser technologies. The complexity of many interconnected processes involved in laser–matter interaction gives rise to elaborated theoretical and computational descriptions of the laser ablation phenomenon, relying on different approaches including atomistic and continuum ones. Atomistic modeling based on molecular dynamics (MD) and Direct Simulation Monte Carlo (DSMC) approaches has a great potential, being however still strongly restricted to the description of a relatively small amount of matter when compared with that usually involved in laser–solid interaction and subsequently developed laser ablation. The continuum models, being properly applied and treated, can represent powerful and effective techniques to study and predict the behavior of such large-sized atomic systems and provide valuable insights into the extremely complicated phenomenon of the interaction of light with matter. An excellent example is a continuum shell model developed by Yakobson et al. [4] to study instability patterns in carbon nanotubes.

Here, we discuss general principles of applying the continuum approaches to small objects or relatively small material amounts and demonstrate these principles on a number of examples.

4.2 Ultrashort Laser–Matter Interaction

By convention, the processes occurring under ultrashort, powerful laser irradiation of matter can be divided into ionizing and nonionizing ones. The kind of interaction depends mainly on the material properties. In metals, the laser light is absorbed predominantly by free electrons in a quasi-linear manner, via the inverse bremsstrahlung process that allows the application of the Beer–Lambert law to describe the laser beam propagation into a metal sample. The Beer-Lambert law is based on the assumption of a linear relationship between light attenuation and the concentration of absorbing species and reads as $I(t,z) = I_0(t)\exp(-\alpha z)$, where I denotes the laser intensity, α the absorption coefficient, and z the beam propagation path through the absorbing media. It is usually assumed that the absorption coefficient is a constant value dependent only on laser wavelength for a given metal that is determined by the free-electron density. This assumption disregards the ionization processes in metals. However, it should be noticed that the excitation of d band electrons in transition and noble metals can possibly contribute to the metal optical response affecting also the electronic and thermal properties [5]. Additionally, for simplicity, the temperature effects on electronic collisional processes which lead to light absorption are neglected as well.

In the laser excitation of bandgap materials the ionizing processes play a key role in generating free-electron populations which, in turn, absorb laser photons similar to free electrons in metals. In wide-bandgap dielectrics absorption of visible or near-infrared radiation can proceed by an interband, multiphoton ionization process that is a simultaneous absorption of several photons with their total energy exceeding the bandgap energy. A comprehensive theory of multiphoton ionization was proposed by Keldysh [6] indicating that the probability of multiphoton ionization depends on the laser field intensity as I^k with k being equal to the number of photons necessary to overcome the ionization barrier. With increasing laser power, the tunneling ionization mechanism (electron tunneling across a barrier in a strong electric field) becomes dominating over multiphoton absorption [6]. Transition from multiphoton absorption to tunneling ionization under ultrafast laser excitation of dielectric materials is widely covered in the literature (see, e.g. [7, 8] and references therein). It should be noted that, in semiconductors with a relatively small bandgap when the energy of a single photon is mostly sufficient to overcome the ionization barrier, the two-photon ionization can be important or even dominating at intensities close to the ablation threshold (e.g. Si [9], InSb [10]).

The electrons excited to the conduction band absorb laser radiation via inverse bremsstrahlung and intraband transitions in a metal-like manner. This results in increasing their kinetic energy and allows them to produce secondary electrons in impact with the lattice atoms, developing the avalanche process. A very simple, but intuitive rate equation describing the evolution of a free-electron density $n_e(t)$ in wide-bandgap dielectrics under intense laser irradiation was proposed by Stuart et al. [11, 12].

$$\frac{\partial n_e(t)}{\partial t} = \sigma_k I^k(t) + \delta I(t) n_e(t). \tag{4.1}$$

Here, σ_k is the constant of multiphoton ionization and the avalanche constant δ is derived from the comparison with experimental data for a particular material. More adapted approaches which take into account the energetic distribution of the electrons were recently proposed [13]. Note that (4.1) yields in the exponential multiplication of the free-electron population in the collisional ionization process called thus avalanche. When the free-electron density exceeds the value of the critical plasma density, the reflectance of the sample surface increases [14], as well as the material absorbance. The excited layer of a semiconductor or a dielectric sample gains optical properties similar, to a certain extent, to metals. Optical response of laser-excited dielectrics and semiconductors at surfaces can be described via the complex dielectric function, which can be seen as a mutual contribution of the unexcited solid and the response of the laser-induced free-electron gas [9]:

$$\varepsilon^*(n_e) \cong 1 + (\varepsilon_g - 1)\left(1 - \frac{n_e}{n_0}\right) - \frac{n_e}{n_{cr}} \frac{1}{1 + i \,/\, \omega\, \tau}. \tag{4.2}$$

Here, ε_g is the dielectric constant of the unexcited material; $n_{cr} = \varepsilon_0 m_e \omega^2 / e^2$ and n_0 are the critical electron density in vacuum and the valence band electron density, respectively; and m_e is the optical electron mass [8].

The thermalization rate within the free-electron subsystem depends on both the material properties and laser fluence. In metals exposed to laser fluences sufficient to melt the excited layer, this process proceeds usually in several femtoseconds [15] whereas in dielectrics electron thermalization may require hundreds of femtoseconds [7]. The free-electron gas transfers energy to the lattice by coupling to the vibration bath [5, 16], which results in heating and triggering a whole range of phase-transformation processes in the material, including its melting, ablation via the different mechanisms such as phase explosion or explosive boiling [17–19], spallation [18, 19], fragmentation [20], plastic deformations [21,22], and, upon cooling, solidification with formation of amorphous and/or polycrystalline phases [23]. These processes are developed in pico- and nanosecond time scales after the laser-pulse termination. In nonmetallic materials, a gradual electron-hole recombination takes place. In semiconductors, the main electron recombination process is Auger recombination [24], which is a three-particle collision process with the recombination energy release carried away by a third particle, usually another electron. Under high excitations, the rate of this process is saturated at the electron densities $\sim 10^{21}$ cm^{-3} due to the plasma screening effect [24,25]. In dielectrics, the main process for a free-electron decay is a trapping-like recombination with creation of excitons and defects of various natures [26,27].

Also other important processes should be mentioned. In semiconducting materials, strong levels of electronic excitation in antibonding states may lead to bond-breaking and premature lattice destabilization which culminate with ultrafast phase transitions to disordered liquid-like phase [10,14,28]. An essential part of laser irradiation of different materials is the process of electron photo- and thermionic emission [29], which can determine important consequences in the dynamics of material ablation [30, 31]. The latter will be considered in more details in Sect. 4.4. In the case of femtosecond-material processing in gaseous environments, femtosecond laser pulses can also cause air breakdown in front of irradiated samples, leading to initiation of a number of additional processes such as shock-wave formation (irrespective to material ablation) [32,33], etching of nonirradiated target surface [32,34,35], and redeposition of ablated material [35]. Additionally, buffer pressure and gas-phase kinetics of cluster growth affect the size distribution and properties of particles emitted from the target during the ablation stage [36]. The listed processes are summarized in Table 4.1. The broken line between the fields indicates a fuzzy temporal boundary between the processes which can also occur on the neighboring time scales. The majority of these processes are discussed in detail in the other chapters of this book. Here, we focus on the applications of continuum approaches for modeling of ultrashort, laser–matter interaction, and, in Sect. 4.3, we discuss their place among the other, more sophisticated theoretical methods and also determine some principles of their application.

4.3 Notes on Continuum Modeling in Application to Ultrashort, Laser–Matter Interactions

The complexity and multiplicity of the processes involved in ultrashort laser–matter interaction and the wide range of the time- and space scales of their manifestation makes it impossible to create a model describing this phenomenon with all its features. Numerous models have been developed to treat different aspects of material evolution under the action of ultrafast laser pulses. One of the oldest but important models is the two-temperature model for the description of laser-induced metal heating. The idea of this model was

Table 4.1. The processes involved in ultrashort laser–matter interaction

<table>
<tr><th></th><th>Metals</th><th>Semiconductors</th><th>Dielectrics</th></tr>
<tr><td rowspan="2">During laser exposure</td><td></td><td colspan="2">One- and/or multiphoton ionization, collisional avalanche multiplication</td></tr>
<tr><td colspan="3">Free-electron absorption, electron photoemission, ambient gas breakdown</td></tr>
<tr><td rowspan="2">Sub-ps</td><td rowspan="2"></td><td>Ultrafast melting</td><td></td></tr>
<tr><td colspan="2">Free-carrier recombination</td></tr>
<tr><td>ps</td><td colspan="3">Free-carrier recombination, electron–phonon coupling, thermionic electron emission</td></tr>
<tr><td></td><td colspan="3">Melting, ablation (phase explosion, fragmentation, spallation)</td></tr>
<tr><td>ns</td><td colspan="3">Plastic deformations, solidification, shock waves in ambient gas and plasma etching, gas phase cluster growth</td></tr>
</table>

proposed by Kaganov et al. [37] even before the first lasers could strike the matter. The authors assumed that, under extremely fast heating, the electron-absorbing laser radiation could gain a temperature T_{e} higher than that of the lattice ions (T_{l}), and the dynamics of thermalization between electrons and lattice could be described by a simple relaxation equation with a characteristic time t_{r}

$$\frac{\partial T_{\mathrm{l}}}{\partial t} = \frac{T_{\mathrm{e}} - T_{\mathrm{l}}}{t_{\mathrm{r}}}. \tag{4.3}$$

In the first decade after the discovery of the laser effect, when the field of laser interaction with solids was rapidly developed, this idea obtained the form of a widely known, two-temperature model (TTM), due to Anisimov et al. [38]. The model is based on the assumption of thermal equilibrium within the electronic and lattice subsystems independently, and is expressed in the form of heat-flow equations for electrons absorbing laser radiation and lattice heated due to heat exchange from electrons:

$$C_{\mathrm{e}}\frac{\partial T_{\mathrm{e}}}{\partial t} = \frac{\partial}{\partial z}K_{\mathrm{e}}\frac{\partial T_{\mathrm{e}}}{\partial z} - g(T_{\mathrm{e}} - T_{\mathrm{l}}) + \alpha(1-R)I_0(t)\exp(-\alpha z), \tag{4.4}$$

$$C_{\mathrm{l}}\frac{\partial T_{\mathrm{l}}}{\partial t} = \frac{\partial}{\partial z}K_{\mathrm{l}}\frac{\partial T_{\mathrm{l}}}{\partial z} + g(T_{\mathrm{e}} - T_{\mathrm{l}}). \tag{4.5}$$

Here, z is the coordinate directed from the target surface to the bulk; C_{e}, C_{l}, K_{e}, and K_{l} are the heat capacities and thermal conductivities of electrons and lattice, respectively; g is the electron–lattice coupling constant; and R is the reflection coefficient of the irradiated sample. Note that the laser energy source term in (4.4) is written on the basis of the Beer–Lambert law. This model treats the material as a *continuum* medium having macroscopic properties, thermal conductivity, and heat capacity, which, however, can be temperature-dependent.

The TTM has proven to be exceptionally important for describing and understanding the dynamics of material heating under the ultrafast laser exposure (e.g. [1,5,14,16,17,39]) and is used as an integral part of more sophisticated approaches, both atomistic (MD simulations [19]) and continuum ones (drift-diffusion approach [31, 40], thermal elastoplastic model [22] or hydrodynamic codes [41–43]). Continuum means a homogeneous medium whose behavior may be characterized by the macroscopic properties and parameters such as density, thermal conductivity, diffusivity, absorption, etc. A most important characteristic of a continuum is the temperature. In consideration of a medium as continuum one, the concept of the local thermal equilibrium is the principal one. In essence, TTM describes the interaction of two continua, electrons and lattice, with broken equilibrium between them; but each system is in local equilibrium within itself.

From above, some principles of application of continuum approaches can be derived. Before undertaking an attempt of modeling, one has to answer the following basic questions:

1. What are the objects and processes under consideration?
2. In which spatial domain and on which time scale can it be or should it be studied?
3. Which approaches may be applied and which of them is most suitable to attain the aims of studies?

What are the objects and the processes under consideration? Answering this question requires an analysis of a medium or a set of objects which is to be studied and the processes that occur within the studied object(s). If this is a collection of particles, it is necessary to determine whether they can be ascribed with a defined temperature and density and if they "behave collectively," i.e., to say they can be described by means of thermodynamic, optical, or kinetic (such as diffusivity) properties. For a gas-phase system, a parameter called the Knudsen number is useful in determining if the continuum approach may be applied or one should use kinetic methods (the Boltzmann equation, DSMC method). The Knudsen number, Kn, is defined as the ratio of the mean free path of the particles to a characteristic size of the ensemble of studied particles [44]. If $Kn < 10^{-2}$ and is kept at such value during the dynamic process in the studied system, the continuum approaches (the Navier–Stokes or Euler equations) may be applied, otherwise, one should use the kinetic methods [44]. Comparison of the simulation results on an argon flow over a cylindrical body obtained with a sophisticated DSMC and with the Navier–Stokes equations for $Kn < 0.01$ demonstrates a high accuracy of the latter in spite of rather crude boundary conditions assuming zero slip at the walls [45]. For the problems of vaporization, the above Knudsen number criterion has also been verified [46]. However, the continuum-flow dynamics is not applicable at

$Kn > 10^{-2}$ because of the breaking local thermal equilibrium, and kinetic modeling approaches should be used to study the gas-dynamic processes in such flows.

If a bulk material or liquid is under investigation, it is important to define if there is a possibility, during the time of interest, of breaking the homogeneous nature of the studied matter (e.g. homogeneous nucleation of vapor phase with subsequent material disintegration to atoms, small clusters, and droplets [18–20], or dismantling of a crystalline solid [47]). This imposes strong restrictions on the application of continuum considerations, and the MD simulations ought to be used to study a matter with structural transformations. For the analysis of strongly nonequilibrium processes and dynamics of thermalization within a system (e.g. energy evolution of electrons in metals and dielectrics absorbing laser radiation [7, 15]), the Boltzmann equation presents the best choice. However, labor-intensiveness and large computational burden limit the applicability of this method for studies, in our cases, of relatively low-laser fluences at the regions where optical response of matter is not significantly disturbed by laser radiation. It should also be mentioned that, for strongly nonequilibrium matter containing electrons as degrees of freedom, the MD technique has to be based on the many-body, potential energy, surface simulations, taking into account the occupation of the electronic levels in the valance band as well as free-electron contribution [48]. So, at high levels of electronic excitations, the models based on an interatomic potential such as Lennard-Jones, Tersoff, or Stillinger–Weber do not allow a reasonable description of ultrashort structural changes [48].

In which spatial domain and in which time scale can it be or should it be studied? This question implies the following:

- Do we have to examine the individual behavior of atoms, molecules, or larger particles, or is it informative enough to consider the overall collective behavior of matter?
- Do we intend to study material that responds to laser light only during the laser pulse action or is it essential to follow the material relaxation, its evolution, and phase transformations on a much larger temporal domain?

The answers to these questions determine in many respects the choice of the modeling approach. Continuum models enable us to consider the behavior of large systems over a long time of their evolution. The limitations on their application are connected with a homogeneous behavior of the studied object(s). Limits imposed on the MD or kinetic approaches relate largely to costs of computer time.

Which approaches may be applied and which of them are most suitable to attain the aims of studies? As it was mentioned above, the TTM model represents an example of a powerful tool which can be independently used for studies of different aspects of material reaction (mainly for metals) to ultrashort laser excitation and it could also be integrated into other sophisticated models, serving as a block module for laser light absorption and material heating. Another group of continuum models, namely the models of continuum elasticity, has proven to have a good applicability for studies of relatively small objects. A continuum shell model used by Yakobson et al. [4] to describe response of carbon nanotubes to axial compression and torsion showed an excellent agreement with observations and MD simulations. This stimulated the applications of the theory of elasticity to micro- and nanosized objects (see [49,50] and references therein). The power of this classic but effective approach to describe the behavior of bulk materials under laser exposure producing strong thermal stresses seems to be underestimated. We have demonstrated its usefulness in revealing the mechanism of microbump and nanojet formation [22] observed experimentally on the nanosized gold films exposed to femtosecond laser radiation [21] and in studying dynamics of refractive index changes and waveguide writing in optical glasses [51]. The third important group of continuum models represents the two-temperature hydrodynamic approaches based on the application of realistic equations of states [41–43]. These models allow elucidating the material evolution from heating to its decomposition, though the decomposition process is crudely described. However, they give valuable information on material response to laser excitation at high laser fluences and, thus, high stresses with the generation of shock-wave features.

With the development of the MD and DSMC methods and computer performance, an opinion spreads that, with these powerful atomistic techniques, needs in continuum modeling will fade. However, up to now, the continuum approaches are still indispensable for a great variety of scientific problems including material science, allowing fast-engineering estimations as well as computations of large systems. On the other hand, as it was demonstrated with the Navier–Stokes equations against the DSMC [45], for relatively dense media kinetic approaches are not advantageous, being time-consuming (note that, for a similar problem solved with the DSMC method, decreasing the Knudsen number implies a significant increasing of computer time). However, in order to get a sophisticated insight into matter behavior on a microscopic level, the atomistic approaches are of benefit.

As a summary of the above mentioned ideas, none of these methods should be disregarded if it can give useful information on an object or a process. Advanced modeling techniques combine continuum and atomistic approaches or different techniques on the same scale level. As examples of successful combinations of different approaches for modeling of ultrashort laser–matter interaction, one can list already mentioned considerations as follows: two-temperature – MD simulations [19], two-temperature – hydrodynamic modeling [41–43], a hydrodynamic (large particles) – DSMC model [52], a combined MD–DSMC model [36], the thermal elastoplastic model [22] and its evolution into optical–thermal–elastoplastic model [51], and many others. In Sect. 4.4, we present a general continuum approach of modeling the laser-induced, charge-transfer effects in materials of different kinds, showing an example of how the modeling process was developing. The model combines the elements of the TTM, kinetics of free-electron generation, and electrodynamics.

4.4 A General Continuum Approach for Modeling of Laser-Induced Surface Charging

The development of this approach was aimed at getting insights into the processes that could be responsible for a "gentle" ablation of dielectric surfaces, which encumber a number of phenomena – among them one mechanism for fast ion emission, otherwise known as Coulomb explosion [30,53–56]. Coulomb explosion (CE) is one of the electronic mechanisms leading to swift ion emission upon laser ablation which was widely discussed for different materials during the last decade [30, 31, 39, 53–59]. The main features of CE are the following: The energetic ions of different species (e.g. aluminum and oxygen in the case of ultrafast laser ablation of sapphire [30]) observed in time-of-flight signals show the same momenta and not the same energy. Also, doubly charged ions have velocities twice as high as singly charged ones (the so-called momentum scaling) in conditions where the ablation rates are low and gas-phase collisions are reduced. This indicates that the ions could be extracted and accelerated in a potential electric field produced inside the target. The electric field can be generated due to intensive electron photoemission, leading to accumulation of positive charge in a superficial target layer. Thus, the fundamental concept of Coulomb explosion is based on the fact that, due to photoemission, the irradiated surface gains a high positive charge, so that, provided that neutralization is not fast enough, the repulsion force between ions exceeds the lattice-binding strength, resulting in surface layer disintegration.

So, the *object* in consideration is a bulk material experiencing a strong charging of surface layer caused by ultrashort laser irradiation. Sapphire, silicon, and gold were chosen as model *objects* of different material classes. The possible processes for studying are listed in Table 4.1 and depend on the material type. The purpose was to study the generation of the electric field inside the target *during the pulse action*. This implies the possibility to limit the range of the considered processes to the following set:

- Free-carrier generation in dielectrics and semiconductors and associated changes in optical response
- Electron photoemission
- Electric-field generation
- Charge-carrier relocation under the action of the electric field

These processes can be presented in the form of a scheme shown in Fig. 4.1 in the oval frames. We address a *macroscopic* Coulomb explosion that is removal of at least several monolayers as observed experimentally for sapphire [30]. Hence, we opted for a *continuum* approach. The existing models which treat most, or at least some of the processes described above may, by convention,

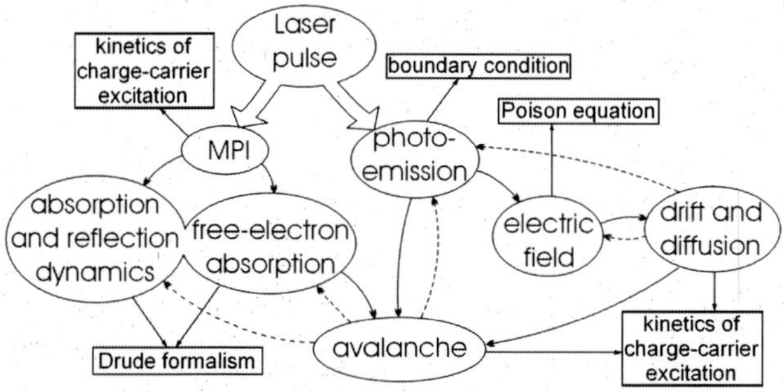

Fig. 4.1. Schematic representation of the laser-induced processes leading to dielectric material charging and the associated equation types. Major interconnections between different processes are shown by the *arrows*. Direct effects are indicated by *arrows* with hard lines and backward (feedback) influences are shown with *broken lines*

be divided into three groups. One of the dominating approaches to describe carrier dynamics in silicon targets is based on the ambipolar diffusion with an implicit assumption of an equal number of electrons and holes in the solid and the preservation of local quasi-neutrality of the sample [24, 60, 61]. Another group of models, developed for semiconductors irradiated by laser pulses [62] and dielectrics under the action of electron beams [63, 64], takes into account the generation of local electric fields inside the target with the assumption that the target remains neutral as a whole. This implies the absence of electron photoemission [62] or relies on a secondary electron emission equal to the absorbed electron flux [63, 64]. A third approach [65, 66] proposed for the case of a dielectric target (MgO) irradiated by a laser pulse of nanosecond duration may be labeled as the drift-diffusion one. The authors studied the self-consistent generation of the electric field, as a result of laser heating and thermionic emission of the electrons excited to the conduction band and their diffusion and drift in the locally established fields. It was found that the self-consistent electric field could reach values exceeding $10^8\,\mathrm{V\,m^{-1}}$ under normal ablation conditions. This third approach perfectly fits the goal of our studies and was taken as a basis for the development of the drift-diffusion model for the case of ultrashort laser irradiation of materials of different classes.

To compose a set of equations describing the laser-interaction phenomenon under study, each process presented in Fig. 4.1 is assigned with a proper equation (shown in the rectangular frames). Therefore, the following equations are assumed to be solved self-consistently:

1. The kinetics of charge-carrier generation can be described in the frames of a rate equation similar to (4.1), taking into account that free carriers may relocate under the action of the electric field and/or density and temperature gradients. Thus, the general continuity equation takes the form:

$$\frac{\partial n_x}{\partial t} + \frac{1}{q_x}\frac{\partial J_x}{\partial z} = S_x + L_x. \tag{4.6}$$

Here, S_x and L_x are the source and loss terms describing the free-carrier populations; q_x and n_x denote the carrier charges and densities with subscript $x = e; h$ representing electrons and ions (holes), respectively; and J_x is the carrier flux density. In metals $S_x = 0$ while the loss term L_e can take into account the electron photoemission from the surface of the irradiated sample. Alternatively, one can set $L_x = 0$ and use the electron photoemission flux density as a boundary condition. For each dielectric or semiconductor material, the source and loss terms are individually constructed taking into account the photoionization mechanism, avalanche multiplication, electron photoemission, and recombination which can proceed either as electron trapping with generation of the defect states or, as well, via the Auger and photo-recombination mechanisms, as mentioned above.

2. The expressions for the electric current densities $J_x = q_x n_x v_x$ with v_x being the directional velocity include drift and diffusion terms [66] and can be considered as the equation of motion:

$$J_\mathrm{e} = -en_\mathrm{e}\mu_\mathrm{e}E - eD_\mathrm{e}\nabla n_\mathrm{e} \quad J_\mathrm{h} = |e|n_\mathrm{h}\mu_\mathrm{h}E - |e|D_\mathrm{h}\nabla n_\mathrm{h}, \tag{4.7}$$

where e and $|e|$ are the electron and hole charges and μ_x is the charge carrier mobility. The time- and space-dependent diffusion coefficient D_x can be calculated according to the Einstein relation as $D_x = kT_x\mu_x/|e|$ with T_x representing the carrier temperature. We assume that the charge-carrier flows are caused by quasi-neutrality violation on and below the target surface due to electron photoemission and strong density and temperature gradients. In semiconductors, the holes are mobile and their current strongly influences the charging dynamics, otherwise one can assume $J_\mathrm{h} = 0$.

3. The electric field generated as a result of breaking quasi-neutrality in the irradiated target (quasi-stationary with respect to the laser oscillating field) is calculated with the Poisson equation:

$$\frac{\partial E}{\partial z} = \frac{|e|}{\varepsilon_r\varepsilon_0}(n_\mathrm{i} - n_\mathrm{e}). \tag{4.8}$$

4. As the diffusion coefficient is dependent on the temperature of charge carriers, the energy conservation equations have to be applied to account for the heating of electronic and lattice subsystems. We assume [40] that laser-excited metals and strongly ionized insulators and semiconductors can be considered as dense plasmas so that TTM [37, 38] may be applied for an energy-balance description:

$$C_\mathrm{e}\left(\frac{\partial T_\mathrm{e}}{\partial t} + \frac{J_\mathrm{e}}{en_\mathrm{e}}\frac{\partial T_\mathrm{e}}{\partial z}\right) = \frac{\partial}{\partial z}K_\mathrm{e}\frac{\partial T_\mathrm{e}}{\partial z} - g(T_\mathrm{e} - T_\mathrm{l}) + \Sigma(z,t), \tag{4.9}$$

$$(C_\mathrm{l} + L_\mathrm{m}\delta(T_\mathrm{l} - T_\mathrm{m}))\frac{\partial T_\mathrm{l}}{\partial t} = \frac{\partial}{\partial z}K_\mathrm{l}\frac{\partial T_\mathrm{l}}{\partial z} + g(T_\mathrm{e} - T_\mathrm{l}). \tag{4.10}$$

Even if complete equilibration does not take place in the electronic system, the value T_e can be considered as a measure for the average energy of electrons. In (4.9) and (4.10), index l refers to the lattice parameters; C_e, C_l, K_e, and K_l are the heat capacities and thermal conductivities of electrons and lattice, respectively; g is the electron–lattice coupling constant; and $\Sigma(z, t)$ is the laser energy source term. All these parameters are specific for each type of material. We introduced into the energy equations several features (compare with (4.4) and (4.5)). The term $L_m\delta(T_l–T_m)$ allows calculations of the liquid–solid interface, having the temperature T_m. L_m is the latent heat of fusion [67]. Also, energy transport provided by free electrons is taken into account.

The source term in (4.9) should be constructed to account for the energy balance of the electrons. In metals this term can be simply described by the Beer–Lambert law as in (4.4) or, additionally, can take into account the ballistic electron transport [16]. In wide-bandgap dielectrics, the processes of free-electron generation via photoionization, the energy expenses used for the development of avalanche multiplication, free-electron absorption, and the energy localized in the strained lattice via trapping processes play a role in the overall electron energy balance. As an example, here, the source term is given for a silicon sample irradiated with an ultrashort laser pulse of 800 nm wavelength:

$$\frac{\partial E_f}{\partial t} = \left((\hbar\omega - E_g)\frac{\sigma_1 I}{\hbar\omega} + (2\hbar\omega - E_g)\frac{\sigma_2}{2}\frac{I^2}{\hbar\omega} - E_g\delta n_e \right)\frac{n_a}{n_a + n_i} + \alpha_{ab}(x,t)I(x,t) + E_g\frac{n_e}{(\tau_0 + 1/Cn_e n_i)} - E_e PE(x,t). \quad (4.11)$$

Here, E_f and E_e are the energy density of the electronic subsystem and the average energy per electron ($E_e = 3kT_e/2$), respectively; E_g is the bandgap energy; σ_1 and σ_2 are the cross-sections for one- and two-photon ionization; δ is the avalanche rate constant; α_{ab} is the absorption coefficient; $C = 3.8 \times 10^{-31}\,cm^6\,s^{-1}$ is the Auger recombination coefficient [61]; and $\tau_0 = 6 \times 10^{-12}\,s$ [25]. The last term in (4.11) takes into account the energy carried away from the target with the ejected photo-electrons.

The details concerning the photoemission description in the particular problem depicted here can be found in [31, 40, 59]. It should be noted that, in studying the possibility of high surface charging which could be realized in Coulomb explosion, we do not limit the electron photoemission term by considering an influence of the generated electric field on the electron work function. Hence, our modeling results most probably overestimate the photoelectron yield from the laser-irradiated materials, though being reasonable by the integral value [31, 40]. For more accurate description of the photoelectron yield, the effective charge potential barrier can be used [68] as discussed in [69]. Concerning the term describing the free-carrier absorption, a comment should be made on the dielectric permittivity. As mentioned, the optical response of laser-excited dielectrics and semiconductors on laser irradiation (the absorption and reflection coefficients) can be calculated via the high-frequency, complex dielectric function presented by (4.2). The dielectric permittivity ε_r in the Poisson equation (4.8) represents the material response on the slow varying (100fs-ps) electric field generated due to electron photoemission and charge separation (quasi-dc electric field) and, thus, it does not relate to (4.2). The dc-dielectric permittivity can be determined from the characteristic time of dielectric (Maxwell) relaxation $t_M = \varepsilon_r \varepsilon_0 / \sigma$, where σ is the electric conductivity. For most metals, the Maxwell relaxation time is below 1 fs [70] that gives for gold $\varepsilon_r \sim 10^3$.

An important feature of modeling laser-induced breakdown of dielectric and semiconductor materials is the time- and space-dependent absorption coefficient that requires calculating the spatial and temporal distribution of the laser intensity in the sample. For a case of silicon presented by (4.11), this reads as

$$\frac{\partial}{\partial x} I(x,t) = -\left(\sigma_1 \frac{n_a}{n_a + n_i} + \sigma_2 \frac{n_a}{n_a + n_i} I(x,t) + \alpha_{ab}(x,t) \right) I(x,t). \quad (4.12)$$

More details on the approach presented in this section can be obtained in [31,40]. The application of this model to metal, semiconductor, and dielectric materials has allowed to make an important conclusion that, for bulk materials at fluences slightly above the ablation threshold, the macroscopic Coulomb explosion can be observed only for dielectrics while in metals and semiconductors this mechanism of ablation is strongly inhibited due to the high mobility of charge carriers. An example is calculated for gold, silicon, and sapphire, representing materials of different classes. The irradiation regimes correspond to the experimental conditions of [30, 56]: 800-nm laser wavelength and 100-fs pulse duration. Laser fluences were chosen to be slightly above the ion emission thresholds for each material (4, 0.8, and 1.2 J cm^{-2} for Al_2O_3, Si, and Au, respectively). The spatial distributions of the self-consistently generated electric field are presented in Fig. 4.2 at time instants corresponding

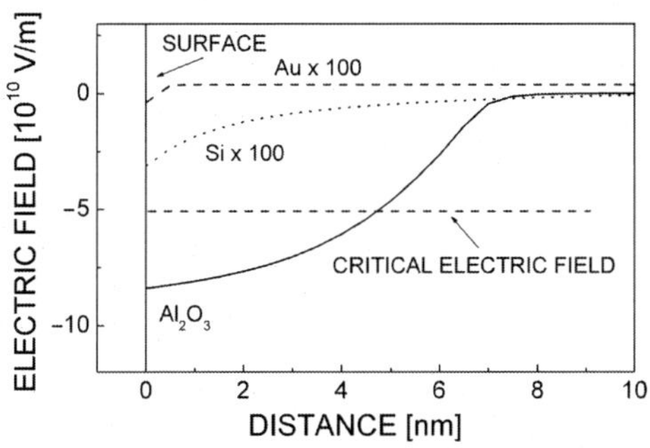

Fig. 4.2. Calculated spatial profiles of the electric field induced in metals, semiconductors, and dielectrics at time moments of reaching the maximum values of the electric field for every material for the regimes employed in the experiments [30,56]. Laser fluences are slightly above the ion emission thresholds for each material. The specific irradiation conditions are given in the text

to reaching the field maximum values for each material, as calculated with the present model together with levels corresponding to the internal material strength. For sapphire, the layer with overcritical electric field where electrostatic disintegration of the lattice should occur is approximately 40 Å wide, in excellent agreement with the experimental estimation of the Coulomb-exploded region [30]. With semiconductors and metals, the higher electron mobility and higher density of available free electrons ensure effective screening and a much smaller, net positive charge accumulated during the laser pulse. This is not sufficient to induce a macroscopic, electrostatic break-up of the outer layers of the substrate. The maximum values of the electric field are only 4.1×10^7 and $3.4 \times 10^8\,\mathrm{V\,m^{-1}}$ for gold and silicon, respectively in the irradiation conditions given above.

At high laser fluences, when thermal ablation and plasma formation take place, the momentum scaling in the fast ion distribution is widely observed. However, it is difficult to state unambiguously the CE origin of these fast ions as the plasma effects can be responsible for ion acceleration in the gas phase [32, 71]. The experimental evidence of Coulomb explosion in metals was demonstrated at much higher laser fluences than usually applied in laser-material processing. Generation of high electric fields in the order of $10^{10}\,\mathrm{V\,m^{-1}}$ was achieved in microscopic metal targets under ultra-intense laser irradiation ($\sim 10^{19}\,\mathrm{W\,cm^{-2}}$) [72]. At low laser fluences, below the melting threshold, nonthermal ion emission can be explained by a field enhancement related to the optical rectification effect [73] at nanoscale protuberances. However, the latter process cannot be described within a continuum approach as it refers to a desorption process of the separate, loosely bonded ions.

A final remark on the application of the drift-diffusion approach concerns the numerical procedures. During the numerical integration of the involved equations, a special care should be taken with respect to energy and particle conservation, controlling the free-electron generation, recombination, their supply through the remote boundary, and photoemission.

4.5 Concluding Remarks

Thus, we have demonstrated that the drift-diffusion continuum approach has allowed elucidating the interrelating processes taking place in materials irradiated with ultrashort laser pulses. The important feature of the model consists of taking into account quasi-neutrality breaking that is difficult to consider with other approaches. Moreover, the modeling results indicating the less-probable occurrence of CE in metal and semiconductors in view of their enhanced transport properties stimulated further studies on the origin of fast ions. New effects were consequently discussed, including field-enhancement and energetic emission by surface optical rectification effect [73] and ion acceleration mechanism in the gas phase [32]. It should be noted that particle desorption by localized charge trapping also leads to fast ion detection in the desorption products [74, 75].

Returning to the importance of the different continuum approaches, the problem of plasma chemistry should be mentioned. Femtosecond laser micromachining is applied mainly under atmospheric conditions. This leads to an unavoidable formation of air plasma in front of the irradiated samples and results in shock-wave generation as shown in [32, 33]. The shocked ambient plasma can serve as an additional factor of modification for the laser-irradiated material [32]. The target surface in contact with the compressed plasma in the shock-wave front is exposed to fluxes of ions, electrons, and neutral atoms. This can result in a substantial modification of the surface via mechanical sputtering and chemical etching. Clear indications of ambient plasma effects are observed in numerous studies (e.g. [34, 35, 76–78]). The involved chemical processes depend on both – the ambient gas composition as well as the type of target material and can be described in the frames adapted for plasma chemistry [79]. The studies of the ambient plasma effects can stimulate an increased interest to the continuum modeling approaches as discussed in [32].

References

1. B.N. Chichkov, C. Momma, S. Nolte, F. Von Alvensleben, A. Tünnermann, Appl. Phys. A **63**, 109 (1996)
2. D. Bäuerle, *Laser Processing and Chemistry* (Springer, Berlin, 2000)
3. J. Koch, F. Korte, T. Bauer, C. Fallnich, A. Ostendorf, B.N. Chichkov, Appl. Phys. A **81**, 325 (2005)
4. B.I. Yakobson, C.J. Brabec, J. Bernholc, Phys. Rev. Lett. **76**, 2511 (1996)
5. Z. Lin, L.V. Zhigilei, V. Celli, Phys. Rev. B **77**, 075133 (2008)
6. L.V. Keldysh, Sov. Phys. – JETP **20**, 1307 (1965)
7. A. Kaiser, B. Rethfeld, M. Vicanek, G. Simon, Phys. Rev. B **61**, 11437 (2000)
8. V.E. Gruzdev, Phys. Rev. B **75**, 205106 (2007)
9. K. Sokolowski-Tinten, D. von der Linder, Phys. Rev. B **61**, 2643 (2000)
10. A. Rousse, C. Rischel, S. Fourmaux, I. Uschmann, S. Sebban, G. Grillon, Ph. Balcou, E. Förstler, J.P. Geindre, P. Audebert, J.C. Gauthier, D. Hulin, Nature **410**, 65 (2001)
11. B.C. Stuart, M.D. Feit, A.M. Rubenchik, B.W. Shore, M.D. Perry, Phys. Rev. Lett. **74**, 2248 (1995)
12. B.C. Stuart, M.D. Feit, S. Herman, A.M. Rubenchik, B.W. Shore, M.D. Perry, Phys. Rev. B **53**, 1749 (1996)
13. B. Rethfeld, Phys. Rev. Lett. **92**, 187401 (2004)
14. C.V. Shank, R. Yen, C. Hirlimann, Phys. Rev. Lett. **50**, 454 (1983)
15. B. Rethfeld, A. Kaiser, M. Vicanek, G. Simon, Phys. Rev. B **65**, 214303 (2002)
16. S.-S. Wellershoff, J. Hohlfeld, J. Güdde, E. Matthias, Appl. Phys. A **69**, S99 (1999)
17. N.M. Bulgakova, I.M. Bourakov, Appl. Surf. Sci. **197**, 41 (2002)
18. L.V. Zhigilei, Appl. Phys. A **76**, 339 (2003)
19. E. Leveugle, D.S. Ivanov, L.V. Zhigilei, Appl. Phys. A **79**, 1643 (2004)

20. D. Bouilly, D. Perez, L.J. Lewis, Phys. Rev. B **76**, 184119 (2007)
21. F. Korte, J. Koch, B.N. Chichkov, Appl. Phys. A **79**, 879 (2004)
22. Y.P. Meshcheryakov, N.M. Bulgakova, Appl. Phys. A **82**, 363 (2006)
23. J. Bonse, G. Bachelier, J. Siegel, J. Solis, H. Sturm, J. Appl. Phys. **103**, 054910 (2008)
24. E.J. Yoffa, Phys. Rev. B **21**, 2415 (1980)
25. J. Bok, M. Combescot, Phys. Rev. Lett. **47**, 1564 (1981)
26. A.M. Stoneham, N. Itoh, Appl. Surf. Sci. **168**, 186 (2000)
27. A.M. Stoneham, Physica B **340–342**, 48 (2003)
28. P. Stampfli, K.H. Bennemann, Phys. Rev. B **49**, 7299 (1994)
29. R. Haight, Surf. Sci. Rep. **21**, 277 (1995)
30. R. Stoian, D. Ashkenasi, A. Rosenfeld, E.E.B. Campbell, Phys. Rev. B **62**, 13167 (2000)
31. N.M. Bulgakova, R. Stoian, A. Rosenfeld, I.V. Hertel, E.E.B. Campbell, Phys. Rev. B **69**, 054102 (2004)
32. N.M. Bulgakova, A.V. Bulgakov, V.P. Zhukov, W. Marine, A.Y. Vorobyev, C. Guo, Proc. SPIE **7005**, 70050C (2008)
33. N.M. Bulgakova, V.P. Zhukov, A.Y. Vorobyev, C. Guo, Appl. Phys. A **92**, 883 (2008)
34. A. De Giacomo, O. De Pascale, Thin Solid Films **453–454**, 328 (2004)
35. A. Weck, T.H.R. Crawford, D.S. Wilkinson, H.K. Haugen, J.S. Preston, Appl. Phys. A **90**, 537 (2008)
36. T.E. Itina, K. Gauriet, L.V. Zhigilei, S. Noël, J. Hermann, M. Sentis, Appl. Surf. Sci. **253**, 7656 (2007)
37. M.I. Kaganov, I.M. Lifshitz, M.V. Tanatarov, Sov. Phys. – JETP **4**, 173 (1957)
38. S.I. Anisimov, B. Kapeliovich, T.L. Perel'man, Sov. Phys. – JETP **39**, 375 (1974)
39. N.M. Bulgakova, I.M. Burakov, Y.P. Meshcheryakov, R. Stoian, A. Rosenfeld, I.V. Hertel, J. Laser Micro/Nanoeng **2**, 76 (2007)
40. N.M. Bulgakova, R. Stoian, A. Rosenfeld, I.V. Hertel, W. Marine, E.E.B. Campbell, Appl. Phys. A **81**, 345 (2005)
41. J.P. Colombier, P. Combis, A. Rosenfeld, I.V. Hertel, E. Audouard, R. Stoian, Phys. Rev. B **74**, 224106 (2006)
42. J.P. Colombier, P. Combis, R. Stoian, E. Audouard, Phys. Rev. B **75**, 104105 (2007)
43. M.E. Povarnitsyn, T.E. Itina, K.V. Khishchenko, P.R. Levashov, Appl. Surf. Sci. **253**, 6343 (2007)
44. G.A. Bird, *Molecular Gas Dynamics and the Direct Simulation of Gas Flows* (Oxford University Press, Oxford, 1994)
45. A.J. Lofthouse, I.D. Boyd, M.J. Wright, AIAA Paper 2006–993 (2006)
46. N.M. Bulgakova, M.Y. Plotnikov, A.K. Rebrov, Fluid Dyn. **32**, 870 (1997)
47. Y. Kubota, R. Matsumoto, M. Nakagaki, Key Eng. Mater. **340–341**, 985 (2007)
48. H.O. Jeschke, M.E. Garcia, in *Nonlinear Optics, Quantum Optics and Ultrafast Phenomena with X-rays*, ed. by B.W. Adams (Kluwer, Boston/Dordrecht/London, 2003) p. 175
49. J. Peddieson, G.R. Buchanan, R.P. McNitt, Int. J. Eng. Sci. **41**, 305 (2003)
50. Q. Wang, V.K. Varadan, Smart Mater. Struct. **14**, 281 (2005)
51. A. Mermillod-Blondin, I.M. Burakov, Y.P. Meshcheryakov, N.M. Bulgakova, E. Audouard, A. Rosenfeld, A. Husakou, I.V. Hertel, R. Stoian, Phys. Rev. B **77**, 104205 (2008)

52. T.E. Itina, J. Hermann, P. Delaporte, M. Sentis, Phys. Rev E **66**, 066406 (2002)
53. H. Varel, D. Wämer, A. Rosenfeld, D. Ashkenasi, E.E.B. Campbell, Appl. Surf. Sci. **127–129**, 128 (1998)
54. J. Reif, M. Henyk, D. Wolfframm, SPIE Proc. **3933**, 26 (2000)
55. F. Costache, J. Reif, Thin Solid Films **453–454**, 334 (2004)
56. R. Stoian, A. Rosenfeld, D. Ashkenasi, I.V. Hertel, N.M. Bulgakova, E.E.B. Campbell, Phys. Rev. Lett. **88**, 097603 (2002)
57. H. Dachraoui, W. Husinsky, G. Betz, Appl. Phys. A **83**, 333 (2006)
58. A. Kaplan, M. Lenner, R.E. Palmer, Phys. Rev. B **76**, 073401 (2007)
59. W. Marine, N.M. Bulgakova, L. Patrone, I. Ozerov, J. Appl. Phys. **103**, 094902 (2008)
60. S.S. Mao, X.-L. Mao, R. Greif, R.E. Russo, Appl. Surf. Sci. **127–129**, 206 (1998)
61. H.M. van Driel, Phys. Rev. B **35**, 8166 (1987)
62. T. Held, T. Kuhn, G. Mahler, Phys. Rev. B **44**, 12873 (1991)
63. A. Melchinger, S. Hofmann, J. Appl. Phys. **78**, 6224 (1995)
64. A. Miotello, M. Dapor, Phys. Rev. B **56**, 2241 (1997)
65. R.M. Ribeiro, M.M.D. Ramos, A.M. Stoneham, J.M. Correia Pires, Appl. Surf. Sci. **109–110**, 158 (1997)
66. R.M. Ribeiro, M.M.D. Ramos, A.M. Stoneham, Thermophys Aeromech **5**, 223 (1998)
67. S.P. Zvavyi, G.D. Ivlev, Inzh.-Fiziol. Zh. **69**, 790 (1996) (in Russian)
68. D.M. Riffe, X.Y. Wang, M.C. Downer, D.L. Fisher, T. Tajima, J.L. Erskine, R.M. More, J. Opt. Soc. Am. B **10**, 1424 (1993)
69. N.M. Bulgakova, A. Rosenfeld, L. Ehrentraut, R. Stoian, I.V. Hertel, Proc. SPIE **6732**, 673208 (2007)
70. Z. Insepov, A. Hassanein, D. Swenson, M. Terasawa, Nucl. Instrum. Meth. B **241**, 496 (2005)
71. R. Stoian, A. Rosenfeld, I.V. Hertel, N.M. Bulgakova, E.E.B. Campbell, Appl. Phys. Lett. **85**, 694 (2004)
72. M. Borghesi, L. Romagnani, A. Schiavi, D.H. Campbell, M.G. Haines, O. Willi, A.J. Mackinnon, M. Galimberti, L. Gizzi, R.J. Clarke, S. Hawkes, Appl. Phys. Lett. **82**, 1529 (2003)
73. A. Vella, B. Deconihout, L. Marucci, E. Santamato, Phys. Rev. Lett. **99**, 046103 (2007)
74. J.T. Dickinson, S.C. Langford, J.J. Shin, D.L. Doering, Phys. Rev. Lett. **73**, 2630 (1994)
75. J. Kanasaki, M. Nakamura, K. Ishikawa, K. Tanimura, Phys. Rev. Lett. **89**, 257601 (2002)
76. J.P. McDonald, S. Ma, T.M. Pollock, S.M. Yalisove, J.A. Nees, J. Appl. Phys. **103**, 093111 (2008)
77. T. Lehecka, A. Mostovych, J. Thomas, Appl. Phys. A **92**, 727 (2008)
78. C.H. Crouch, J.E. Carey, J.M. Warrender, M.J. Aziz, E. Mazur, F.Y. Génin, Appl. Phys. Lett. **84**, 1850 (2004)
79. H.F. Winter, J.W. Coburn, Surf. Sci. Rep. **14**, 161 (1992)

5

Cluster Synthesis and Cluster-Assembled Film Deposition in Nanosecond Pulsed Laser Ablation

Paolo M. Ossi

Summary. Pulsed laser ablation in an ambient gas, using nanosecond pulses, allows producing films whose elementary building blocks are atomic or molecular clusters. When such clusters are grown in the expanding plasma, their size, energy at landing and mobility onto the substrate surface affect film morphology and structure. The deposition parameters, such as laser wavelength and energy density, target to substrate distance, nature and pressure of the ambient gas influence the expansion of the ablation plasma and consequently cluster size and kinetic energy, together with the related distributions. In this chapter, the phenomenology of plasma expansion through an ambient gas is presented and the most popular models that describe plume propagation are critically reviewed. A phenomenological model of mixed-propagation that aims at capturing the main features of the collisional processes behind the formation of nanoclusters during plume attenuation and thermalisation in the gas is discussed and the growth of clusters nucleated in the plume is modelled.

Model predictions are compared to representative experiments including ablation of elemental and compound targets, as well as deposition of nanometer-sized cluster-assembled films, illustrating the complex dependence of cluster size on ambient gas properties and on plume energetics. Plasma chemistry and its relevance in the presence of a reactive background gas are discussed with attention to the growth rate of nanostructured films.

5.1 Introduction

The presence of clearly recognisable particles in materials synthesized by pulsed laser deposition (PLD) is rather common and it was observed since the first scanning electron microscopy (SEM) observations, both of the surface morphology of films prepared by this technique and of laser irradiated targets [1]. Big particles, with size ranging from hundreds of nanometers to a few micrometers are classified as particulates and are considered a kind of debris whose accidental occurrence constitutes a potentially severe drawback of a deposited film. The surface analysis of metallic targets supports one of the pictures of particle formation in PLD when nanosecond pulses are used: accordingly droplets are directly ejected from the irradiated surface by a hydrodynamic mechanism [2]; in turn, photomechanical effects driven by the relaxation of laser-induced stresses [3] can give rise to the formation/expulsion of large liquid droplets and solid particles.

When ultra-short laser pulses, in the *fs* range, are used, efficient cluster-nanoparticle (the two words are used as equivalent throughout this chapter) production is observed. In the latter, experimental configuration clusters are ejected *directly* from the target following localised target disintegration by a laser-induced explosion-like process [4]. The associated phenomenology and the related mechanisms are discussed in Chaps. 6 and 3, respectively.

In the past, most attention was paid to obtain compact films, with controlled thickness, density and surface roughness, besides being well adherent to the substrate. Such requisites are important when manufacturing surface microstructures and in protective films with elevated mechanical properties, resistant to wear and corrosion, for application in severe conditions.

In the last 10 years, however, several experiments have investigated the mechanisms by which a laser pulse of duration in the nanosecond range, impinging onto a solid surface generates a non negligible number of clusters, besides neutral and excited atoms, electrons and ions with different charge states. In particular, even just above the ablation threshold (of the order of $1\,\mathrm{J\,cm^{-2}}$), given a fixed laser power density deposited at the target surface, with decreasing laser wavelength an increase of the total mass, of the number density and of the kinetic energy of the species ejected from the target is observed. Pulses from excimer lasers predominantly produce vapours consisting of high-energy individual atoms, either excited, or ionised. On the opposite side, laser pulses with wavelength in the IR region produce vapours formed mostly by clusters with large mass number and relatively low kinetic energy.

The above studies were at least in part motivated by the interest in cluster-assembled (CA) films that are an example of the bottom–up strategy to obtain nanostructured materials. Indeed, when the number of atoms per particle is progressively reduced from several tens of thousands atoms to a few atoms, strong variations in the surface to bulk atom ratio occur and dramatic changes are expected in the physico–chemical properties of the material that depend on such a ratio [5]. In fact, vibrant research activity was driven by the predicted and in part observed unusual transport (electronic, optical), magnetic, chemical properties that characterise matter when its typical sizes are pushed to the nanometer scale; in particular, attention was focussed on noble metal nanoparticles (NPs), exploiting their optical properties, of interest in surface enhanced spectroscopies and plasmonics, while their catalytic behaviour finds application in the growth of nanotubes and nanorods [6].

Even limiting our attention to quasi two-dimensional systems like films are propedeutical to any meaningful application of CA materials is a careful control of cluster size, composition, number density on the supporting substrate in the *initial* stage of deposition, besides a knowledge and control of the mode of mutual cluster aggregation when an extended film is formed.

It was soon realised that laser ablation in an ambient gas strongly favours cluster formation [7]. Yet, the characterisation of NPs produced by PLD is difficult and complex due to the interplay among thermal and electronic processes, which are considerably more entangled than in vacuum depositions. Understanding how clusters are synthesized by laser ablation in an ambient gas requires knowing the time and space scales for nanocluster formation, besides the mechanisms of their transport and deposition onto a substrate. A key issue to control cluster production is the knowledge of *when* and *where* the clusters are formed. This is the objective of the present contribution. Behind these questions is a picture of nanoparticle formation by PLD that assumes that they are synthesized in gas phase [7–9].

A further difficulty in this study stems from a traditional separation between two approaches to analyse thin films grown by PLD. On the one side, much attention is dedicated to establishing useful correlations between film properties and deposition parameters, such as ambient gas nature and pressure, target-substrate distance, laser wavelength, power density deposited at the target surface, number of pulses [10]. On the other side, attention is driven by plasma expansion dynamics; in this family of studies, time and space resolved plasma diagnostics provide us with details on the dynamics of the ejected species, mainly by excimer laser ablation in ambient gases of different compositions. The most used diagnostics are optical emission spectroscopy (OES) [11], optical time of flight measurements (TOF) [12], laser-induced fluorescence (LIF) [13], Langmuir probes [14] and fast photography, using an intensified charge coupled device (ICCD) [7]. The latter allows obtaining two-dimensional pictures of the expanding plasma plume from which plasma front position and velocity are obtained, besides the length and shape of the expanding plasma. After solving for conservation of energy, momentum and mass [15], vapour temperature, pressure and density are obtained. The above classes of analyses complement each other because the latter can provide detailed information on the dynamics of the relevant species in the deposition process, yet only in the last few years the first plasma plume studies were carried out at conditions suitable to have a feedback on film deposition.

In this chapter, a review is offered of the phenomena associated to the propagation through an ambient gas of a plasma plume produced by a nanosecond laser pulse. The main features of such phenomena are critically discussed in terms of the most popular models adopted to interpret plasma expansion, before introducing a phenomenological model of mixed-propagation for plume dynamics through the ambient gas. Recent results on the deposition of CA films of carbon, silicon, tin, $LaMnO_3$ and tungsten under different conditions are interpreted. It is shown that the average asymptotic sizes of the *deposited* NPs, as deduced by mixed-propagation model, nicely agree with those measured by transmission electron microscopy (TEM) on suitable substrates. Plasma chemistry and its relevance in presence of a reactive background gas are discussed, with particular attention to the attainment of specific NP stoichiometries and to the growth rate of a nanostructured film, with defined size of the constituent NPs.

5.2 Phenomenology of Plume Expansion through an Ambient Gas

The target surface upon absorbing the laser pulse undergoes strong heating at a rate of the order of $10^{12}\,\mathrm{Ks}^{-1}$ in most materials. As a consequence, intense matter evaporation occurs, with the concurrent formation of a dense, highly anisotropic (the area is comparatively large, the thickness being very shallow) vapour cloud, lying just above the irradiated target surface. In the early expansion stage, the pressure inside the vaporised matter, of the order of several MPa, is orders of magnitude higher than the surrounding ambient gas pressure. Such a high density, strongly collisional volume (typical particle number density $n \sim 10^{19}\,\mathrm{cm}^{-3}$, particle mean free path $\lambda \sim 1\,\mu\mathrm{m}$) behaves like a high temperature fluid that after further interaction with the laser radiation forms an isothermal expanding plasma up to the end of the pulse. Both forward and lateral expansion is observed, due to the high recoil pressure from the target surface until the plume is transparent to the incident laser beam. Strong laser-plasma interaction, following intense ionisation of plasma species, creates a further high-pressure kinetic energy region that stimulates additional plume expansion. At the end of the laser pulse, particle ejection from the target surface ceases.

Direct observation [16] indicates that in the range of ambient pressure from vacuum to 1 Pa the plume undergoes *free* propagation whose features are described theoretically as an adiabatic self-similar expansion of an elliptical gas cloud in the vacuum [17]. Basically, expansion is driven by the pressure gradients of the plume that experiences a dominant acceleration in the direction normal to the target surface, where the initial plume size is smaller. Thus the plasma quickly elongates in the forward direction, as observed experimentally. A fast conversion of thermal energy into kinetic energy corresponds to high expansion velocities of the plume, around $10^4 - 10^5\,\mathrm{ms}^{-1}$. With increasing plume size, the acceleration of the plume front decreases asymptotically to zero, so that the front quickly reaches a constant asymptotic velocity. In turn, plasma temperature shows an initial (within 10^2 ns) quick drop, followed by a smoother one at longer times. At last, in vacuum the plasma propagates in the same way as a supersonic expansion with a *linear* relation between the delay time and the position of the plume front. Only a weak fluorescence is visible close to the target, due to the collisions between the plasma species that occur just after the end of the laser pulse. Conceptually, ablation can be considered as an extension of thermal desorption, thereby monolayers consecutively evaporate from the target surface in a quasi-equilibrium way. The presence of NPs in the plume, although as a minor constituent, is explained by a condensation model [15]. Just to offer a few examples, free plume propagation is observed when aluminium is ablated in atmospheric air between 1.3×10^{-4} and 1.3 Pa, using pulses of 8 ns duration at 532 nm, depositing a power density of 3×10^7 Wmm^{-2} per pulse [18]. Plume shows free propagation between vacuum and a few tens of pascal (pictures are displayed for 0.3 Pa) when $LaMnO_3$ is ablated in O_2 with pulses of 20 ns duration at 351 nm, the deposited power density per pulse being 7.5×10^5 Wmm^{-2} [19].

With increasing gas pressure the collisions between species ejected from the target and ambient gas decelerate the expanding plume and lead to formation of *shock waves*. Given the high density of the ambient gas, the ablated material experiences maximum braking in the direction normal to the target as compared to the expansion in radial directions. This results in the observed *spherical* shape of the propagating plume. For ambient gas pressures relatively low (below 10 Pa), the initial plume expansion is similar to that in vacuum [19, 20, for Al ablation in N_2], but at times longer than 1 μs at most, plume front slows down due to the confinement effect of the background gas. At larger times the plume sharpens and its front shows an oscillatory behaviour that persists over a range of ambient gas pressures (up to a few tens of pascal), usually occurring at earlier times for higher pressures. As an example, in Fig. 5.1 are shown pictures taken at different delay times of plumes ablated from a SnO_2 target, freely propagating in vacuum (Fig. 5.1a), and of the formation of shock waves during expansion in O_2 at a pressure of 67 Pa (Fig. 5.1b). At ambient gas pressures above about 10^2 Pa, such oscillations disappear. At intermediate gas pressures, about 30–50 Pa, plume sharpening is meaningful and is associated with increasing confinement of the emission to the plume front; at these pressure values the deceleration of plume front begins after a few microseconds. Such a slowing down continues until a stationary behaviour is reached. At the same time, the rear edge of the plume moves *backwards* to the target. This behaviour marks a transition to a *diffusion-like* propagation of plume species through the ambient gas typical of longer times, for pressures in tens of pascal range [21]. During this stage, the plume is characterised by strong interpenetration of plasma species and ambient gas that leads to plume *splitting*, besides sharpening. In such a process, the ions and neutrals split into *two* velocity populations; the faster group, that moves practically at the same velocity as in a vacuum, consists of particles that cross the ambient gas almost collision less. The slower, delayed population results from the interaction between ablated species and ambient gas atoms. The effect was observed by TOF distribution analysis of the ablated species and affects both ions and neutrals, as demonstrated by ion probe [22] and OES [23]. Along this propagation regime of mutual penetration of the laser generated plasma and ambient gas, a consistent fraction of kinetic energy is converted into heat that in turn increases both gas and radiation temperature. At pressure values around 10^2 Pa, *turbulence* has been observed in the decelerating plume front [18]. A further pressure increase causes a contraction of the mutual penetration zone and the plasma front becomes compressed.

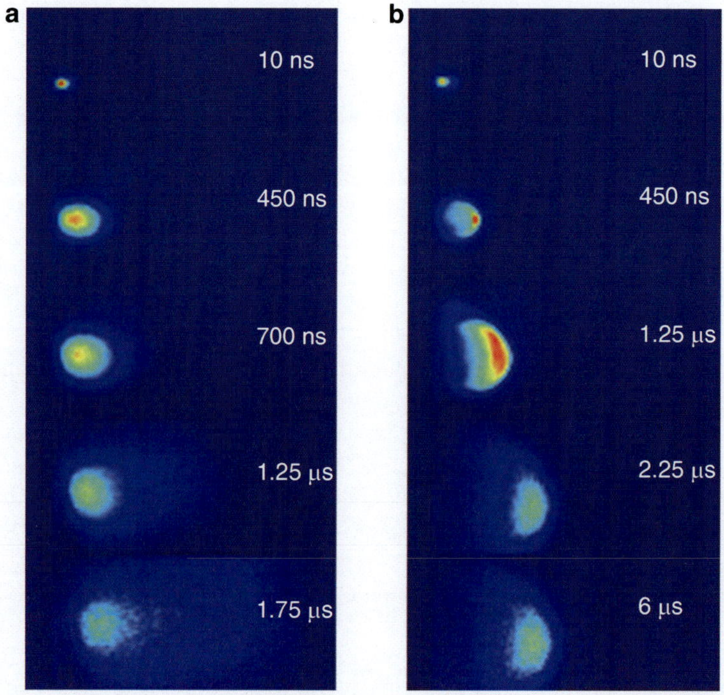

Fig. 5.1. ICCD fast photography pictures of ablation plumes expanding from a SnO_2 target irradiated with pulses from a KrF excimer laser (wavelength 248 nm, pulse width 25 ns, repetition rate 10 Hz, energy density 1.0 J cm^{-2}). The laser beam was focused at an incident angle of 45° onto the target, placed on a rotating holder. Ablation was carried out. (**a**): in vacuum (residual pressure better than 1.00×10^{-4} Pa). (**b**): in high purity O_2 ambient gas at 67 Pa. Notice the different evolution of plasma size and shape, the development of a shock wave and plasma confinement (courtesy of Dr. S. Trusso, CNR-Istituto per i Processi Chimico-Fisici, Sez. di Messina, Italy)

Moving from a combination target-ambient gas-process conditions to another one basically the above-illustrated sequence of evolution steps experienced by the ablation plume is found with increasing ambient gas pressure. Yet the pressure ranges characterising the different propagation regimes are quite broad and not always all of the phenomenology just discussed is observed.

To sum up, compared to an expansion into vacuum, the interaction of the plume with a background gas is a much more complex gas dynamic phenomenon. A wealth of physical processes are observed: they include scattering, slowing down, thermalisation, diffusion, recombination of the ablated particles, formation of shock waves, particle clustering. They give rise to increased visible fluorescence both in the plume body and in the expansion front, to a shape change of the plume itself, to a better definition than in the vacuum of plume edge, to a spatial confinement of the plasma.

5.3 Analytical Models for Plume Propagation through an Ambient Gas

The forward directed flow with weak scattering of the vaporised particles typical of the vacuum-like plume propagation regime is followed by a transition state with strong momentum transfer to the ambient gas and weak scattering of the ablated species and lastly by a diffusion regime at high pressure. Thus background gas affects both plume dynamics and the spatial distribution kinetic energy and kinetic energy distribution of its constituents at the same time. A condition to observe meaningful deviations of plume behaviour from a free expansion is that the mass of snowploughed ambient gas, whose density is ρ_{g}, at the plume frontier be comparable to plume mass M_{p}. The radius r_{p} of a supposed hemispherical plume is obtained by the equality

$$(2/3)\,\pi\rho_{\mathrm{g}}r_{\mathrm{p}}^{3} \cong M_{\mathrm{p}}. \tag{5.1}$$

Gas pressure p_{g} is related to r_{p} as

$$p_{\mathrm{g}}^{1/3} = \left[(3\,M_{\mathrm{p}}k_{\mathrm{B}}T_{\mathrm{g}})\,/\,(2\pi m_{\mathrm{g}})\right]^{1/3} r_{\mathrm{p}}^{-1}, \tag{5.2}$$

where k_{B} is the Boltzmann constant and T_{g} and m_{g} are the temperature and atomic mass of the ambient gas [24].

A few phenomenological analytical models are available in the literature to describe plume propagation in a gas environment; their intrinsic simplicity allows applying them to interpret the full expansion regimes for a generic combination ablation plume/ambient gas at the expenses of a partial or limited reproducibility of experimental results beyond narrow time-pressure intervals. Apart from these models, some numerical studies were performed in selected cases; a scattering-hydrodynamical numerical model for Si ablation in helium and in argon [25] accurately describes the expansion of laser-produced plumes through low and moderate pressure inert gases, where the initial particle mean free path may be long enough to allow meaningful plume-ambient gas interpenetration. Plume splitting is explained *quantitatively* and the differences between plume propagation in He and in Ar are put into evidence. Such a study was motivated by the interest in controlling the size of Si NPs for application in microelectronics.

Other gas-dynamical numerical approaches [26, 27] provide good fits to specific experiments, but the degree of complexity of the mathematical treatment and the required approximations limit their extensive applicability. We now address the most popular analytical models for ablation plume expansion, namely the drag, the shock wave and the diffusion models, discussing their strengths and limitations.

At *low* pressure and in the *initial* expansion stages, plasma dynamics is well fitted by the *drag* model [28]. In a strictly phenomenological picture, the observed trends of the distance travelled by the plume, i.e. the position of the front edge of the plume, in ablation experiments on different classes of materials, including YBCO, C, Al, $BaTiO_3$ [18, 28–31] are described considering the ejected species as an ensemble that feels a viscous force proportional to its average velocity v through the ambient gas. The corresponding equation of motion is

$$(\mathrm{d}v\,/\,\mathrm{d}t) = -bv \tag{5.3}$$

with solution

$$x\,(t) = x_{st}\left(1 - \mathrm{e}^{-bt}\right). \tag{5.4}$$

In (5.4), x_{st} and b are numerical coefficients determined by fitting experimental data to (5.4); they are known as stopping distance and slowing down coefficient, respectively [28]. The estimated x_{st} values [30] can be more than one order of magnitude larger than the calculated inelastic mean free paths λ. Such large differences are presumably due to the fact that x_{st} is a complex function of different experimental parameters, including nature and pressure of the ambient gas, plasma mass and energy and atomic mass ratio of the target. With increasing gas pressure the viscous force increases too, the internal expansion pressure of the plasma drops and the backward pressure on the plasma towards the target increases, leading to a decrease of the plume expansion velocity. At ambient gas pressure higher than 10^2 Pa and times longer than 4 μs the predicted distances travelled by the plume are *shorter* than observed [30]; the plasma will eventually be arrested by collisions with background gas atoms. Thus, high gas pressures result in a non-linear dependence of the position of the plasma front edge on the distance from the target.

In the *delayed drag* model [30]

$$x\,(t) = x_{st}\left(1 - \mathrm{e}^{-bt}\right) - x_0, \tag{5.5}$$

where x_0 is a boundary condition to take into account the delay before emission starts at $x = 0$. In selected experiments ($BaTiO_3$ ablation in O_2 atmosphere), fits to data on *initial* stages of plume propagation are satisfactory [30].

The increase in plasma emission associated with increased ambient gas pressure is due to collisions among gas atoms and plasma particles at plume–gas interface, as well as to particle–particle collisions within the plume body. UV radiation by laser-target interaction in turn energizes the ambient gas, giving rise to a density increase in a narrow region that propagates as a shock wave through the ambient atmosphere with speed higher than the ion sound velocity $v_{s,i}$

$$v_{s,i} = \left(< Z_\mathrm{i} > k_\mathrm{B} T_\mathrm{e} m_\mathrm{i}^{-1}\right)^{1/2}, \tag{5.6}$$

with $< Z_\mathrm{i} >$ and m_i ion average charge and mass, k_B the Boltzmann constant, T_e the absolute electron temperature. In the *shock wave* model, that was introduced [15] to describe the propagation of a shock wave through the atmosphere after the explosive release of an amount of energy E_0, just after the arrival of a laser pulse at a point on the target surface, a plasma ball develops and expands along the normal to the target surface. In a mechanistic picture [32], the propagating plasma acts like a piston, compressing and accelerating ahead of it the gas molecules to a supersonic velocity. A shock wave forms ahead of the plasma-ambient gas contact surface and propagates away from the target, being immediately followed by an expansion wave that progressively smoothes out the shock strength. The ambient gas confines the plasma, thus generating a rise up of the density of the species in the plume. Such behaviour is consistent with a slowing down of the plume both at large distances from the target and at high gas pressure. The position x of the plasma front edge, as a function of time t, is given by

$$x(t) = c\left(E_0/\rho_0\right)^{1/5} t^{2/5}; \tag{5.7}$$

ρ_0 is the unperturbed ambient gas density, E_0 the plasma energy and the constant c ($c \cong 1$) is given by

$$c = \left[(75/16\pi)(\gamma - 1)(\gamma + 1)^2 / (3\gamma - 1)\right]^{1/5}, \tag{5.7a}$$

γ being the ratio between the gas specific heats. The model can be applied when the mass of the gas set in motion by the shock wave is larger than the ablated mass (a situation *not* typical of PLD) and up to distances from the target at which the pressure driving the shock front is higher than the ambient gas pressure p_0 [15]. Thus the shock wave can be observed only in a spatial region X defined by the inequalities

$$\left(\frac{3M\mathrm{p}}{4\pi\rho_0}\right)^{\frac{1}{3}} << X << \left(\frac{E_0}{p_0}\right)^{\frac{1}{3}}, \tag{5.8}$$

where M_p is the mass of the expanding plume and p_0 is the pressure ahead of the shock wave front [33]. At a distance from the target surface where the pressure driving plasma motion becomes comparable to ambient gas pressure (5.7) is no longer valid [15]. In the shock wave region, the temperature attains values in the range of tens of thousands of Kelvin degrees and leads to increased optical emission from excited species in the plasma. The temperature in this region can be evaluated from the equation of state for a polytropic gas

$$T_\mathrm{sw} = T_0\ \left[(2\gamma)/(\gamma + 1)\right]\ \left[1 + \mathfrak{M}^2(\gamma - 1)/(\gamma + 1)\right] \tag{5.9}$$

with T_0 the temperature of the unperturbed gas and $\mathfrak{M}$ the Mach number, that is the ratio of shock wave velocity to sound velocity in the ambient gas. Here the temperature values can significantly affect the physical and chemical evolution of the species, although chemical reactions take place only if meaningful plasma-ambient gas mixing occurs. The latter is driven by inter diffusion across the contact surface between the plasma and the ambient gas; it is relevant enough if the thickness of the shock wave region is comparable to the gas diffusion length $l_{d,g} = (4D_g t)^{1/2}$ in the same region. $l_{d,g}$ can be evaluated from the expression for gas diffusion coefficient

$$D_{\mathrm{g}} = D_0 \left(T_{\mathrm{sw}}/T_0\right)^{3/4} \left(\rho_0/\rho_{\mathrm{sw}}\right), \tag{5.10}$$

where D_0 is the diffusion coefficient of the ambient gas at atmospheric pressure and room temperature; the density in the shock wave region ρ_{sw} is

$$\rho_{\mathrm{sw}} = \rho_0 \left(\gamma + 1\right) / \left(\gamma - 1\right). \tag{5.11}$$

In the *delayed shock* model [30]

$$x\left(t\right) = \left(k/p_{\mathrm{g}}\right)^{1/5} \left(t - t_{\mathrm{s}}\right)^{2/5} - x_0, \tag{5.12}$$

with k a constant proportional to the laser energy density, p_{g} the gas pressure, t_{s} the boundary condition for the delayed shock wave and x_0 a boundary condition analogous to that in (5.5). The model fits well experimental results on plasma expansion from a $BaTiO_3$ target at *large* times and ambient gas (O_2) pressures [30].

Again at *low* buffer gas pressures, assuming diffusion as *the* transport mechanism, the distance x travelled by the plume is

$$x\left(t\right) = \left(Dt\right)^{1/2}. \tag{5.13}$$

The equation defines the classical *diffusion* model [34]. In the gas theory $D = \left(\lambda v_{\mathrm{th}}\right)/3$ [35] with v_{th} the thermal velocity of the colliding particles and $\lambda = \left(n_{\mathrm{g}}\sigma\right)^{-1}$ the mean free path, where n_{g} is the number density of the ambient gas and σ the appropriate scattering cross section.

The limitation of the above-discussed analytical models is that, given their dependence on numerical fitting parameters, they allow interpreting a posteriori experimental data, but have *no* predictive capability about plume dynamics. In the following section, a two-stage phenomenological analytical approach that accounts for plume expansion under rather general conditions is introduced and tested against experimental data on the propagation through various ambient gases of plumes ablated from elemental and compound targets. Model parameters depend on easily available process parameters such as ambient gas nature, pressure and temperature, as well as number density of ablated species. Thus it becomes possible to *predict* plume propagation for reasonably wide ranges of experiments moving from knowledge of the chosen process parameter values.

5.4 Mixed-Propagation Model

The established [24] high values of temperature and particle number densities in the first stage of plume life, with the associated quasi-explosive initial plume expansion result in a Knudsen layer [36], where the leading contribution is the particle flux velocity. It is assumed that the particles ejected from the target, with a strongly dominant velocity component in the direction normal to the surface, yet undergo a diffusive motion. This corresponds to diffusion through an ambient gas whose *effective* number density n_{eff} is considerably reduced with respect to the value n_g deduced from gas pressure p_{g}. In the resulting *modified diffusion* model, the diffusion coefficient [37] is

$$D' = K\lambda v_0 = K v_0 \left(n_{\text{g}}\sigma\right)^{-1}, \tag{5.14}$$

where v_0 is the ejection velocity from the target of the *fastest* group of ablated particles. v_0 is obtained from the initial slope of the measured distance-time curve for plumes produced and propagating under given conditions. The choice of the particle group with *maximum* flux velocity v_0 instead of the usual thermal velocity v of the particles is an *ansatz* of the model to enhance the relevance of flux velocity. Although plume front slows down during expansion (see Figs. 5.2–5.5 below) modified diffusion model fits reasonably well experimental data, particularly beyond the initial plume expansion stage. This means that flux velocity indirectly influences plume propagation also at comparatively *large* distances from the target.

Considering elemental targets of C, Si, Sn, Ag, Ta, W, literature data on the expansion velocities of UV laser-generated ablation plumes through different gases were analysed [38]. The K values that better fit the data scale with the target atomic mass, from $K = 2$ (light elements: C; Si) to $K = 6$ (intermediate mass elements: Ag; Sn) to $K = 8$ (heavy elements: Ta; W). Thus, given the target mass, the K value is *uniquely* defined.

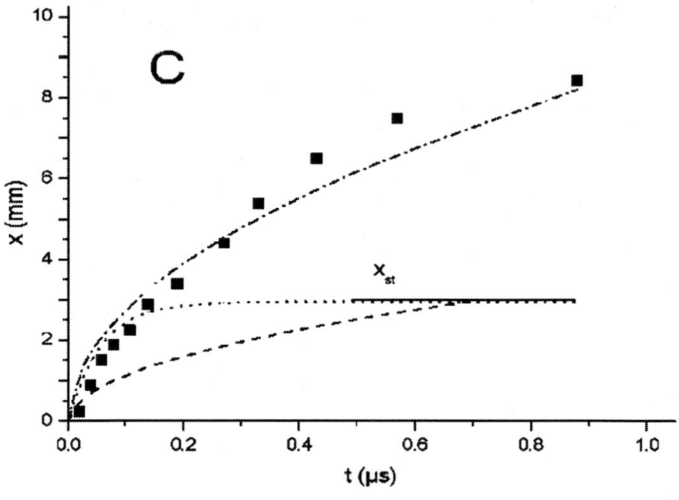

Fig. 5.2. position x (*full squares*) of the front edge of carbon plumes propagating through N_2 [29] as a function of time. *Dashed curve*: diffusion model; dashed-dotted curve: modified diffusion model; *dotted curve*: modified drag model. x_{st}: stopping distance

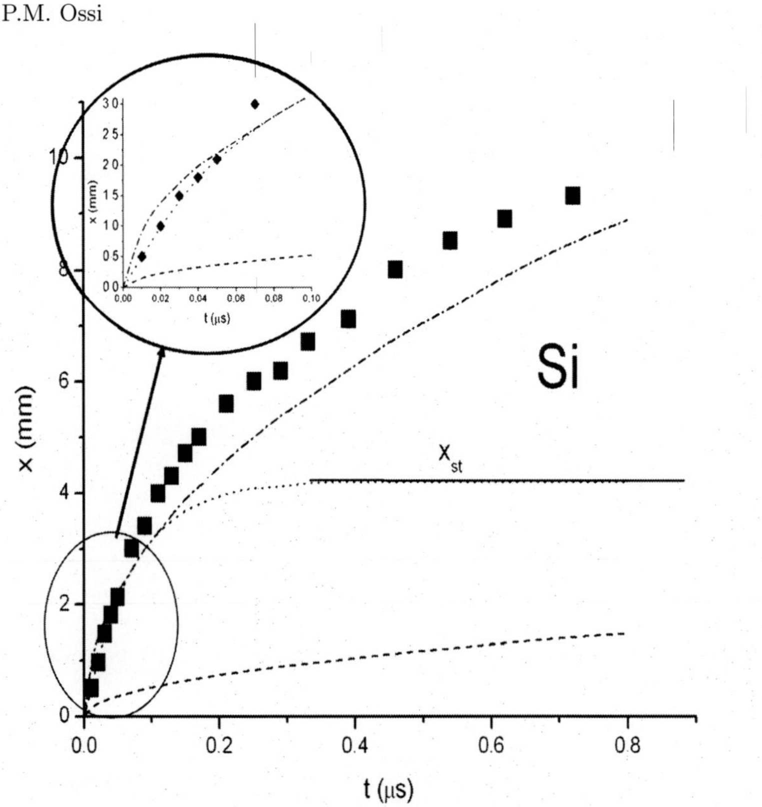

Fig. 5.3. Model analysis of Si plume propagation through He vs. time [8]. General information as for Fig. 5.2

To reproduce the experimentally observed initial linear behaviour of plume expansion (see e.g. [8] for Si and [29] for C) drag model is most suited. To this task a one-dimensional plume expansion along the principal plume axis x is considered. Moving from (5.4), taking into account the diffusion dynamics of a fluid of classical particles in the presence of viscosity, whose effect is embodied in a coefficient ξ [39] (ξ is a diffusion coefficient multiplied by time), with the initial conditions $x(0) = 0$ and $(\mathrm{d}x/\mathrm{d}t)(0) = v_0$, the position $x(t)$ of an atom is

$$x(t) = v_0 D' \xi^{-1} \left[1 - \exp\left(-\xi D'^{-1} t\right)\right] . \tag{5.15}$$

Both the slowing down coefficient $b = \xi D'^{-1}$ and the stopping distance $x_{\mathrm{st}} = v_0 D' \xi^{-1}$ of this *modified drag* model have a clear physical meaning.

The initial number density n_{a} of the ablated particles is larger than n_{g}, but the violent plume expansion leads to the condition $n_{\mathrm{a}} = n_{\mathrm{g}}$, when a stable shock wave front forms, the expansion regime changes and the inequality $n_{\mathrm{a}} < n_{\mathrm{g}}$ is established in the body of the plume. Ambient gas atoms are strongly scattered by fast plume constituents (normally, positive ions) that

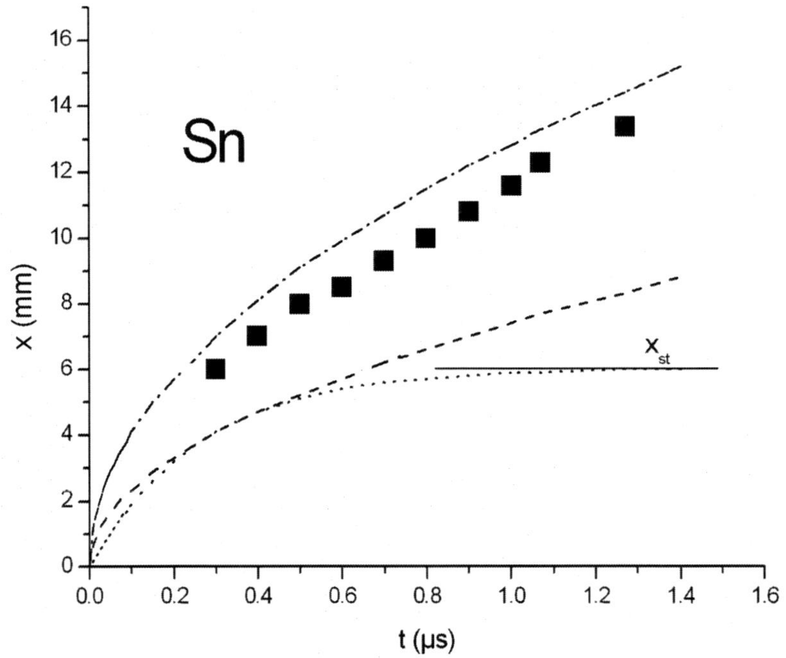

Fig. 5.4. Model analysis of Sn plume propagation through O_2 vs. time [45]. General information as for Fig. 5.2

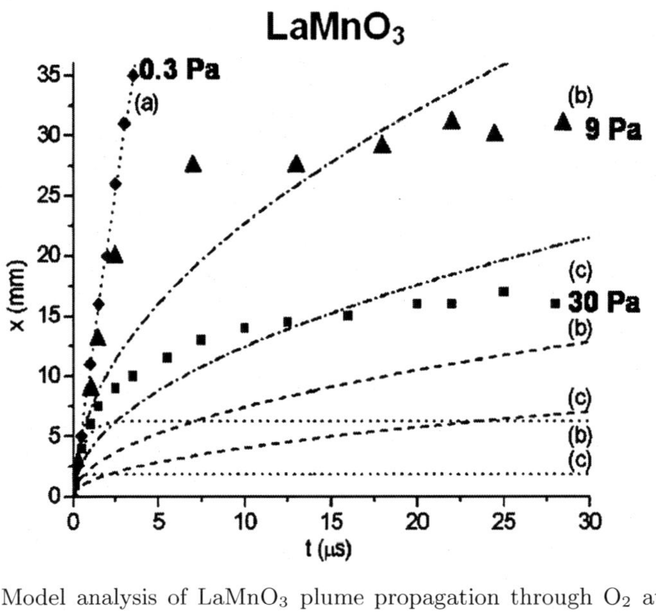

Fig. 5.5. Model analysis of $LaMnO_3$ plume propagation through O_2 at different pressures vs. time. Diamonds (**a**): 0.3 Pa; triangles (**b**): 9 Pa; squares (**c**): 30 Pa [19]. General information as for Fig. 5.2

are slowed down and *aggregate* themselves with the slower particles initially grouped in the body and tail of the plume.

The sequence of such complex phenomena is mimicked by the combination of modified diffusion and modified drag models to give *mixed-propagation* model. The unphysical discontinuity of plume propagation at x_{st} corresponds to the region where the viscous slowing down of plume front leads to the formation of the stable shock wave front; at distances $x < x_{st}$ modified drag model holds, while for $x \geq x_{st}$ modified diffusion model holds. The x_{st} value is chosen according to the estimate $x_{st} \cong 4\lambda$ [25] and it is calculated using the n_g and pertinent σ values.

At low ambient gas pressure (less than 1 Pa), x_{st} can be much larger than the usual values of target-substrate distance x_{T-S} (a few centimetres), so that modified drag model is enough to describe plume dynamics. Mixed-propagation model takes into account the limitations both of modified drag model that, at variance with experiment predicts plume stopping at x_{st}, and of modified diffusion model that provides propagation distances larger than the experimental values both in the early expansion stages and at low ambient gas pressure. Although mixed-propagation model offers a strongly simplified description of plume expansion, its easily accessible input parameters v_0, σ, n_g and n_a make it useful to *predict* general trends of plume behaviour.

Mixed-propagation model has been tested against literature data on the dynamics of ablation plumes from different targets, propagating in different ambient gases at various pressures [40–44]. The expansions of C in molecular nitrogen [29], Si in helium [8], Sn [45] and $LaMnO_3$ [19] in molecular oxygen are discussed here; the model is then applied to study the propagation of W ablation plumes in helium and in argon, for which data on plasma expansion are lacking. Tables 5.1 and 5.2 display data collected from ablation experiments and parameter values used in mixed-propagation model, respectively. In Fig. 5.2, data on the propagation of the front of carbon plumes in N_2 (full squares) [29] are compared with predictions of diffusion model (dashed curve), modified diffusion model (dashed-dotted curve) and modified drag model (dotted curve). The latter fits well the first stage of plume expansion, but it suffers from predicting that plume is stopped at x_{st}. Diffusion model appears inaccurate; its modified version well fits the data at intermediate and delayed times, but slightly overestimates plume expansion velocity in the neighbourhood of the target. From a fit on available data [46, 47] on carbon ablation with excimer lasers in vacuum, or at low ambient gas pressure, over the energy density interval between 1 and 10^2 J cm^{-2} the initial velocity is $v_0 = 5.2$ cm µs^{-1}. Mixed-propagation model successfully fits experimental data on carbon plume expansion for different gas pressures and laser energy densities (50 Pa, 12 J cm^{-2} [29]; 66 Pa, 6 J cm^{-2} [48]). Taking $x_{st} = 4l$, an estimate for the slowing down coefficient is obtained from $b = v_0 \ (4l)^{-1}$. For C, $K = 2$.

In Fig. 5.3, data [8] on the propagation of Si plume fronts in helium (full squares) obtained by recording ICCD pictures of the plume self-emission are

Table 5.1. Collection of data from ablation experiments on C [29], Si [8], Sn [45], LaMnO3 [19] and W [63,64]

Target	λ (nm)	E ($J\,cm^{-2}$)	τ (ns)	x_{T-S} (cm)	p_g (Pa)	T(K)	$n_g(cm^{-3})$
C	248	38	20	6	30; N_2	300	7.2×10^{15}
Si	532	4.0	8	5	65; He	300	1.6×10^{16}
Sn	532	4.0	8	5	65; O_2	300	1.6×10^{16}
$LaMnO_3$	351	1.5	20	6	0.3; O_2	300	7.2×10^{13}
					9; O_2		2.2×10^{15}
					30; O_2		7.2×10^{15}
W	248	4.5	20	5	40; Ar	300	9.7×10^{15}
					60; Ar		1.5×10^{16}
					100; Ar		2.4×10^{16}
					40; He		9.7×10^{15}
					60; He		1.5×10^{16}
					100; He		2.4×10^{16}

Table 5.2. Collection of parameter values for mixed-propagation model (see text)

Target	$<n_a>$ (cm^{-3})	σ_{a-g} (cm^2)	σ_{a-a} (cm^2)	$<v>$ ($cm\,\mu s^{-1}$)	t_f (μs)	x_{aggr} (cm)
C	1.7×10^{17}	7.4×10^{-16}	1.5×10^{-15}	4.5	1.8	2.5
Si	3×10^{16}	7.5×10^{-16}	2.2×10^{-15}	3.0	4.6	2.6
Sn	3×10^{16}	7.5×10^{-16}	2.2×10^{-15}	3.0	4.6	2.6
$LaMnO_3$	7×10^{12}	3×10^{-15}	4.6×10^{-15}	0.55	0.9	7.6
$LaMnO_3$	1×10^{13}	3×10^{-15}	4.6×10^{-15}	0.55	14.6	3.9
$LaMnO_3$	1.3×10^{13}	3×10^{-15}	4.6×10^{-15}	0.55	30.1	3.0
W_{Ar}	6.7×10^{13}	1.6×10^{-15}	3.2×10^{-15}	0.77	14.7	2.9
W_{Ar}	2.4×10^{14}	1.6×10^{-15}	3.2×10^{-15}	0.77	18.7	2.7
W_{Ar}	1.5×10^{15}	1.6×10^{-15}	3.2×10^{-15}	0.77	24.6	2.4
W_{He}	6.7×10^{14}	1.04×10^{-15}	3.2×10^{-15}	0.80	6.9	2.9
W_{He}	2.4×10^{14}	1.04×10^{-15}	3.2×10^{-15}	0.80	8.8	2.7
W_{He}	1.5×10^{15}	1.04×10^{-15}	3.2×10^{-15}	0.80	18.1	2.1

fitted to the same models as for carbon; also for this element $K = 2$ and the value of v_0 used in the fits is $6\,cm\,\mu s^{-1}$ [27]. Looking at Fig. 5.3 modified diffusion model reproduces in an accurate enough way all stages of plume expansion. The extent of accuracy is comparable to that of the original data analysis [8], yet, if *ad hoc* experiments were performed focusing on the very initial plume propagation, within $0.1\,\mu s$ (see the inset of Fig. 5.3), the linear behaviour predicted by modified drag model would be most suited.

Figure 5.4 shows the result of the fitting procedure to the expansion of Sn plumes [45]. All data fall beyond the stopping distance $x_{st} = 6.05\,mm$, so modified diffusion model is used to describe plume dynamics, with $v_0 = 2.3\,\mathrm{cm\,\mu s^{-1}}$ [45] and $K = 6$; plume velocity is slightly overestimated.

Figure 5.5 shows data collected for the expansion of plumes ablated from a $LaMnO_3$ target in O_2 at three pressures (full symbols), obtained from ICCD pictures of plasma self-emission [19]; the figure displays fits to the data with the same models as discussed for elemental targets. Again, modified drag model (dotted lines) better reproduces the linear trend of initial plume expansion, but it predicts plume stopping at $x_{st} \rightarrow \infty$ for $p_g = 0.3\,\mathrm{Pa}$, $x_{st} = 6.1\,\mathrm{mm}$ for $p_g = 9\,\mathrm{Pa}$ and $x_{st} = 1.9\,\mathrm{mm}$ for $p_g = 30\,\mathrm{Pa}$. Diffusion model is clearly inaccurate; its modified version gives a reasonable fit to the data at intermediate and delayed times, but it slightly underestimates propagation distances near the target. The value of v_0 used in the fit is $1.1\,\mathrm{cm\,\mu s^{-1}}$ [19]. In the case of this compound, Mn and O are light elements, while La belongs to intermediate mass elements; with the above condition on K values, an average value $\langle K \rangle = 3$ is used. For such a massive target modified diffusion model is sufficient to reproduce all stages of plume expansion, taking into account that no plume velocity data are available at times shorter than $0.5\,\mathrm{\mu s}$ [19]. The comparatively poor fit to data points at $p_g = 9\,\mathrm{Pa}$ could be related to the unusual "knee" in the experimental data set, possibly associated to an overestimate of plume velocity.

In the case of tungsten $v_0 = 1.5\,\mathrm{cm\,\mu s^{-1}}$ [49]; the datum refers to ablation in vacuum, but v_0 is nearly independent of gas pressure [19]; plasma velocity was not measured both for expansions in Ar and in He, yet, given the similarity between thermophysical properties of Ta and W [50] the expansion of W plumes is taken to be comparable to that of Ta plumes [51, 52] under comparable irradiation conditions.

In summary, for all examined systems mixed-propagation model offers fits to available experimental data of reasonable accuracy, comparable, or slightly better than existing approaches. Its main advantage is that it does not require using fitting parameters, apart from K, whose value can be easily chosen, as discussed. The model appears suitable for a wide range of ablation conditions. The results of mixed-propagation model are specifically useful as input parameters to model nanoparticle growth in the expanding ablation plasma.

5.5 Nanoparticle Growth

When the landing of ablation plasma onto the substrate is a high-energy deposition process, with an associated energy of several eV at.$^{-1}$, the NPs possibly present in the plume either undergo penetration, or pinning [53, 54], resulting in well adherent, compact, smooth films. Low energy depositions, characterised by fractions of eV at.$^{-1}$, lead to cluster diffusion and aggregation on the substrate surface [55], followed by coalescence in large clusters beyond a critical degree of surface coverage. Such a deposition is attractive to synthesize CA materials retaining a memory of the properties of their precursor building blocks. The kinetic energy of the deposited species was evaluated only in very few cases; in particular, the kinetic energy of Au species was changed from several tens to fractions of eV $at.^{-1}$ by adjusting gas pressure and target–substrate distance in PLD experiments [56], but no evidence of cluster fragmentation, or soft landing on impact was reported. STM observations of Pt NPs at different degrees of substrate coverage and different plume kinetic energies [57] indicate that the average particle diameter increases with increasing film thickness, according to a power law with an exponent comparable to that observed for metals grown by MBE [58]. This is taken as evidence for the fact that NPs grow *on the substrate* by surface diffusion of the deposited material. Again, mechanisms operating at the substrate surface, thereby nucleation and growth up to a critical particle size are followed by coalescence and coarsening at large degrees of substrate coverage are reported [59] to drive the growth of Cu nanocrystals with size less than 10 nm, being synthesized in nanocomposite films deposited by PLD in argon up to a pressure of about 0.66 Pa. For Ar pressures between 6 Pa and 13 Pa, nanocrystal morphology suggests that nucleation and growth mainly occur at the substrate, but the reduced surface mobility inhibits the coarsening stage, so that highly anisotropic crystals are observed. These findings were recently confirmed [60] with Au NPs grown in amorphous Al_2O_3 layers. A different picture emerges from the growth of silicon NPs with narrow size distribution that was analysed in depth in the framework of a search for new optoelectronic devices [61]; plasma spectroscopy shows that cluster nucleation and growth occurs *in* the expanding ablation plume. A more recent deposition and modelling study confirms that Si nanoclusters are formed in the plume and are strongly affected by ionisation processes that occur when plume propagates through an ambient gas, resulting in very high nucleation rates and small cluster critical radii [8]. The synthesis of C nanoclusters in plumes propagating through He and Ar atmospheres up to pressures of 1 kPa was reported and modelled [40], again keeping into account the relevant role of ionisation phenomena occurring in the interface region between the shock wave front and the ambient gas [62]. Cluster-assembled W films were deposited in different atmospheres and pressure ranges [63,64]; the surface morphology, bond coordination and oxidation path of the deposited films, both when exposed to ambient atmosphere and when synthesized in dry air were systematically studied and complemented a detailed HREM analysis of structure, size and morphology of the deposited NPs [65]. In situ STM observations of the first stages of film formation upon ablation of a W target in Ar atmosphere, selecting different combinations of gas pressure and target-substrate distance [66] allowed identifying deposition regimes ranging from cluster deposition-diffusion-aggregation, to cluster melting and coalescence, to cluster implantation. From the body of the above analyses the picture of NP formation in the expanding ablation plume is strengthened.

In the framework of this picture, a further step of mixed-propagation model using the parameters calculated from the analysis of plasma expansion is to provide an evaluation of the average cluster asymptotic size that is of the number N of constituent atoms in a cluster that reached a steady state during plume propagation.

It is generally accepted that clusters form through the steps of nucleation, growth and cooling [61]; in the presence of an ambient gas a set of hydrodynamic equations for plume expansion should be solved, taking into account vapour condensation. This approach is out of the present calculation capability. We assume an initial seed population of tiny clusters [67] in the propagating plume. This is justified by the high ionisation degree of the plume and by the ion tendency to be jacketed by surrounding neutral atoms [8]. The hypothesis implies that plume evolution is not affected by the mechanisms of cluster formation [61]. For a given set of ablation conditions (see Table 5.1 for the examples discussed here) the average *asymptotic* number N of atoms in a cluster that attained a steady state during plume flight is now calculated. Although the plume experiences a range of internal pressures and is spatially inhomogeneous, so that growing clusters experience the above steps at different times, averages over long times are considered.

In the ideal gas approximation, N is given by

$$N = (<n_{\mathrm{a}}>\sigma_{\mathrm{a-a}}<v>t_{\mathrm{f}})\ (n_{\mathrm{g}}\sigma_{\mathrm{a-g}}<v>t_{\mathrm{f}})\, x_{\mathrm{T-S}}x_{\mathrm{aggr}}^{-1}, \tag{5.16a}$$

when the target-substrate distance $x_{\mathrm{T-S}}$ is shorter than the distance x_{aggr} over which clusters grow in the flying plume and

$$N = (< n_{\mathrm{a}} >\sigma_{\mathrm{a-a}}<v>t_{\mathrm{f}})\ (n_{\mathrm{g}}\sigma_{\mathrm{a-g}}<v>t_{\mathrm{f}})\,, \tag{5.16b}$$

when $x_{\mathrm{T-S}}$ is larger than x_{aggr}.

In both equations, the distance x_{aggr} travelled by the plume during cluster growth is deduced from optical emission data, considering the relevant species in the plume (e.g., LaO, in the case of $LaMnO_3$ [68] and SnO, for Sn ablation [45]). When such data are lacking

$$x_{\mathrm{aggr}} = 6p_{\mathrm{g}}^{-1/5}. \tag{5.17}$$

This relation gives an upper limit for the value of x_{aggr}; the estimate was deduced from an analysis [38] of TaO [52] and CN [48, 69] bands during Ta and C ablation in O_2 and N_2 respectively, over wide ranges of laser energy density (from 4 to 95 J cm^{-2}) and of ambient gas pressure p_{g} (from 1 to 1.32×10^5 Pa). The monotonic dependence of cluster size on p_{g} in (5.17) is in qualitative agreement with a number of results on Si [10, 70]; mixed-propagation model well fits [9] the measured monotonic increase of Si cluster size [8]. It is to be noticed, however, that the size of Si NPs grown in helium at very high pressure shows a peaked dependence on gas pressure [71] that was also reproduced numerically [72]. The choice of the value of x_{aggr} is a critical issue of mixed-propagation model, as was observed when the measured sizes of silver NPs deposited directly on a-C supported TEM grids was compared to the calculated values obtained by the model. The experiments consisted of ablation of an Ag target in Ar atmosphere at different pressures, ranging from 10 Pa to 10^2 Pa, keeping fixed all other deposition parameters. Using (5.17) NP sizes comparable to the measured ones were obtained, although slightly overestimated. From an analysis of fast photography pictures of the propagating plumes, $x_{\rm aggr}$ was identified as the distance corresponding to a decay of absolute plasma luminosity by three orders of magnitude with respect to the initial value. It was assumed that at such a point any meaningful collisional plume behaviour actually came to an end. Afterwards, the propagating plume consists of steady-size NPs. The plasma temperature dropped from $\sim 1.2 \times 10^5$ K in the initial expansion stage to around 10^4 K at $x_{\rm aggr}$ (see (5.9)), the distance $x_{\rm aggr}$ being travelled in 580 ns. On a temperature-time plot, such an instant corresponds to an inflection point, where the slope of the temperature profile changes significantly, in coincidence with a likely change of the plume propagation regime. Adopting the $x_{\rm aggr}$ value, deduced from a more physically sound estimate, mixed-propagation model provides average diameter values for isolated sphere-like Ag NPs in excellent agreement with the sizes directly measured by TEM observations [73].

In (5.16a) and (5.16b) the cluster formation time t_f is the time required by the plume to travel the distance $x_{\rm aggr}$.

$<n_{\rm a}>$ is the average number density of ablated atoms; it is given by the ratio between the number of ablated atoms per pulse and the volume of the plume, obtained from fast imaging pictures of the plume taken in proximity of the target and at a distance from the target around $x_{\rm aggr}$. Increasing $<n_{\rm a}>$, the number of collisions between ablated atoms increases, while increasing $n_{\rm g}$ plume confinement is enhanced. Both mechanisms favour cluster formation and growth. $\sigma_{\rm a-a}$ and $\sigma_{\rm a-g}$ are the geometric cross sections for ablated particle-ablated particle and ablated particle-gas atom binary collisions. A unit sticking coefficient is assumed. While the contribution of elastic collisions to cluster growth is negligible, they play a role to spread the kinetic energy of plume species. Thus both for ambient gas atoms and for ablated species, velocity distributions should be considered. The former is a Boltzmann distribution while the latter is non-equilibrium at least until the plume becomes non-collisional, as discussed in the section concerning the choice of v_0 value. Yet a single value for both families is assumed, namely v_0 for ablated particles and the average velocity $v_{\rm g}$ for gas atoms, as deduced from the gas temperature. The average between $v_{\rm g}$ and v_0 is taken as the representative average velocity $<v>$ of plume particles; it represents the impact velocity in a binary collision between an ambient gas atom (slow) and a plume particle (fast). This choice corresponds to assign a leading role in cluster formation to the fastest group of ablated particles. When $<v>$ increases the time interval between two subsequent collisions decreases, thus enhancing the rate of cluster growth.

In both (5.16a) and (5.16b) the first term ($<n_{\mathrm{a}}> \cdot \sigma_{\mathrm{a-a}} \cdot <v> \cdot t_f$) is associated to cluster growth and is proportional to the scattering probability between ablated particles, while the second term ($<n_{\mathrm{g}}> \cdot \sigma_{\mathrm{a-g}} \cdot <v> \cdot t_f$) represents the slowing down and confinement of the plume, which is proportional to the scattering probability between ablated particles and gas atoms.

In the first stage of plume propagation, atoms mainly aggregate together and NPs grow; beyond the distance x_{aggr}, cluster growth is *balanced* by cluster cooling both by a leading evaporative and by a less relevant collisional mechanism. The competition between growth and cooling mechanisms in a cluster is taken into account by the term $x_{\mathrm{T-S}} x_{\mathrm{aggr}}^{-1}$ in (5.16a) and 1 in (5.16b). Indeed, this allows avoiding an unphysical, indefinitely persisting cluster growth in the limit of large distances flown by the plume, like in (5.16a).

The above phenomenological model of NP growth in the expanding ablation plasma is highly simplified. The dependence of x_{aggr} on laser energy density and on ambient gas nature and pressure is complex as illustrated by the contrasting experimental results discussed for Si, the only material whose behaviour was explored in depth, although not exhaustively, till now. This is an indication that our understanding of the combined effect on cluster formation of the parameters that drive plume propagation is far from complete.

The model of NP growth was applied to evaluate the average asymptotic size of NPs grown in ablation plumes of C, Si, Sn, $LaMnO_3$ and W, propagating in the ambient conditions listed in Table 5.1. Parameter values for the mixed-propagation model from Table 5.2 have been used in the calculations. Except for $LaMnO_3$ ablation in 0.3 Pa oxygen atmosphere, x_{T-S} is larger than x_{aggr} and (5.16b) was used. The average number of atoms per cluster N is reported in Table 5.3 together with the average diameter of spherical NPs; the latter was calculated using packing efficiency $\eta = 0.67$ for close-packed non-crystalline structures, or $\eta = 0.74$ for cubic crystals, depending on NP structure from electron diffraction patterns taken on samples observed by HREM.

Looking at Table 5.3 a reasonable agreement is found between calculated and observed NP sizes, when available. The different efficiency of different inert gases to favour cluster growth is put into evidence in the case of W films pulsed laser deposited under otherwise identical conditions.

Summarising, although blind with respect to the detailed interaction mechanisms among particles in the ablation plasma that propagates through an ambient gas, mixed-propagation model appears to have a degree both of interpretative and of predictive ability concerning NP growth.

Yet it is unable to take into account plasma *chemistry*. Nonetheless, when a film is deposited at high temperature in a reactive ambient gas and/or when the plume moves through and interacts with a plasma fed by e.g. a rf discharge dramatic effects on the growth kinetics of the film are observed. A specific study was conducted on the radio frequency assisted PLD of WO_x

Table 5.3. Asymptotic number of atoms per NP, N and average NP diameters, d calculated with mixed-propagation model; for comparison available NP diameters measured by TEM are reported

Target	p_g (Pa)	η	N	d_{th} (nm)	d_{exp} (nm)
C	30; N_2	0.67	5.5×10^4	10	$5 \div 10$ [40]
Si	65; He	0.67	1.55×10^4	13.5	~10 [8]
Sn	65; O_2	0.67	30	1.1	–
$LaMnO_3$	0.3; O_2	0.67	0	0	–
$LaMnO_3$	9; O_2	0.67	27	1	–
$LaMnO_3$	30; O_2	0.67	455	2.8	–
W_{Ar}	40; Ar	0.67	432	2.3	$5 \div 10$ [65]
W_{Ar}	60; Ar	0.67	3.77×10^3	4.8	$5 \div 10$ [65]
W_{Ar}	100; Ar	0.67	8.9×10^4	13.8	$5 \div 10$ [65]
W_{He}	40; He	0.74	7	1.3	~10 [44]
W_{He}	60; He	0.74	585	2.7	~10 [44]
W_{He}	100; He	0.74	5×10^4	11.8	~10 [44]

nanostructured films [74]. A W target was ablated either in pure O_2 atmosphere, or in a mixture of O_2 and He, in the same relative proportion, at a total pressure of 900 Pa, with rf power fixed at 150 W, and films were deposited on substrates maintained at a fixed temperature $T_s = 873$ K. TEM observations, as shown in Fig. 5.6, indicate that, irrespective of the ambient gas composition, the same kind of spherical NPs are present in the deposited films, being in part uniformly scattered, in part assembled together to give chain-like structures. On average, larger NP sizes are found when the plume suffers from more severe scattering from the ambient gas: the size is about 25 nm for films deposited in O_2 (Fig. 5.6a) and about 15 nm for films deposited in O_2 + He (Fig. 5.6b). However, on a larger scale, as shown by SEM pictures, dramatic differences are found in film morphology. Pure O_2 atmosphere results in a complete coverage of the substrate by snowflake-like agglomerates with average size around 40 nm, visible in Fig. 5.7a. In mixed O_2 + He atmosphere, under otherwise identical process conditions isolated big particles with size around 1.5 μm are distributed on the bare substrate, as shown in Fig. 5.7b. The features of Raman spectra from the two films show that in pure O_2 a nanocrystalline film was obtained, while in mixed gas atmosphere a nanostructured non-crystalline sub-oxide was deposited.

The substitution of a chemically reactive gas species, such as O_2 with an inert gas (He) appears to have a twofold effect: the average number of W–O collisions, assuming W atoms as the relevant plume constituent and O atoms as the relevant ambient gas species, is much lowered than in a pure O_2 atmosphere, so that the plume is expected to land onto the substrate with higher kinetic energy than when it flies through O_2 at the same total pressure. Yet, to attain an extended nanostructure, this larger kinetic energy contribution is insufficient to compensate for the loss of chemical reactivity of the process. Such a reactivity is associated both to *gas-phase* reactions

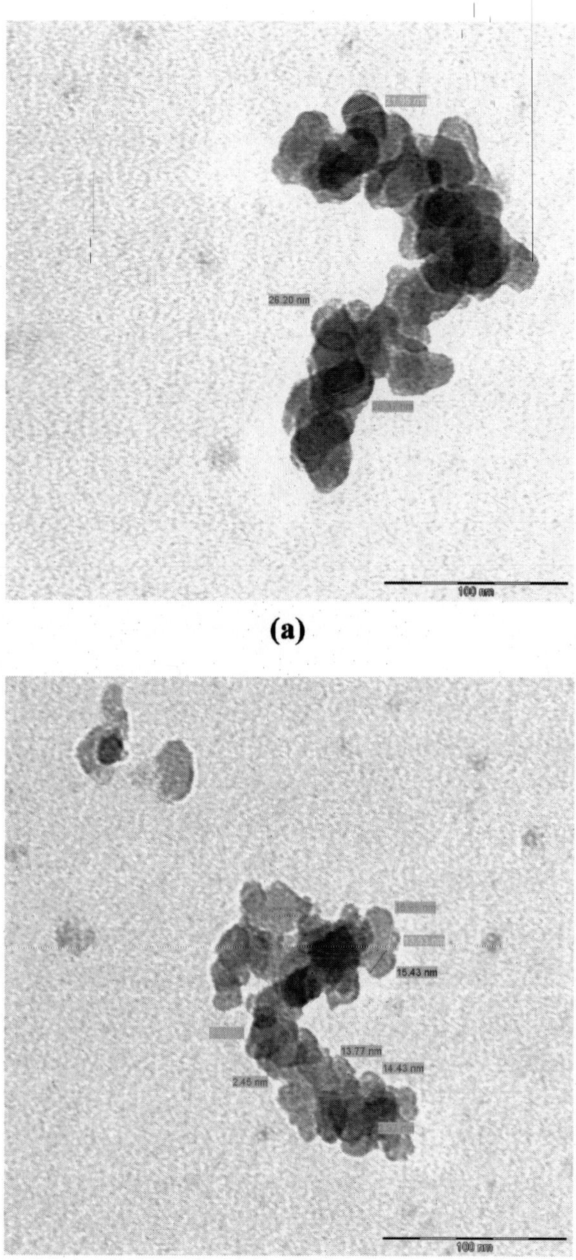

Fig. 5.6. Representative TEM pictures showing the structure of films deposited at $p_g = 900$ Pa and $T_s = 873$ K in (**a**) pure O_2 atmosphere; average NP size: 25 nm. (**b**) mixed $O_2 + He$ atmosphere; average NP size: 15 nm

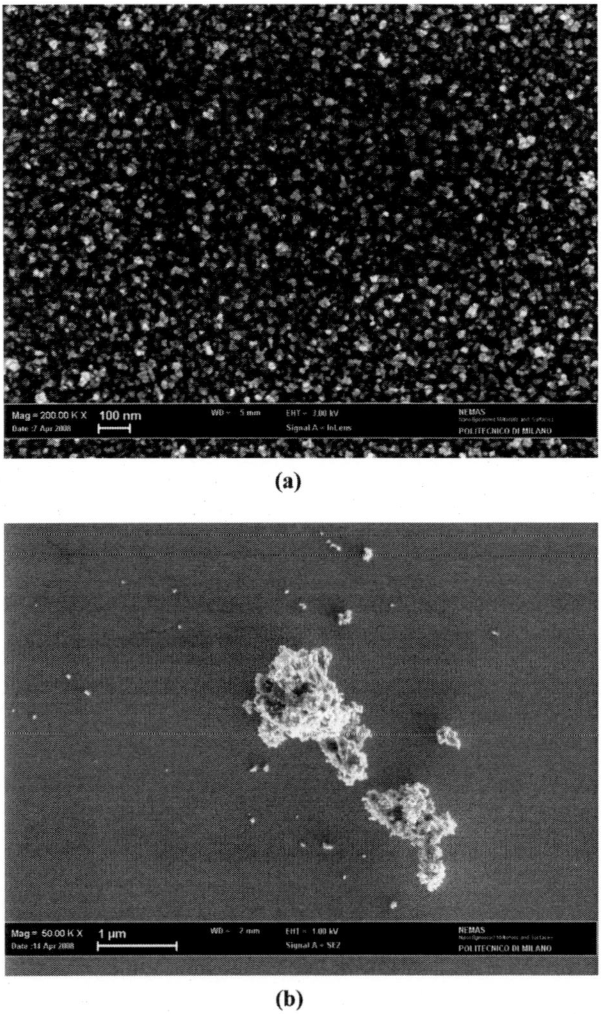

Fig. 5.7. Representative surface microstructures (SEM) of films deposited at $p_g =$ 900 Pa and $T_s = 873$ K in (**a**) pure O_2 atmosphere; substrate completely covered by agglomerates (average size: 40 nm). (**b**) mixed O_2 + He atmosphere; isolated large (1.5 µm) particles on the substrate

involving excited/ionised oxygen molecules and atoms produced by the rf discharge with which the ablation plasma interacts during its flight from the target to the substrate, and to reactions at the *surface* of the growing film, in turn continuously bombarded and activated by the high pressure chemically and electrically non-neutral oxygen species from the ambient gas. The kinetics of both families of reactions is dramatically increased by the presence of high temperature, non-equilibrium electron populations in the two plasmas. A number of such reactions are essentially unknown, being irrelevant, or kinetically prevented in thermodynamic equilibrium.

5.6 Concluding Remarks

Ablation of a solid target in an ambient gas using nanosecond laser pulses results in the synthesis of clusters/NPs that can self-assemble together after landing onto a substrate. Energy exchanges occurring both intra-plasma and between plasma and ambient gas during plume propagation critically affect cluster growth, besides determining the energy available to cluster mutual interaction on the substrate. All these factors contribute to define the morphological and structural features of CA films. In particular, plume interaction with background gas is a complex gas dynamic phenomenon including scattering, slowing down, thermalisation, diffusion, recombination of the ablated particles, formation of shock waves and particle clustering. By modelling plasma propagation, it is possible to calculate the average size of clusters grown in the plume up to their steady size. A comparison with available experimental data appears satisfactory, showing that the comprehension and control of the basic mechanisms underlying the synthesis of nanometer-sized clusters by PLD is a steadily advancing research field. However, the ability to design realistic nanostructured films with ad hoc tailored properties that requires a detailed understanding of the mechanisms involved in the assembly of clusters on a suitable substrate appears to be still in its infancy.

References

1. L. Chen, in *Pulsed Laser Deposition of Thin Films*, ed. by D.B. Chrisey, G.K. Hubler (Wiley, New York 1994), p. 167
2. J.E. Rothenberg, R. Kelly, Nucl. Instrum. Methods Phys. Res. B **1**, 291 (1984)
3. E. Borsella, S. Botti, R. Giorgi, S. Martelli, S. Turtu, G. Zappa, Appl. Phys. Lett. **63**, 1345 (1993)
4. T.E. Itina, K. Gouriert, L.V. Zhigilei, S. Noël, J. Hermann, M. Sentis, Appl. Surf. Sci. **253**, 7656 (2007)
5. M.F. Jarrold, in *Clusters of Atoms and Molecules*, ed. by H. Haberland (Springer, Berlin, 1994) p. 315
6. G. Bertoni, C. Cepek, F. Romanato, C.S. Casari, A. Li Bassi, C.E. Bottani, M. Sancrotti, Carbon **42**, 440 (2004)
7. D.B. Geohegan, A.A. Puretzky, G. Duscher, S.J. Pennycook, Appl. Phys. Lett. **73**, 438 (1998)
8. M.S. Tillack, D.W. Blair, S.S. Harilal, Nanotechnology **15**, 390 (2004)
9. A. Bailini, P.M. Ossi, Europhys. Lett. **79**, 35002 (2007)
10. T. Yoshida, S. Takeyama, Y. Yamada, K. Mutoh, Appl. Phys. Lett. **68**, 1772 (1996)
11. A.A. Voevodin, J.C. Jones, J.S. Zabinski, J. Appl. Phys. **88**, 1088 (2000)

12. X.T. Wang, B.Y. Man, G.T. Wang, Z. Zhao, B.Z. Xu, Y.Y. Zia, L.M. Mei, X.Y. Hu, J. Appl. Phys. **80**, 1783 (1996)
13. S.J.P. Laube, A.A. Voevodin, Surf. Coating. Tech. **105**, 125 (1998)
14. Z. Zhang, P.A. VanRompay, J.A. Nees, P.P. Pronko, J. Appl. Phys. **92**, 2867 (2002)
15. Y.B. Zel'dowich, Y.P. Raizer, *Physics of Shock Waves and High-temperature Hydrodynamic Phenomena* (Academic, New York, 1966)
16. P.R. Willmott, J.R. Huber, Rev. Mod. Phys. **72**, 315 (2000)
17. S.I. Anisimov, D. Bäuerle, B.S. Luk'yanchuk, Phys. Rev. B **48**, 12076 (1993)
18. S.S. Harilal, C.V. Bindhu, M.S. Tillack, F. Najmabadi, A.C. Gaeris, J. Appl. Phys. **93**, 2380 (2003)
19. S. Amoruso, A. Sambri, X. Wang, J. Appl. Phys. **100**, 013302 (2006)
20. A.K. Sharma, R.K. Thareja, Appl. Surf. Sci. **243**, 68 (2005)
21. T.E. Itina, J. Hermann, P. Delaporte, M. Sentis, Phys. Rev. E **66**, 066406 (2002)
22. R.F. Wood, J.N. Leboeuf, D.B. Geohegan, A.A. Puretzky, K.R. Chen, Phys. Rev. B **58**, 1533 (1998)
23. S. Amoruso, B. Toftmann, J. Schou, R. Velotta, X. Wang, Thin Solid Films **453–454**, 562 (2004)
24. S. Amoruso, B. Toftmann, J. Schou, Phys. Rev. E **69**, 056403 (2004)
25. R.F. Wood, K.N. Chen, J.N. Leboeuf, A.A. Puretzky, D.B. Geohegan, Phys. Rev. Lett. **79**, 1571 (1997)
26. A. Bulgakov, N.M. Bulgakova, J. Phys. Appl. Phys. **31**, 693 (1998)
27. V. Gusarov, I. Smurov, J. Phys. Appl. Phys. **36**, 2962 (2003)
28. D.B. Geohegan, Appl. Phys. Lett. **60**, 2732 (1992)
29. S. Acquaviva, M.L. De Giorgi, Appl. Surf. Sci. **186**, 329 (2002)
30. J. Gonzalo, C.N. Afonso, I. Madariaga, J. Appl. Phys. **81**, 951 (1997)
31. S. Acquaviva, M.L. De Giorgi, Appl. Surf. Sci. **197–198**, 21 (2002)
32. N. Arnold, J. Gruber, J. Heitz, Appl. Phys. A **69**, S87 (1999)
33. W.K.A. Kumuduni, Y. Nakayama, Y. Nakata, T. Okada, M. Maeda, J. Appl. Phys. **74**, 3340 (1993)
34. A.V. Rode, E.G. Gamaly, B. Luther-Davies, Appl. Phys. A **70**, 135 (2000)
35. F. Llewellyn-Jones, *Ionization and Breakdown in Gases* (Science Paperbacks, London, 1957)
36. A. Peterlongo, A. Miotello, R. Kelly, Phys. Rev. E **50**, 4716 (1994)
37. A. Bailini, P. M. Ossi, Appl. Surf. Sci. **252**, 4364 (2006)
38. A. Rivolta, "*Modello per la sintesi di cluster durante deposizione a laser pulsato in gas ambiente*," Laurea Thesis, Politecnico di Milano (2006) (in Italian)
39. A. Quarteroni, *Modellistica Numerica per Problemi Differenziali* (Springer, Milano, 2003) p. 105
40. D. Bolgiaghi, A. Miotello, P. Mosaner, P.M. Ossi, G. Radnoczi, Carbon **43**, 2122 (2005)
41. A. Bailini, P. M. Ossi, A. Rivolta, Appl. Surf. Sci. **253**, 7682 (2007)
42. P.M. Ossi, A. Miotello, J. Non-Cryst. Solids **353**, 1860 (2007)
43. A. Bailini, P.M. Ossi, Rad. Eff. Def. Sol. **163**, 497 (2008)
44. P.M. Ossi, A. Bailini, Appl. Phys. A **93**, 645 (2008)
45. F. Barreca, E. Fazio, F. Neri, E. Barletta, S. Trusso, B. Fazio, Rad. Eff. Def. Sol. **160**, 647 (2005)
46. J.C. Orlianges, C. Champeaux, A. Catherinot, Th. Merle, B. Angleraud, Thin Solid Films **453–454**, 285 (2004)

47. Y. Yamagata, A. Sharma, J. Narayan, R. M. Mayo, J. W. Newman, K. Ebihara, J. Appl. Phys. **88**, 6861 (2000)
48. C. Vivien, J. Hermann, A. Perrone, C. Boulmer-Leborgne, A. Luches, J. Phys. Appl. Phys. **31**, 1263 (1998)
49. J.M.D. Coey, D. Cole, R. Jordan, J.C. Lunney, J. Magn. Magn. Mater. **165**, 246 (1997)
50. M. Winter, Dept. of Chemistry, University of Sheffield, UK: http://www.webelements.com
51. M.F. Zhou, Z.W. Fu, Q.Z. Qin, Appl. Surf. Sci. **125**, 208 (1998)
52. Z.W. Fu, Q.Z. Qin, M.F. Zhou, Appl. Phys. A **65**, 445 (1997)
53. S.J. Carroll, P.D. Nellist, R.E. Palmer, S. Hobday, R. Smith, Phys. Rev. Lett. **84**, 2654 (2000)
54. M. Di Vece, S. Palomba, R.E. Palmer, Phys. Rev. B **72**, 073407 (2005)
55. P. Jensen, Rev. Mod. Phys. **71**, 1695 (1999)
56. E. Irissou, B. Le Drogoff, M. Chaker, D. Guay, J. Appl. Phys. **94**, 4796 (2003)
57. R. Dolbec, E. Irissou, M. Chaker, D. Guay, F. Rosei, M.A. El Khakani, Phys. Rev. B **70**, 201406(R) (2005)
58. M. Siegert, M. Plischke, Phys. Rev. Lett. **73**, 1517 (1994)
59. C.N. Afonso, J. Gonzalo, R. Serna, J.C.G. de Sande, C. Ricolleau, C. Grigis, M. Gandais, D.E. Hole, P.D. Townsend, Appl. Phys. A **69**, S201 (1999)
60. J. Gonzalo, A. Perea, D. Babonneau, C.N. Afonso, N. Beer, J.-P. Barnes, A.K. Petford-Long, D.E. Hole, P.D. Townsend, Phys. Rev. B **71**, 125420 (2005)
61. W. Marine, L. Patrone, B. Luk'yanchuk, M. Sentis, Appl. Surf. Sci. **154–155**, 345 (2000)
62. A. Bailini, P.M. Ossi, Carbon **44**, 3049 (2006)
63. F. Di Fonzo, A. Bailini, V. Russo, A. Baserga, D. Cattaneo, M.G. Beghi, P.M. Ossi, C.S. Casari, A. Li Bassi, C.E. Bottani, Catal. Today **116**, 69 (2006)
64. A. Baserga, V. Russo, F. Di Fonzo, A. Bailini, D. Cattaneo, C.S. Casari, A. Li Bassi, C.E. Bottani, Thin Solid Films **515**, 6465 (2007)
65. P.M. Ossi, A. Bailini, O. Geszti, G. Radnoczi, Europhys. Lett. **83**, 68005 (2008)
66. D. Cattaneo, S. Foglio, C.S. Casari, A. Li Bassi, M. Passoni, C.E. Bottani, Surf. Sci. **601**, 1892 (2007)
67. D.B. Geohegan, A.A. Puretzky, Appl. Surf. Sci. **96–98**, 131 (1996)
68. S. Amoruso, A. Sambri, M. Vitiello, X. Wang, Appl. Surf. Sci. **252**, 4712 (2006)
69. R.K. Thareja, R.K. Dwivedi, K. Ebihara, Nucl. Instrum. Methods Phys. Res. B **192**, 301 (2002)
70. T. Makimura, Y. Kunii, K. Muratami, Jpn. J. Appl. Phys. **35**, 4780 (1996)
71. D.H. Lowndes, C.M. Rouleau, T. Thundat, G. Duscher, E.A. Kenik, S.J. Pennycook, Appl. Surf. Sci. **127–129**, 355 (1998)
72. T. Ohkubo, M. Kuwata, B. Luk'yanchuk, T. Yabe, Appl. Phys. A **77**, 271 (2003)
73. E. Fazio, F. Neri, P.M. Ossi, N. Santo, S. Trusso, Laser Part. Beams **27**, 281 (2009)
74. M. Dinescu, M. Filipescu, P.M. Ossi, N. Santo, Appl. Surf. Sci. **255**, 9699 (2009)

6

Nanoparticle Formation by Femtosecond Laser Ablation

Chantal Boulmer-Leborgne, Ratiba Benzerga, and Jacques Perrière

Summary. Ultra short femtosecond (fs) pulses for the laser ablation of materials lead to deposited films which are very different from those obtained by the well-known classical nanosecond (ns) pulsed laser deposition (PLD). In very specific cases, epitaxial thin films can be obtained, whereas in the majority of materials, the films formed by fs PLD are constituted by the random stacking of nanoparticles (Nps) in the 10–100 nm size range. As a result, fs PLD has been rapidly considered as a viable and efficient method for the synthesis of Nps of a wide range of materials presenting interesting physical properties and potential applications. The Np synthesis by fs laser ablation has been studied, and theoretical investigations have been reported to establish their formation mechanisms. Two possibilities can be assumed to explain the Np synthesis: direct cluster ejection from the target or collisional sticking and aggregation in the ablated plume flow.

6.1 Introduction

The use of ultra short femtosecond (fs) pulses for the laser ablation of materials leads to results in terms of morphology, composition and structure of the deposited films which are very different from those obtained by the well-known classical nanosecond (ns) pulsed laser deposition (PLD) [1]. In very specific cases, i.e. SnO_2 [2] or ZnO [3], epitaxial thin films can be obtained, while in the majority of materials, the films formed by fs PLD are constituted by a random stacking of nanoparticles (Nps) in the 10–100 nm size range [4]. As a result, fs PLD has been rapidly considered as a viable and efficient method for Np synthesis of a wide range of materials presenting interesting physical properties [5] and potential applications. Owing to the increasing interest in nanosciences and nanotechnology, the formation of Nps by a dry and clean physical method has been the subject of further investigations. The Np synthesis of semiconductors (Si [6] or GaAs [7]) or metals (Ti [8], Al [9,10] or Ni [11]) by fs laser ablation has been studied, and theoretical investigations have been reported to establish their formation mechanisms during ultra short pulsed laser irradiation of materials [12,13].

Two possibilities can be assumed to explain the Np synthesis: direct cluster ejection from the target or collisional sticking and aggregation in the ablated plume flow. The generation of Nps by fs laser ablation is quite different from the formation of Nps by ns PLD under high pressure [14] by condensation in the confined plasma (up to a few 10^2 Pa), as discussed in Chap. 6. Np synthesis by fs laser ablation occurs under vacuum (in the 10^{-5} to 10^{-1} Pa pressure range) without gas phase condensation, and theoretical investigations were carried out to explain this formation. The different models and simulations are based on two important points: the fs laser pulses heat the target without changing its density, and then the further rapid expansion and cooling of this solid density matter result in Np synthesis. Various mechanisms have been considered for the Np synthesis [12, 15]. Currently, the most admitted hydrodynamic model suggests that the Nps are formed via mechanical fragmentation [16, 17] of a highly pressurized fluid undergoing rapid quenching during expansion.

Such models are based on the studies of fs PLD of single element targets, in which the sole relevant parameter governing Np emission is the laser fluence. In this work, we report on fs PLD of various materials (single element and polyatomic targets with various physical properties), under a broad range of experimental conditions of ultra short laser irradiation. Our results cannot be explained by Np formation via the mechanical fragmentation of the pressurized fluid following laser irradiation. The influence of a geometrical parameter (laser beam spot size) on Np emission has been also evidenced. This parameter is not taken into account in the above-mentioned theoretical approaches.

The aim of this chapter is thus to review the results obtained on the formation of clusters by fs laser irradiation of various targets (especially multielement targets), and to extract the most relevant parameters governing their emission. The results are compared with the proposed models and simulations of the formation of clusters by fs PLD already published. It can be concluded that a different approach is needed in order to explain all the phenomena taking place during ultra short laser pulse irradiation of a material, leading to the emission of nanoparticles.

6.2 Experimental

The PLD experiments were carried out using a laser beam operating at 620 nm with 90 fs pulse duration and 10 Hz repetition rate [8], at the femtosecond laser facility in LOA laboratory (ENSTA) in Palaiseau, France. About 1–3 mJ laser energy from an amplifier femtosecond colliding mode locking (CPM) dye laser was focused onto the target, leading after focalisation to power densities on the target in the 10^{16} to 10^{18} W m^{-2} range. In order to vary the laser fluence, the laser beam energy was modified by using optical densities on the beam path (from 2×10^{-3} to 3 mJ) and the laser spot size was varied by defocusing the beam, displacing the lens from its focal length position. The laser spot size on the target was determined by viewing its CCD camera image on a computer (from 10^{-8} to $1.5 \times 10^{-6}\,\mathrm{m}^2$).

Various materials were used as targets: metals (Ti, W, Al), semiconductor (Si) and insulators (AlN, $BaTiO_3$, MgO). These targets were irradiated by the fs laser with a 45° incidence under a 10^{-3} Pa pressure. Some experiments were carried out as a function of oxygen pressure up to 10 Pa. The particles emitted by the targets were collected onto Si single crystal substrates located in front of the target at a distance of $5 \times 10^{-2}\,\mathrm{m}$; the substrate was maintained at room temperature. The dynamics of the plasma expansion was studied using a charged-coupled device (CCD) camera to collect the luminescence of the plume. The image intensifier gate was triggered by the laser and the delay between the laser pulse and the image recording could be varied to obtain a time-resolved plasma imaging study. The camera gate duration time (minimum 5 ns) was adjusted in order to obtain sufficient signal intensity at each delay. The shape and evolution of the plasma plume at different times, for different experimental conditions (laser fluence, beam spot size) were studied. A time resolved spectroscopy experiment was also performed using this camera coupled to a spectrometer operating in the 300–900 nm spectral range. An optical filter was used to suppress the laser wavelength from the optical emission spectroscopy (OES) results. The different kinds of spectra (discrete or continuous) led to the determination of the nature of the emitted species in the plasma, i.e. atomic species through well defined emission peaks, or nanoparticles through blackbody emission.

In order to investigate the kinetics of the plasma emission, the CCD camera was replaced by a photomultiplier to record the time dependent luminous signal. This device records the signal of the different types of emitted species at different times, leading to the determination of their velocity.

The morphology of the deposited films was studied by scanning electron microscopy (SEM) and their crystalline structure by X-ray diffraction (XRD) analyses. The layer thickness and their chemical composition were deduced from Rutherford backscattering spectrometry (RBS) analyses, using the RUMP simulation program [18]. Nuclear reaction analyses (NRA) were used to quantitatively determine the absolute amounts of light elements (such as oxygen) in the films.

6.3 Results

Following the hypotheses and main characteristics of the models and simulations previously reported on fs laser irradiation of a target, the absorption of the laser energy and its transfer to the lattice result in a very fast heating of the near surface region of the target that can reach a very high temperature. Due to the ultra short pulse duration (fs), no significant expansion of the absorbing volume of the target can occur during the pulse, and at the end of the fs pulse, the density remains very close to that of the solid. This

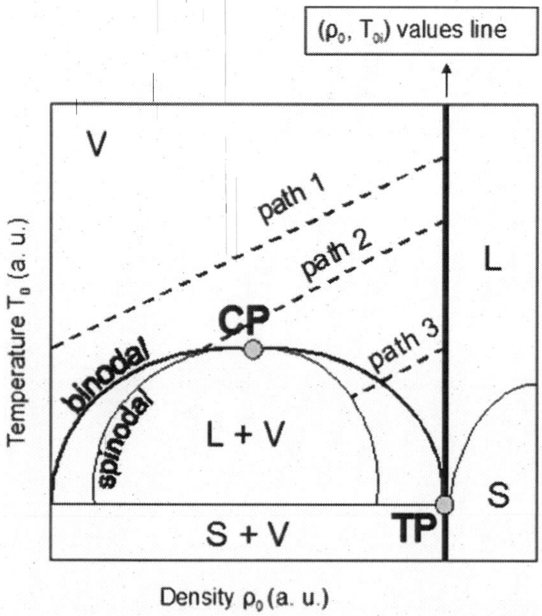

Fig. 6.1. Schematic (ρ, T) diagram with the various domain boundaries (solid S, liquid L and vapour V). CP and TP are the critical and triple points. The 1–3 lines show the various symbolic adiabatic trajectories for various initial temperature T_0 conditions

situation is depicted in a schematic density–temperature (ρ, T) diagram in Fig. 6.1 based on [9]. The vertical trajectory to the (ρ_0, T_{0i}) points represents the thermodynamic conditions of the irradiated target immediately after the fs pulse, i.e. the initial conditions for the further expansion of the ablated material in the vacuum chamber which is currently assumed to be adiabatic [9, 12, 13, 16, 19], which will lead to the formation of different species in the ablated plume.

The schematic and very simplified (ρ, T) diagram presented in Fig. 6.1 shows different symbolic thermodynamical trajectories (paths 1 to 3) of the adiabatic expansion depending on various T_{0i} conditions for different layers, located at different depths into the target. Detailed descriptions and precise discussions of these complex phenomena can be found in the original papers [9, 12, 16, 19]; here, let us only recall that the initial conditions (ρ_0, T_{0i}) following the laser pulse exclusively determine the nature of the plume species. For example, trajectories in the region over path 1 never reach the binodal. This corresponds to material that is converted into plasma recombining into atoms (i.e. vapourisation), while nanoparticles are preferentially formed in the path 3 region that reaches the spinodal at supercritical or near critical densities [9].

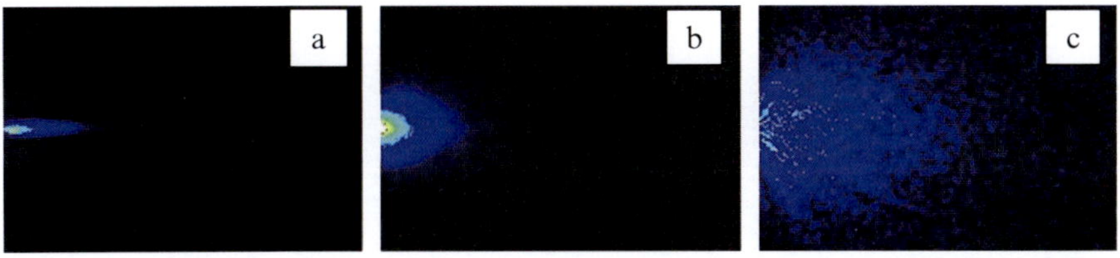

Fig. 6.2. CCD camera images of the plasma plume induced by fs laser irradiation of a Ti target under vacuum. The images were recorded at various delays after the fs laser pulse for different camera gates depending on the intensity of the signal [fluence: 7.2×10^4 J m^{-2}, laser spot size: 3.5×10^{-8} m^2] (**a**) Delay 3.25×10^{-7} s, CCD gate 2×10^{-8} s; (**b**) Delay 5×10^{-6} s, CCD gate 2.5×10^{-5} s; (**c**) Delay 5×10^{-5} s, CCD gate 10^{-4} s

6.3.1 Nature of the Species Emitted During fs PLD

Various species are emitted by the target during fs laser irradiation. As shown in Fig. 6.2, the images recorded by a CCD camera from the ablation of a Ti target show the presence of species with marked differences in velocity, making it possible to distinguish them as functions of time. First, at a very short time delay (Fig. 6.2a), a plume expanding along the normal to the target with very low angular aperture is observed. OES indicates that this plume is composed only of atomic species (ions and neutrals). Precise measurements have been achieved [8] and led to the determination of their velocity around 3×10^4 ms^{-1} for atoms and 6×10^4 ms^{-1} for ions. These values are in accordance with those from literature [6]. This plume is observed whatever the nature of the target, and only weak differences are observed in the species velocity when the target material is changed. The classical explanation of this ion-neutral emission is based on the Coulomb explosion phenomenon [20, 21]. The ion-neutral species are emitted with almost the same velocities from all the studied targets, whatever their nature (insulating, semiconducting or metallic). In the case of metallic targets the ambipolar diffusion is proposed as possible mechanism giving rise to the observed high-energy plasma plume component [22].

A second plume can be observed after a few μs delay (Fig. 6.2b), with a significantly higher angular aperture and lower velocity (a few 10^3 ms^{-1}). The nature of the species present in this plume was studied by OES, and the observed typical blackbody signal leads to the conclusion that this plume is mainly composed of clusters [8, 23]. This conclusion was further checked by SEM analyses of the film surface (Fig. 6.3) which show the presence of a random stacking of nanoparticles, whose size depends upon the material and laser irradiation conditions. Such SEM images are observed whatever the nature of the target material (metal, semiconductor or insulator), indicating that Np emission is a characteristic phenomenon resulting from fs PLD of materials.

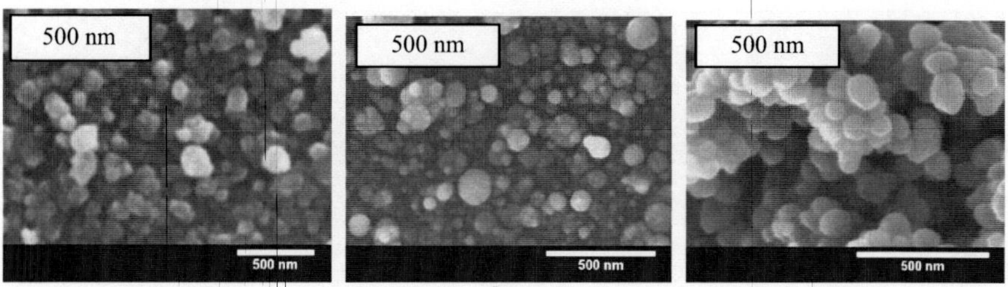

a b c

Fig. 6.3. SEM images of the surface of the films grown by fs laser ablation on (**a**) Ti; (**b**) Si; and (**c**) AlN targets, [fluence: 9.6 10^4 J m^{-2}, deposition time: 1 h]

It can also be deduced from these SEM images that the relevant fraction of the matter emitted by the target during fs PLD corresponds to clusters, i.e. atomic species represent a negligible amount of the emitted matter, in agreement with previous observations [24]. This conclusion has been checked by a careful examination of the angular distribution of film thickness and of the distributions of the various species deduced from the CCD images. A good agreement was found between the angular distribution of the Np plume deduced from the CCD images (Fig. 6.2b) and the angular distribution of film thickness. It appears that only a few percent of the matter emitted by the target corresponds to atomic species.

The third population which can be observed (Fig. 6.2c) at longer delay (a few tens of μs) corresponds to droplets which are easily recognized thanks to their luminous trajectories due to their low velocity and the high integration time for the image recording (i.e. a few tens of μs). From such trajectories an estimation of their velocity can be obtained (a few tens of 10 ms^{-1}). Such droplets, which are present whatever the nature and properties of the target material, are only observed at very high laser fluences (1.5 10^5 Jm^{-2}). The SEM images of the films recorded in these cases show the presence of micrometer-sized particles at the surface of the films [4]. It is important to note that these droplets do not correspond to "big" clusters. Indeed, their size and velocities are very different; their origin and formation mechanisms should therefore also be different. A possible explanation of the formation of droplets based on target stress confinement due to the ultrashort pulse laser irradiation has been proposed [15].

The influence of the laser fluence on the emission of these various species was studied, and three laser fluence thresholds were determined for the emission of atomic species (E_a), clusters (E_c) and droplets (E_d), with the following relationship which holds whatever the target material

$$E_a \approx E_c < E_d \tag{6.1}$$

meaning that it is always possible to find experimental fs PLD conditions leading to the formation of films composed only by the stacking of nanoparticles or clusters, without macroscopic droplets.

6.3.2 Nature of the Nanoparticles Formed During fs PLD

The generation of Nps during fs PLD has been mainly studied on elemental targets and the results lead to the conclusion that their formation and emission directly occurs from the target during a very short time (i.e. 50 ps after the laser pulse) [24], without any interactions with the atmosphere of the chamber (residual vacuum or injected gas). Precise RBS and NRA analyses of the deposited films however showed the incorporation of oxygen in the films, these oxygen atoms coming from the residual pressure in the chamber (10^{-3} Pa). Such oxygen incorporation could be due to Np oxidation during plume expansion. Further analyses on films grown by fs PLD led to conclude that oxygen atoms could take part in the formation process of Nps. Figure 6.4 represents the XRD diagram of a W film grown by fs PLD. The presence of two tungsten phases is deduced from the reflection peaks: the stable α-W phase and the metastable β-W phase. The α-W is the well-known bulk W phase, while the β-W, also called W_3O phase, is a metastable phase which is stabilized by the incorporation of oxygen (less than 10%). Previous studies indicate that the β-W phase is observed only in thin films grown in the presence of oxygen atoms, but is not obtained by the direct oxidation of thc α-W phase [25]. The presence of the β-W phase in the film indicates that the synthesis of W Nps occurs in the presence of oxygen atoms.

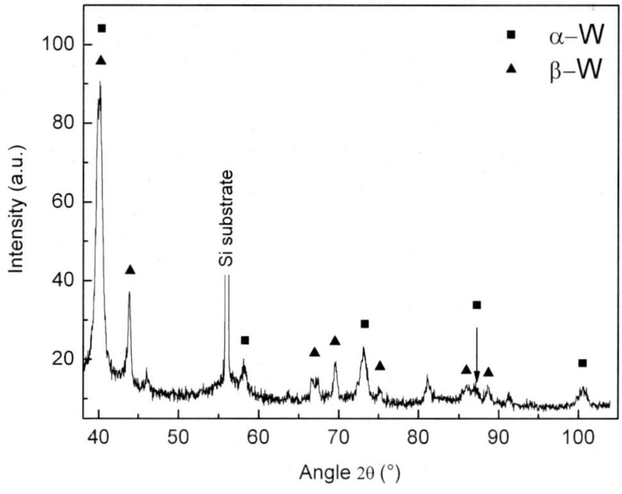

Fig. 6.4. XRD diagram for a fs PLD W film grown under vacuum (10^{-3} Pa) at a laser fluence of 7.1 10^4 J m^{-2} and for a deposition time of 1 h

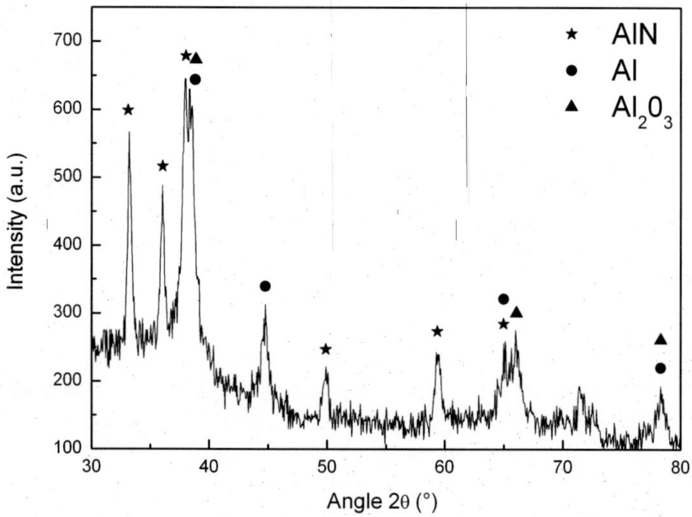

Fig. 6.5. XRD diagram for a fs PLD AlN film grown under vacuum (10^{-3} Pa) at a laser fluence of 7.2 10^4 J m^{-2}, and a deposition time of 1 h

In addition fs laser ablation of a polyatomic target does not inevitably lead to the formation of nanoparticles whose composition respects the target composition. A typical example is given by the fs laser ablation of an aluminum nitride target, for which AlN Nps are not the only species formed. The XRD diagram presented in Fig. 6.5 recorded on a film grown by fs PLD of a stoichiometric AlN target shows reflection peaks of Al and Al_2O_3 crystallites in addition to AlN ones. The RBS and NRA analyses of such films evidenced noticeable amounts of oxygen atoms in the deposited material, while the SEM analyses of these films showed the classical random stacking of Nps characteristic of fs laser ablation (see Fig. 6.3c). This result cannot be explained by a pure fragmentation mechanism as proposed in [16, 17]. The formation of Al and Al_2O_3 Nps from an AlN target needs first the separation of the chemical phases, chemical reactions and finally the Al oxidation by interaction with oxygen atoms of the residual gas in the chamber.

For baryum titanate ($BaTiO_3$) fs PLD, ICCD images of the plume expansion showed the characteristic signal of the clusters, and SEM analyses revealed their presence in the films. RBS analyses of the baryum titanate films grown in these conditions evidenced deviations with respect to the target composition. A clear titanium enrichment (i.e. Ba/Ti ratio lower than 0.9) with respect to the ideal stoichiometric target composition was observed in the films whose position faces towards the target in the normal direction [26]. Moreover, the composition of the films was studied as a function of their angular position with respect to the normal to the target. From these RBS analyses, the variation in the Ba to Ti concentration ratio with the angular position was deduced and is plotted in Fig. 6.6. This figure shows an important

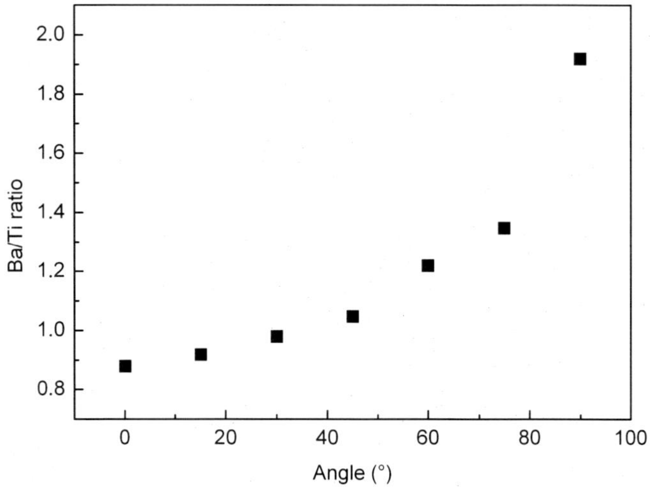

Fig. 6.6. Evolution of the Ba to Ti concentration ratio in the fs PLD films as a function of their angular position with respect to the normal to the target

variation in composition as a function of the angle, i.e. Ti enrichment in the normal direction and a corresponding depletion at large angular aperture. These composition deviations are significantly higher than those observed during ns PLD of the same material [27], in which only atomic species are emitted by the target during ablation. This non uniform angular distribution means that fs PLD leads to the formation of Nps of different compositions, i.e. for instance BaO, TiO_2, and $Ba_xTi_yO_z$, and that the various Nps present different angular distributions.

Moreover, the influence of the oxygen pressure on the Ba/Ti ratio indicates that the gas phase plays an important role in Np synthesis [26], and this has not been yet explained. How the gas phase influences the precise cationic composition of the Nps is still a matter of discussion. A pure condensation in the gas phase like for ns PLD at high pressure with plasma confinement can be excluded [14]. The dependence of Np composition upon the gas pressure in the ablation chamber seems to be a general phenomenon. The fs PLD of InP, CoPt and other compounds [28] evidences variations in the film composition depending upon the gas pressure during growth. Moreover, the crystalline nature of the Si Nps formed during fs laser irradiation was dependent upon the nature and pressure of the gas in the ablation chamber [29].

It can be concluded that phase separation, atomic movements, and chemical reactions have to occur during Np synthesis, and this cannot be envisaged in the framework of a homogeneous nucleation, or a pure fragmentation-based mechanism for fs laser ablation. All these phenomena have never been taken into account in the various models and simulations of fs PLD.

6.3.3 Relevant Parameters of Nanoparticle Formation

The laser fluence plays an important role in Np formation and emission during fs laser irradiation of materials. A laser fluence threshold exists for Np formation (E_c). At very high laser fluences, the emission of Nps decreases while the emission of droplets is observed. This behaviour is illustrated in Fig. 6.7 representing photomultiplicator (PM) signals as a function of time obtained for different laser fluences with the same laser beam spot size. At the lowest laser fluence the signal is composed of 2 peaks (atom-ion emission and Nps one). The highest laser fluences lead to the 3-peak characteristic signal due to macroscopic droplets produced at the longest time (as presented in Fig. 6.2). The Nps are thus formed and emitted in a laser fluence range limited by E_c and E_d, the thresholds for cluster and droplet emission respectively (see [4]). Such a fluence range for nanoparticle formation has already been reported in the case of fs PLD of aluminum [9]. Thus Fig. 6.7 characterizes a bimodal (Nps in the 10 to 100 nm range and droplets in the μm range) distribution. Such a bimodal distribution for particle production is also described in literature for fs PLD at atmospheric pressure [30] and can be related to Np and droplet emission.

In order to show a characteristic point on Np formation, Fig. 6.8 represents the intensity evolution of the PM signal of ions/neutrals and nanoparticle emission as a function of the laser spot diameter for a constant laser energy. For a low beam spot diameter, the Np signal is low. Then by increasing the beam spot size, the Np intensity signal increases and reaches a maximum

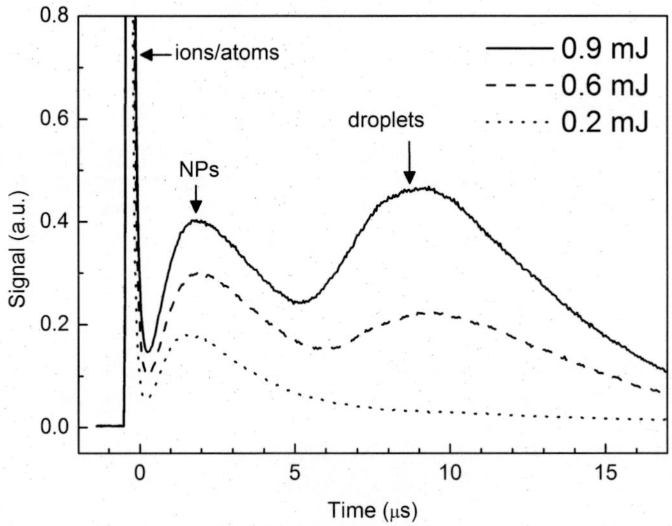

Fig. 6.7. PM signals of the plasma plume obtained by fs laser ablation of MgO for various laser beam energies with the same laser beam spot size (9.4 10^{-9} m^2) versus time

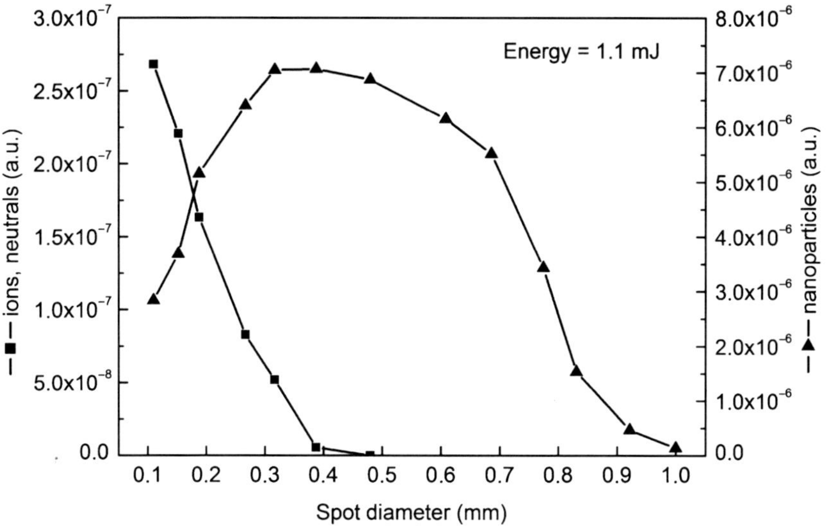

Fig. 6.8. Evolution of the PM signal of the plasma plume obtained by fs PLD of a Ti target, as a function of laser spot diameter and a laser beam energy of 1.1 mJ (signal intensity for ■ ions, neutrals and for ▲ nanoparticles)

value. A further increase in beam spot size leads to a decrease in the Np signal which completely disappears for large beam spot sizes corresponding to the laser fluence threshold for Np emission (E_c). The signal for ions and neutrals presents a very different behaviour, with a clear maximum when the laser is focused (highest laser fluence), followed by a continuous decrease with increasing beam spot size (decreasing laser fluences). It should be noted that the intensity signal of the droplets is characterized by a behaviour that is strictly identical to that of the ions and neutrals. It can thus be concluded that the mechanisms of formation and emission of droplets and nanoparticles are different, i.e. the macroscopic droplets are neither large nanoparticles nor an agglomeration of nanoparticles; the origin of the droplet could be explained by target stress confinement as proposed in [15].

Another important parameter playing a role in the formation of nanoparticles is a geometrical parameter: the *laser beam spot size*. Figure 6.9 shows two different ICCD images of the plume recorded for two different beam spot sizes with almost the same laser fluence. In the case of the smaller beam spot size, the emission of macroscopic droplets is clearly evidenced through the presence of their luminous trajectories. Then for larger beam spot sizes, the image of the plume is solely characteristic of nanoparticle emission. This means that for the same initial thermodynamic conditions (ρ_0, T_{0i}) related to the laser fluence, very different behaviours can be observed depending on the laser beam spot size: Np, or droplet emission. This geometrical effect which has been observed whatever the nature (metal, semiconductor or insulator) of

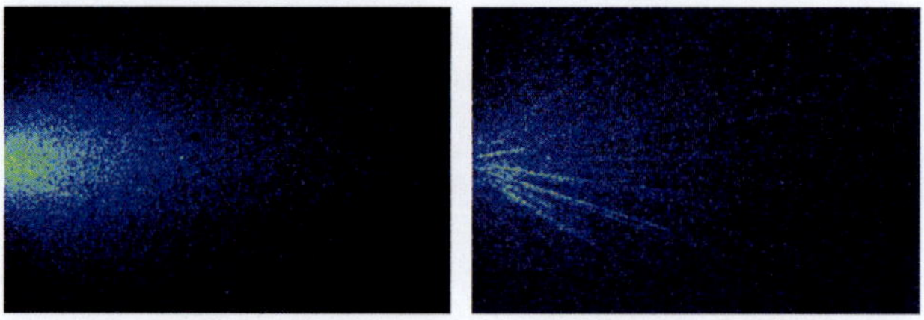

Laser fluence: 2.5 10^4J/m^2, Laser spot size: 1.10^{-7} m^2

Laser fluence: 2.4 10^4 J/m^2, Laser spot size: 2.3.10^{-8} m^2

Fig. 6.9. CCD camera images of the plasma plume obtained by fs PLD of Ti for the same laser fluence (around 2 $10^4\,\mathrm{J\,m^{-2}}$) and two different laser beam spot sizes

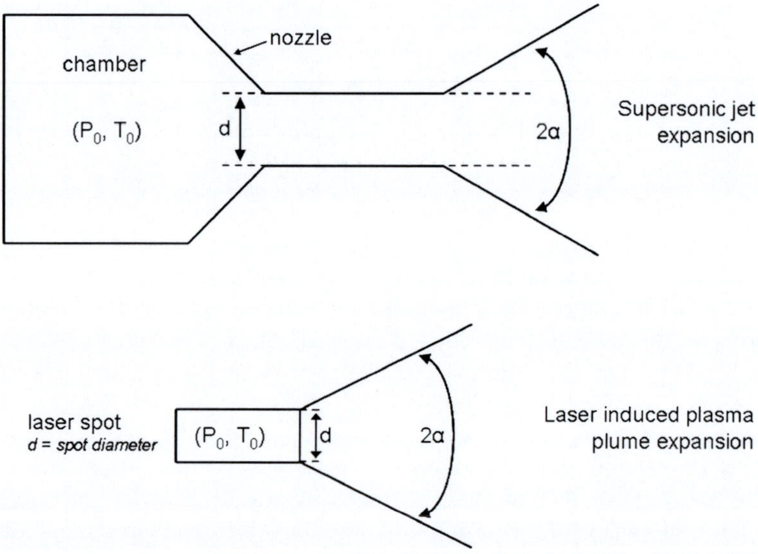

Fig. 6.10. Schematic of the experimental set up for a supersonic jet expansion (*top*) compared to the laser induced plasma plume expansion (*bottom*)

the target is not envisaged in the theoretical models and simulations of the fs PLD of materials.

An analogy of this geometrical effect can be made considering the formation of clusters in a supersonic jet [31,32]. When high-pressure gas flows into vacuum through an orifice, clusters can be formed in a supersonic jet. More precisely, clusters from gases as well as from metal vapours can be obtained from an expanding nozzle flow (see Fig. 6.10) with the appropriate set of flow field conditions, characterized by a condensation scaling parameter Γ^*. This empirical parameter varies with the experimental conditions of the gas jet and was found to be [31]

$$\Gamma^* = k\left[(d/\tan\alpha)^{0.85}\cdot P_0\right]/T_0^{2.29}, \tag{6.2}$$

where k is a constant depending on the material, P_0 and T_0 are the initial gas pressure and temperature in the chamber, d is the jet throat diameter and α the jet expansion half angle (see Fig. 6.10). It was shown that clustering begins when Γ^* exceeds about 300 [32]. It is interesting to note that Γ^* depends on two different kinds of parameters, (P_0, T_0) describing the thermodynamic state of the gas and (d, α) the geometrical parameters of the expansion. For identical initial P_0 and T_0 conditions for the gas, the formation of clusters will depend on gas expansion, as imposed by the shape and dimension of the nozzle.

A clear analogy can thus be made with the phenomena observed in this work, with the (ρ_0, $T_{0\mathrm{i}}$) initial conditions and the geometrical effect related to the laser beam spot size. This analogy is more complete when one takes into account the angular aperture of the plume depending on the beam spot size as α increases with decreasing laser spot size. It can be assumed that the Nps are formed during thc first step of plume expansion, and that the main parameter governing this stage of plasma expansion is the laser beam spot size.

The expansion of the plasma induced by laser irradiation of a material is an adiabatic expansion which first occurs in the axial dimension (along the normal to the target), and then in a three dimension expansion (axial and radial) [33,34], as schematically represented in Fig. 6.11. The strong forward direction of the initial expansion is caused by strong differences in pressure gradients in axial and radial directions. The analysis of the expansion of a

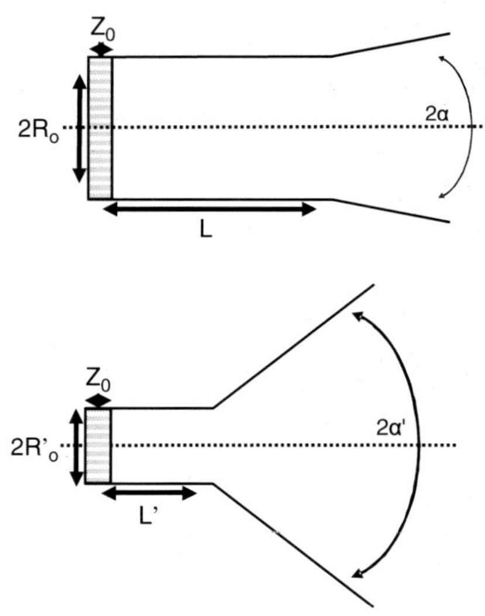

Fig. 6.11. Schematic description of the adiabatic plasma expansion of a fs PLD induced plasma plume, for two different laser beam spot sizes (diameter $2R_0 > 2R'_0$)

PLD plasma [34] is based on the solution of gas-dynamic equations assuming an adiabatic expansion of the plasma plume in vacuum. The evolution of the axial Z(t) and radial R(t) plume components has been determined as a function of time, and it was established that for a long time interval ($t \rightarrow \infty$), both Z(t) and R(t) are linear functions, meaning that the expansion of the plume becomes inertial. In this inertial three dimensional expansion, the ratio $k = Z(t)/R(t)$ is constant.

One important parameter governing the plasma expansion was found to be the value of $\sigma = Z_0/R_0$, with Z_0 and R_0 the initial axial and radial dimensions of the plasma. σ roughly characterizes the initial pressure gradient, i.e. the driving force of the expansion. k was found to increase with decreasing values of σ [34]. The extension of the one dimensional plasma expansion depends on the initial conditions, i.e. the Z_0 and R_0 values, R_0 being the radius of the laser beam spot, and Z_0 being the thickness of the ablated volume of the target. As the laser pulse duration time is ultra short it can be assumed that Z_0 corresponds to the absorption depth.

Assuming the same laser fluence in the two cases presented in Fig. 6.11, the initial conditions (ρ_0, T_{0i}) should be identical and then the same value of Z_0 could be considered. In that case the σ values (i.e. Z_0/R_0) will be different, a larger value being obtained with the larger spot size, meaning that the initial gradient pressure will be larger.

For the larger spot sizes $R_0 > R_0'$ (Fig. 6.11), the length of the one dimensional expansion L will be larger than the corresponding value L'. The situation for Np formation favoured by the greater length of the one dimensional expansion is described in Fig. 6.11. On the contrary a small laser spot size R_0' corresponds to a small L' value limiting Np formation and favouring droplet emission. By comparison with the evolution in the (ρ, T) diagram (Fig. 6.1), it can be concluded that the path followed by the ablated matter in this diagram will be a function of the beam spot size $2R_0$, via the different length L and L' of the one dimensional expansion of the plume.

6.4 Conclusions

By the complementary study of the plasma plume dynamics and nature (composition, morphology and structure) of films grown by fs PLD of various materials, results on the formation of nanoparticles by fs PLD have been obtained.

First, the fs PLD of polyatomic materials does not necessarily lead to the formation of Nps presenting the same composition as that of the target. Complex phenomena occur in this case like separation of phases, atomic movements and chemical reactions, leading to various kinds of Nps with different structures and/or composition. Some interactions between the species emitted by the target and the gas phase in the ablation chamber should also occur during the fs PLD. In addition, the influence of the laser beam spot size on the formation and emission of Nps at constant laser fluence has been evidenced. All these results raise some questions on the relationships between the laser beam spot size and photon absorption and on the first stage of plasma expansion during fs PLD.

To our knowledge all these points have never been considered in the theoretical models and simulations currently proposed to describe Np formation during fs PLD. A new approach taking these various aspects into account has to be envisaged in order to find an agreement with experimental observations and to contribute to our knowledge of fs laser ablation phenomena.

Acknowledgements

The authors would like to thank O. Albert and J. Etchepare for the use of the femtosecond laser facility in LOA laboratory (ENSTA), Palaiseau, France and their fruitful collaboration in this work.

References

1. D.B. Chrisey, G.K. Hubler (eds), *Pulsed-Laser Deposition of Thin Films* (Wiley, New York, 1994), p. 229
2. Z. Zhang, P.A. Van Rompay, P.A. Nees, J.A. Stewart, C.A. Pan, X.Q. Fu, P.P. Pronko, SPIE Proc. **3935**, 86 (1999)
3. R. Eason (ed.), *Pulsed-Laser Deposition of Thin Films, Application Led-Growth of Functional Materials* (Wiley-Interscience, Hoboken, NJ, 2006), p. 261
4. C. Boulmer-Leborgne, B. Benzerga, J. Perrière, SPIE Proc. **6261**, 20 (2006)
5. S. Amoruso, G. Ausanio, R. Bruzzese, M. Vitiello, X. Wang, Phys. Rev. B **71**, 033406 (2005)
6. S. Amoruso, R. Bruzzese, N. Spinelli, R. Velotta, M. Vitiello, X. Wang, Europhys. Lett. **67**, 404 (2004)
7. T. Trelenberg, L. Dinh, C. Saw, B. Stuart, M. Balooch, Appl. Surf. Sci. **221**, 364 (2004)
8. O. Albert, S. Roger, Y. Glinec, J.C. Loulergue, J. Etchepare, C. Boumer-Leborgne, J. Perrière, E. Millon, Appl. Phys. A **76**, 319 (2003)
9. S. Eliezer, N. Eliaz, E. Grossman, D. Fisher, I. Gouzman, Z. Henis, S. Pecker, Y. Horovitz, S. Fraenckel, M. Maman, Y. Lereah, Phys. Rev. B **69**, 144119 (2004)
10. S. Amoruso, R. Bruzzese, M. Vitiello, N. Nedialkov, P. Atanasov, J. Appl. Phys. **98**, 044907 (2005)
11. B. Liu, Z. Hu, Y. Che, Y. Chen, X. Pan, Appl. Phys. Lett. **90**, 044103 (2007)
12. D. Perez, L. Lewis, Phys. Rev. B **67**, 184102 (2003)
13. S. Amoruso, R. Bruzzese, X. Wang, N. Nedialkov, P. Atanasov, J. Phys. Appl. Phys. **40**, 331 (2007)
14. J. Perrière, E. Millon, M. Chamarro, M. Morcrette, C. Andreazza, Appl. Phys. Lett. **78**, 2949 (2001)
15. L. Zhigilei, Appl. Phys. A **76**, 339 (2003)

16. B. Holian, D. Grady, Phys. Rev. Lett. **60**, 1355 (1988)
17. D. Perez, L. Lewis, Phys. Rev. Lett. **89**, 255504 (2002)
18. L. Doolittle, Nucl. Instrum. Methods Phys. Res. B **9**, 344 (1985)
19. F. Vidal, T.W. Johnson, S. Laville, O. Barthelemy, M. Chaker, B. Le Drogoff, J. Margot, M. Sabsabi, Phys. Rev. Lett. **86**, 2573 (2001)
20. R. Stoian, D. Ashkenasi, A. Rosenfeld, E.E.B. Campbell, Phys. Rev. B **62**, 13167 (2000)
21. R. Stoian, A. Rosenfeld, D. Ashkenasi, I. Hertel, N.M. Bulgakova, E.E.B. Campbell, Phys. Rev. Lett. **88**, 097603 (2002)
22. S. Amoruso, X. Wang, C. Altucci, C. de Lisio, M. Armenante, R. Bruzzese, N. Spinelli, R. Velotta, Appl. Surf. Sci. **186**, 358 (2002)
23. S. Amoruso, R. Bruzzese, N. Spinelli, R. Velotta, M. Vitiello, X. Wang, G. Ansiano, V. Ianotti, L. Lanotte, Appl. Phys. Lett. **84**, 4502 (2004)
24. T. Glover, G. Ackerman, A. Belkacem, P. Heimann, Z. Hussain, R. Lee, H. Padmore, C. Ray, R. Schoenlein, W. Steele, D. Young, Phys. Rev. Lett. **90**, 236102 (2003)
25. I. Weerasekera, S. Ismat Shah, D. Baxter, K. Unruh, Appl. Phys. Lett. **64**, 3231 (1994)
26. E. Millon, J. Perrière, R.M. Defourneau, D. Defourneau, O. Albert, J. Etchepar, Appl. Phys. A **77**, 73 (2003)
27. J. Gonzalo, R. Gomez-San Roman, J. Perrière, C. Afonso, R. Perez-Casero, Appl. Phys. A **66**, 487 (1998)
28. T. Trelenberg, L. Dinh, B. Stuart, M. Baboch, Appl. Surf. Sci. **229**, 268 (2004)
29. B. Tull, J. Carey, M. Sheehy, C. Friend, E. Mazur, Appl. Phys. A **83**, 341 (2006)
30. J. Koch, A. von Bohlen, R. Hergenröder, K. Niemax, J. Anal. Spectrom. **19**, 267 (2004)
31. O. Hagena, Rev. Sci. Instrum. **63**, 2374 (1992)
32. T. Ditmire, T. Donnely, A. Rubenchik, R. Falcone, M.D. Perry, Phys. Rev A **53**, 3379 (1996)
33. R.K. Singh, J. Narayan, Phys. Rev. B **41**, 8843 (1990)
34. S. Anisimov, D. Bäuerle, B. Luk'yanchuk, Phys. Rev. B **48**, 12076 (1993)

7

UV Laser Ablation of Polymers: From Structuring to Thin Film Deposition

Thomas Lippert

Summary. UV laser ablation of polymers is a versatile method to structure polymers with high resolution. The mechanism of ablation is often discussed controversially, but it is necessary to keep in mind that polymers are complex systems with a wide variety of properties that can influence the ablation mechanism. Analyzing the data, it appears that the ablation mechanism is a complex interrelated system, where photochemical and photothermal reactions are very important. The pressure jump, which is associated with the creation of small molecules and originates from both types of reaction, is also important for ablation. The importance of each effect is strongly dependent on the type of polymer, the laser wavelengths, the pulse length, and the substrate. UV laser ablation can also be utilized to deposit directly thin polymer films by PLD, but this is limited to certain polymers. Alternative laser-based techniques (LIFT) utilize the decomposition of a thin layer to transfer complete layers with high spatial resolution. This approach can be used to transfer pixels of sensitive materials to a substrate with a minimal thermal and UV load.

7.1 Introduction

7.1.1 Laser Ablation of Polymers

Laser ablation of polymers was first reported by Srinivasan et al. [1] and Kawamura et al. [2] in 1982. Since then, numerous reviews on laser ablation of a large variety of polymers and the different proposed ablation mechanisms have been published [3–11]. There is still an ongoing discussion about the ablation mechanisms, e.g., whether it is dominated by photothermal or photochemical processes.

Since its discovery, laser polymer processing has become an important field of applied and fundamental research. The research can be separated into two fields, the investigation of the ablation mechanism and its modeling and the application to produce novel materials or structures. Laser ablation is used as an analytical tool in *matrix-assisted laser desorption/ionization (MALDI)*

[12,13] and *laser-induced breakdown spectroscopy (LIBS)* [14] or as preparative tool for the deposition of thin films, e.g., by *pulsed laser deposition (PLD)* of synthetic polymers [15–17] (of inorganic films [18,19]), *matrix-assisted pulsed laser evaporation (MAPLE)*, which is a deposition technique that can be used to deposit highly uniform thin films [20], or *laser-induced forward transfer (LIFT)* [21,22].

There are several industrial applications for polymers in laser ablation, mainly for structuring, i.e., for the production of nozzles for inkjet printers [23] and to prepare via-holes in multichip modules through polyimide by IBM [24]. Laser ablation for other applications, e.g., fabrication of micro-optical devices [25] and microfluidic channels [26–29], are under development.

7.1.2 Polymers: A Short Primer

Polymers are macromolecules, which are synthesized from one or more different monomers using different types of polymerization, i.e., radical or ionic polymerization, polycondensation, polyaddition, and special cases such as copolymerization. To start the polymerization reaction, starters have to be applied in many cases, e.g., molecules that form a radical upon reaction that is initiated by temperature or light or even complex initiators and enzymes. The polymerization type has also a direct influence on the characteristics of the polymer, e.g., molecular weight and distribution, impurities, polymer structure (tacticity), or molecular form, and on the decomposition mechanism. The molecular weight, M_w, of the polymer has a direct influence on the state of the polymer, i.e., low molecular weight polymers may still be liquids, while high molecular weight polymers are solids, which may even be insoluble in all solvents if the molecular weight is too high. The M_w subsequently influences the viscosity of the polymer (in the melt or solution), the glass transition temperature, T_g, which is the temperature at which the polymer changes from the glass to rubber state, and possibly the melting and decomposition temperature. The M_w of a polymer is not one well-defined number, but a range of molecular weights is obtained from the synthesis, and normally an average is quoted. To be more precise, the polydispersity is used, which is the ratio of the weight average molecular weight to the number average molecular weight and an indication for the distribution of the molecular weights. In polymer chemistry, a Schulz–Flory distribution is often used to describe the variation of molecular weights. The polymer synthesis and structure of the monomer have a direct influence on the chain regularity/conformation of the polymer, which is also called tacticity. A polymer can have an atactic (random), iso or syndiotactic (ordered, see Fig. 7.1) structure, which again influences properties such as the T_g. In the case of optical active monomers, optical active polymers may be obtained as pure D-, L-, or D-L (racemic) structure, which is common for biopolymers. Another aspect that is specifically important for the photon–polymer interactions is the possibility of polymers to be partially crystalline (never really complete, even if they are called single crystals), which results in

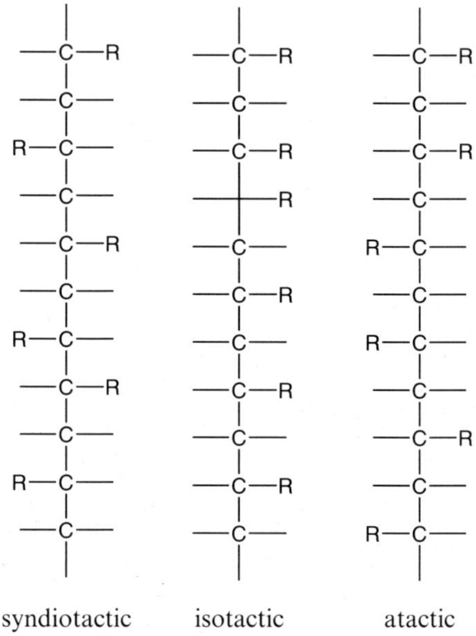

Fig. 7.1. Possible conformations of a polymer chain

light scattering (not absorption) in the polymer. Finally, it is also necessary to consider that most polymers cannot be vaporized (sublimed) intact and that many do not have a melting point prior to decomposition, which is the case for cross-linked polymers or many polyimides.

Most of these polymer characteristics can have, as described below, an influence on the ablation behavior of polymer, while the decomposition type is important for the ablation mechanism and the possibility to form thin films.

Classification of the Decomposition Behavior

The decomposition mechanism of a polymer is a reasonable way to classify polymers for their behavior upon UV laser irradiation. Polymers which decompose into fragments are for example polyimides or polycarbonates (see Figs. 7.2 and 7.3). This method of classification is closely related to the synthesis of the polymers. Polymers that are produced by radical polymerization from monomers, which contain double bonds, are likely to depolymerize into monomers, while polymers that have been formed by reactions such as polycondensation will not depolymerize into monomers upon irradiation, but will be decomposed into different fragments. The second group cannot be used to produce films with the same structure or molecular weight as the original material with methods such as PLD.

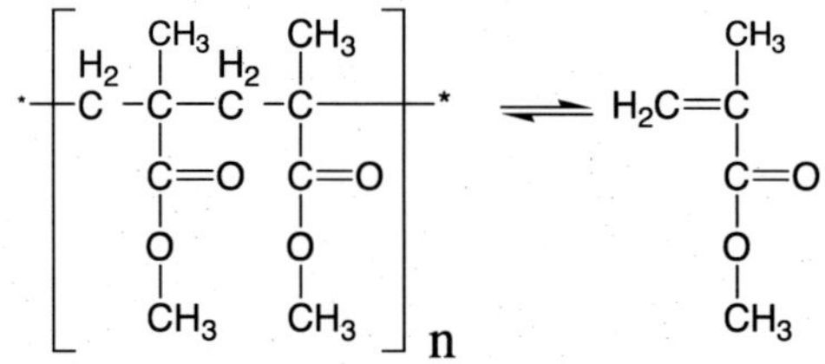

Fig. 7.2. Chemical structure of PMMA and its monomer

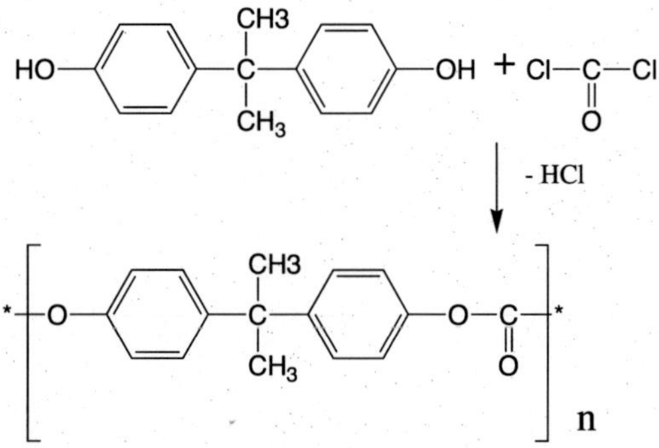

Fig. 7.3. Typical polycondensation reaction to form a polycarbonate

The different mechanisms may be described as:

Depolymerization

Some of these polymers show unzipping reactions (one radical on the polymer main chain yields several monomers) and have a ceiling temperature (T_c, above which the equilibrium between polymer and monomer is totally on the side of the monomer). A typical example is poly(methylmetacrylate) (PMMA, see Fig. 7.2), which has a ceiling temperature of 550 K and the zip length (the number of monomers originating from every chain end radical) is 6 at room temperature and ~200 above the glass transition temperature (378 K) of PMMA. Other examples of unzipping polymers are polystyrene and Teflon.

Decomposition or Fragmentation

Polymers that decompose into fragments are for example polyimides or polycarbonates. The reactions which are used to form these polymers are shown in Figs. 7.3 and 7.4. It is obvious that the monomers cannot be produced during decomposition, because one reaction product, e.g., H_2O or HCl, is removed during polymerization.

These polymers show in the case of decomposition (thermally or photochemically) a tendency for a pronounced fragmentation into various small molecules, as shown for polyimide in Fig. 7.5. All the fragments shown have been detected by various analytical methods [30, 31].

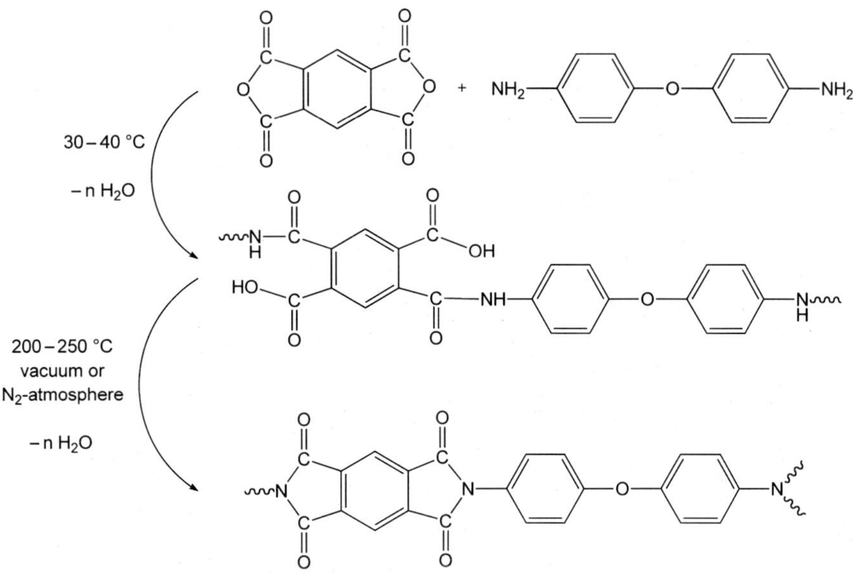

Fig. 7.4. Typical polycondensation reaction to form a polyimide

The large number of small decomposition products will results in a pronounced pressure increase inside the polymer matrix, which is important for ablation, as discussed in detail below.

7.2 Polymer Properties and Ablation

The influence of only some of the polymers properties, as discussed above, on ablation has not been studied in detail and only for the molecular weight several studies have been performed [32–35]. A clear influence of the molecular weight on the ablation rate was detected for doped PMMA (see Fig. 7.6) and has been assigned to the increased viscosity of the higher molecular weight polymer, which is clearly important for an ablation mechanism that shows clear indications of melting (see the ablation crater in Fig. 7.7).

A pronounced influence of the M_w on the ablation behavior has also been detected for doped PMMA and polystyrene doped with Iodo-naphthalene. The formation of Nap_2 (=1,1-binaphthalene) as a product of irradiation has been analyzed by fluorescence spectroscopy, and a complex M_w dependent behavior was detected that cannot be simply explained by the expected increase of viscosity for the higher molecular weight polymers. It seems that additional effects, e.g., higher ablation rates for lower M_w, the T_g, and bubble formation influence the rate of product formation [34, 35].

Another important parameter, which is especially important for technical polymers, is the presence of polymer additives or impurities that originate from the reaction (e.g., catalysts, starters). Additives to technical

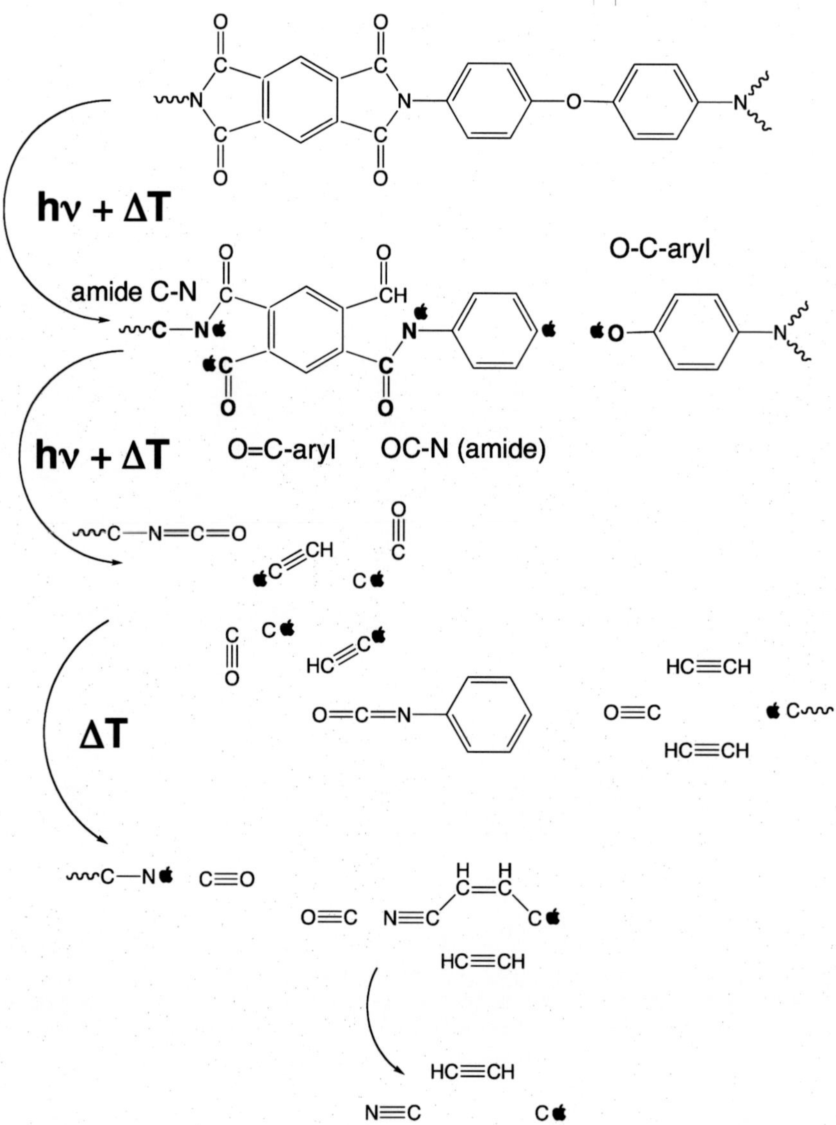

Fig. 7.5. Laser-induced decomposition/fragmentation of Kapton. All shown species have been detected. The ➠ denotes a radical, ion or broken bond

polymers such as antioxidants, UV absorbers, HALS (hindered amine light stabilizers), process and heat stabilizers for the stabilization of polymer recyclates, antistatics/antistatic agents, flame retardants, nucleating agents, oxygen absorbers, slip agents, carbon nanotubes/nanofilled thermosetting resins, optical brighteners/fluorescence indicators, plasticizers, silanes, silanes as bonding agents, silanes as cross-linking additives, antimicrobials, hydrophilic

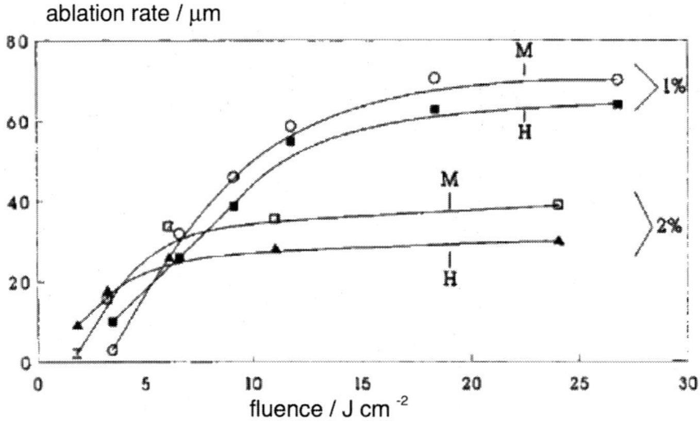

Fig. 7.6. Ablation rate for PMMA with an M_w of 97,000 and 500,000 doped with different amounts of a triazene compound. Irradiation wavelength 308 nm

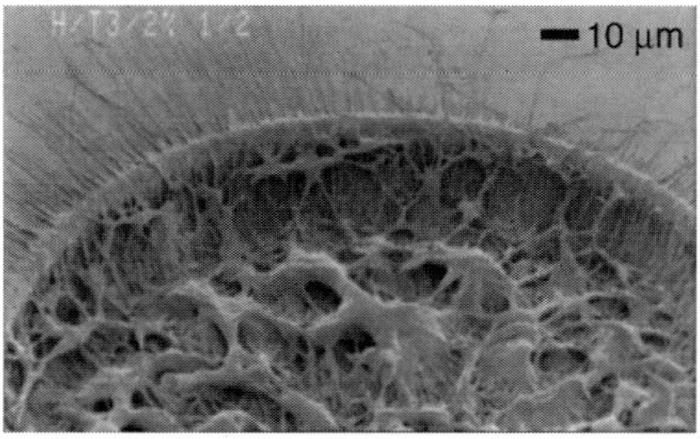

Fig. 7.7. Ablation crater in a triazene doped PMMA, with clear indications for melting during ablation. Irradiation wavelength 308 nm

additives, additives for content protection, photoselective additives, UV-Titan, titanium dioxide, and catalysts (not a complete list) are very common, and they may be inorganic or organic compounds. It is noteworthy that a common UV absorber (stabilizer), i.e., Tinuvin, can be used as dopant to induce effective ablation of PMMA at 308 and 350 nm [36, 37].

One possible effect, which can be observed for impurities in polymers, is the formation of microstructures, e.g., cones in a well-defined fluence range. The cone formation is due to the higher threshold fluence of ablation compared to the pure polymer, while the apex angle of the cones (Θ) varies with the applied fluence (F) and ablation threshold (F_0) according to equation (7.1):

$$\Theta = 2 + \sin^{-1}\left[\frac{F_0\,(1 - R_0)}{F\,(1 - R\,(\Theta))}\right], \tag{7.1}$$

where R_0 and $R(\Theta)$ are the surface reflectivities for incidence angles of 90° (normal to the surface) and Θ degrees, respectively [38, 39]. Typical examples

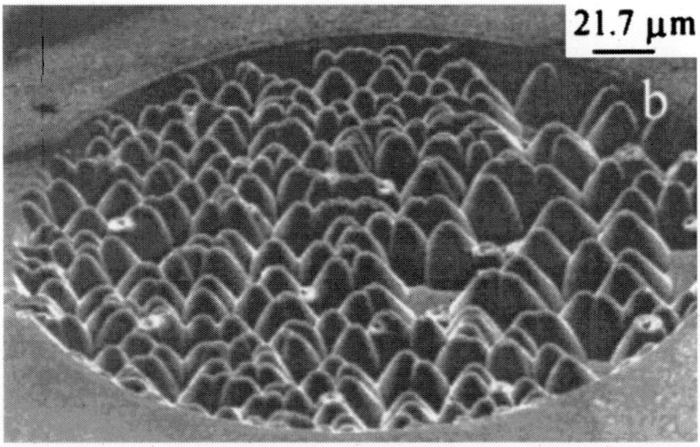

Fig. 7.8. Cone structures in a triazene polymer with a mixture of Si, O, and Cl on top of each cone. Irradiation wavelength 308 nm

Fig. 7.9. Cone structures in a triazene polymer with a Ca-species on top of each cone. Irradiation wavelength 308 nm

of these cone structures are shown in Figs. 7.8 and 7.9. In Fig. 7.8 cone structures produced in a triazene polymer are shown, where on top of each cone Si, O, and Cl were detected, which indicates impurities from the synthesis which have not been removed completely during purification from the polymer [40].

In the case of Fig. 7.9 calcium was detected on top of each cone inside the ablation crater in polyimide sheets (Kapton^{HN}) [41]. According to the manufacturer Ca-stearate is used as antifriction compounds for the Kapton sheets.

The influence of chain end groups (see for example the end groups of the monomers in Figs. 7.3 and 7.4) on the ablation characteristics has not yet been analyzed in detail. End groups can/will influence the surface properties especially for low Mw polymers and may even change the absorption properties. Whether chiral polymers show specific features for ablation is also not known but is possible, if we consider that different microstructures have been observed for ablation with differently polarized light [42].

7.2.1 Polymer Names

It is of utmost importance not only to consider the methods of analysis for the data (e.g., single pulse vs. multi pulse, gravimetric vs. volumetric methods, such as AFM or profilometry) but also know which polymer has been used and whether it has been "prepared," e.g., purified to remove additives, and which polymer is really used. A good example for the latter is polyimide, which is/are probably the most studied polymer for ablation (due to its broad absorption which allows to use wavelength up to 355 nm for ablation). Polyimide is not one single polymer, but a class of polymers that consists of hundreds of different types. Even Kapton is not one polymer, but additionally letters such as HN, describe it in more detail, as almost a hundred different Kaptons exist. The properties of polyimides can even range from photosensitive to "photostable," which has a strong influence on the ablation characteristics (shown in Fig. 7.10). The ablation rates of two different polyimides have been analyzed by a quartz microbalance, and much higher ablation rates and lower threshold fluences have been detected for the photosensitive polyimide (Durimid) as compared to PMDA (a polyimide very similar to Kapton) [43, 44].

7.2.2 Polymers and Photochemistry

Photochemistry of polymers is a well-established field of research that also explains many features of laser ablation, especially in the low fluence range. Incubation of PMMA for example is based on the same photochemical processes, which result in photoyellowing of PMMA. This originates from the formation of double bonds in the polymer chain (chain end and in-chain). The formation of the double bonds is due to a classical photochemical reaction, i.e., the Norrish type I or α-cleavage, which can be described as the homolytic breaking of a bond next to a double bond with a heteroatom (C=O). This reaction creates several small reaction products, i.e., CO, CO_2, CH_4, CH_3OH, and $HCOOCH_3$, which have been all detected for photodecomposition and laser ablation of PMMA [30,31]. Subsequent reactions after this reaction create the double bonds and the monomer (shown in Fig. 7.11). It is also noteworthy to mention that the monomer is the exclusive product from thermal decomposition of PMMA ($T >$ ceiling temperature) which is detected for CO_2 laser ablation, while only a small amount of monomer, i.e., $\approx$1% for 248 nm irradiation and $\approx$18% for 193 nm irradiation [45,46] is detected for UV laser ablation. The rest of the products are the small products and polymer fragments.

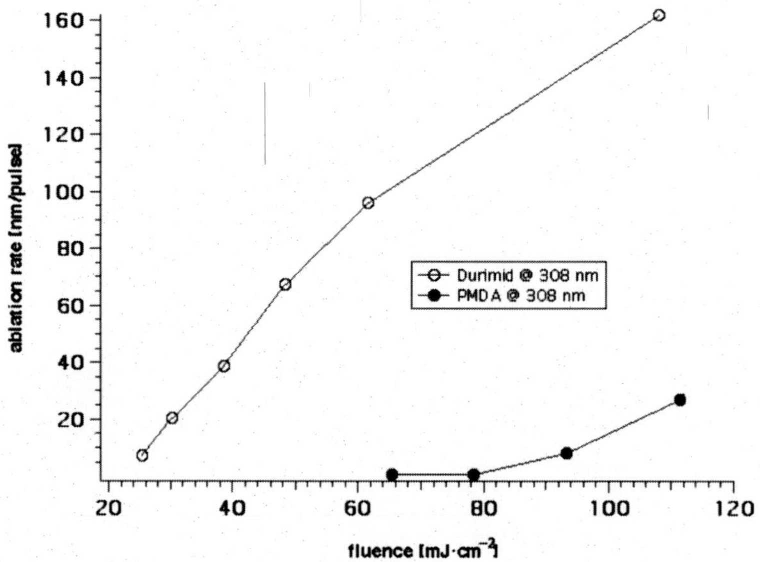

Fig. 7.10. Ablation rates of a photosensitive polyimide (Durimid) and PMDA (Kapton like). Irradiation wavelength 308 nm

7.2.3 Fundamental Issues of Laser Ablation

For an understanding of polymer ablation the main ablation parameters have to be explained and their definition have to be discussed in detail. Also the most frequently proposed ablation mechanisms and models will be discussed.

Ablation Parameters

The main parameters that describe polymer ablation are the *ablation rate*, $d(F)$, and the *ablation threshold fluence* F_{th}, which is defined as the minimum fluence where the onset of ablation can be observed. A third important parameter is the *effective absorption coefficient*, α_{eff}, which yields information on the mechanisms that take place in the ablation process when compared to the *linear absorption coefficient*, α_{lin}, that is measured on thin un-irradiated polymer films.

The ablation process is often described by the following equation [47, 48]:

$$d(F) = \frac{1}{\alpha_{\mathrm{eff}}} \ln\left(\frac{F}{F_{\mathrm{th}}}\right) \tag{7.2}$$

Also the method as to how the ablation parameters are acquired can have a pronounced influence on the results. The ablation rate can be defined either as the depth of the ablation crater after one pulse at a given fluence, or as the slope of a linear fit of a plot of the ablation depth versus the pulse

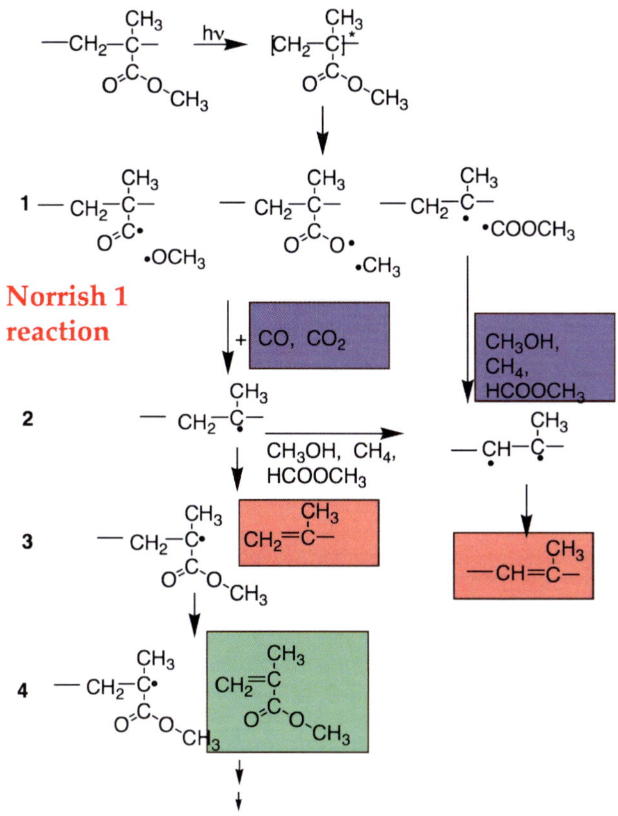

Fig. 7.11. Photochemical decomposition pathway of PMMA

number for a given fluence. Very different ablation rates can result from the two different measurement methods. This is especially the case for materials where ablation does not start with the first pulse, but after multiple pulses, or if the ablation crater depth after one pulse is too small to be measured. The process that occurs if ablation that does not start with the first laser pulse, is called *incubation*. It is related to the physical or chemical modifications of the material by the first few laser pulses, which results often in an increase of the absorption at the irradiation wavelength [49, 50], e.g., the formation of double bonds in *poly(methylmetacrylate) (PMMA)*. Incubation is normally only observed for polymers with low absorption coefficients at the irradiation wavelength. The typical appearance of incubation in a plot of the ablation depths vs. pulse number is shown in Fig. 7.12.

The method applied to measure the depth of the ablated area or the removed mass can also have an influence on the ablation parameters. If profilometric measurements (optical interferometry, mechanical stylus [51] or atomic force microscopy [52]) are used to calculate the ablation rate, a sharp ablation

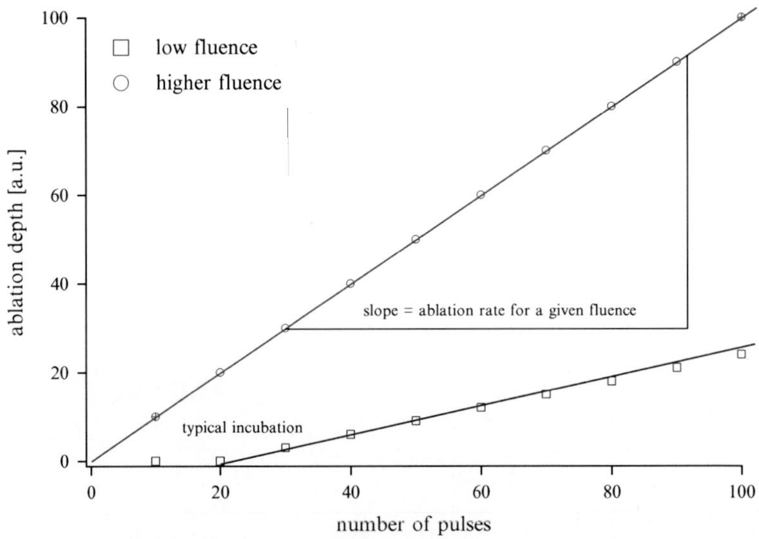

Fig. 7.12. Plot of ablation depth vs. pulse number, which is used to determine the ablation rate for a given fluence. The typical feature of incubation, i.e., ablation starts only after a certain number of pulses is shown for the lower curve

threshold can be defined. This is also supported by reflectivity [53] and acoustic measurements [54]. In mass loss measurements, such as mass spectrometry or with a *quartz crystal microbalance (QCM)*, a so called *Arrhenius tail* [55] has been observed for certain conditions. The Arrhenius tail describes a region in the very low fluence range, where a linear increase of detected ablation products is observed, which is followed by a much faster increase, that coincides with the removal rates of the profilometric measurements [43].

Even if these different approaches are taken into account, it is often the case, that the ablation rate cannot be defined with a single set of parameters. Therefore, one set of parameters has to be defined for each fluence range in which different processes dominate the ablation process and thereby influence the ablation rate. In Fig. 7.13 the dependence of the ablation rate on the irradiation fluence is illustrated as a generic scheme, which is typical for most polymers. The intersection of the gray extensions of the schematic ablation rates (*black lines*) with the x-axis of the ablation rate vs. irradiation fluence plot is the threshold fluence and varies for each fluence range. Also a different effective absorption coefficient can be defined for each region.

Three fluence ranges are visible, which can be characterized as follows:
Low fluence range:

- From this fluence range, the ablation threshold fluence is normally defined
- Incubation can be observed at these low fluences

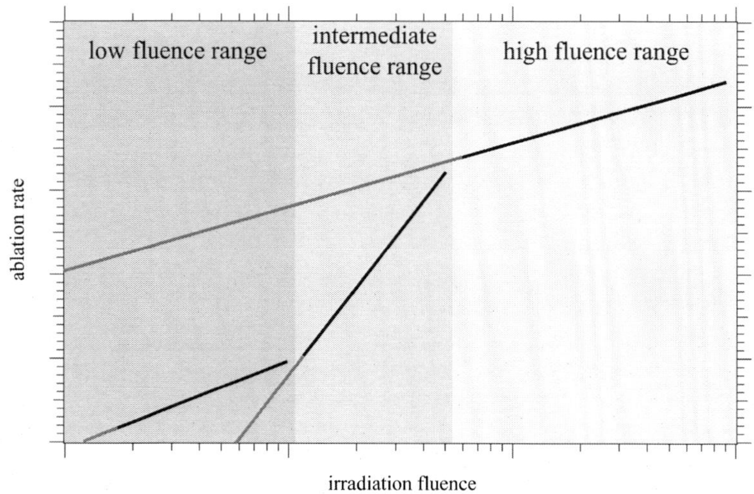

Fig. 7.13. Schematic plot of the three fluence ranges which are typically observed for polymers. The three ranges are indicated with different *shades of gray*

Intermediate fluence range:

- Increase of the slope of the ablation rate, which is caused by a more efficient decomposition of the polymer. Energy that has been gained from an exothermic decomposition of the polymer can also increase the ablation rate

High fluence range:

- The incident laser light is screened by solid, liquid, and gaseous ablation products and the laser produced plasma. This leads to similar ablation rates for many polymers [5] at high fluences

7.2.4 Ablation Mechanism

It is therefore of great importance not only to consider the values for the different ablation parameters, but also information about the technique of analysis and for which fluence range they are valid. An interpolation to values beyond the measurement range is also not advisable, as not all three ranges have to be present for all polymers and irradiation condition.

Even after 25 years of research in the field of laser polymer ablation, there is still an ongoing discussion about the ablation mechanisms, e.g., whether in addition to these mechanisms, photothermal processes, photochemical reactions, or even photophysical and mechanical processes are important.

If we summarize the experimental data and known reactions and products, then the following trends can be established:

- Absorption of the UV laser photons can and will result in direct bond breaking with a certain quantum yield (<1). The photon energy not resulting in bond breaking is transferred to the polymer matrix ➠ increase of T.

- Direct bond breaking is very fast (fs-ps) compared to the time scale of typical UV laser ablation (excimer lasers with ns pulse duration) ➠ primary decomposition products can also be decomposed directly by the laser photons (also possible for secondary products, etc.)
- The reactions create different structures with different absorption properties and different quantum yields (QY) for their further decomposition. Energy can be released from exothermic decomposition reactions (up to $\mathrm{kJ\,g^{-1}}$) ➠ increase of T.
- Small reaction/decomposition products are formed ➠ increase of p.
- The increase of T results in thermal decomposition of the polymer (can be quite fast as high temperatures may be reached in short time scale) ➠ increase of p.
- The fast increase of T can also increase the QY and subsequent reactions (unzipping: for PMMA 6 monomers at RT and over 200 above T_g) ➠ increase of p.
- Ablation starts when a certain number of bonds are broken in a volume element (before: decomposition/incubation) ➠ delayed ablation of a *modified* (not original) polymer.
- All of these processes are dependent on the polymer ➠ very complex ablation mechanism.

This complex behavior explains why the ablation mechanism has been and is still controversial. It is very difficult to separate the products of thermal and photochemical decomposition (they are often very similar), but ablation products must be considered (remember very different products for thermal/photochemical reaction of PMMA). Many different reactions are possible, which makes it difficult to model the complete process. Many constants of the polymer are temperature dependent and partly unknown for the possible high temperatures (and high heating rates present for laser irradiation). It is also difficult to establish an energy balance, because for many reactions the parameters are not known, and energy is carried away by the ablation products, which is difficult to measure (model correctly).

Modeling of Laser Ablation

It is generally accepted that for ns laser pulses, the energy of the laser photons is used for electronic excitation in a first step. The following steps are still under discussion and the different *models* can be summarized as follows:

Photothermal: The electronic excitation is thermalized on a ps timescale, which then results in thermal bond breaking [56–60].

Photochemical: Electronic excitation results in direct bond breaking [5,47, 61–63].

Photophysical: Both thermal and nonthermal processes play a role. Two independent channels of bond breaking [64, 65] or different bond breaking energies for ground state and electronically excited states chromophores are applied [66,67] in this model. It is most adequate for short laser pulses in the ps and fs range [68].

Another way to distinguish the different models is by separating them into *surface* and *volume* models. In the volume models, the different processes that eventually result in ablation take place within the bulk of the material. In the surface models, the processes that are responsible for the material removal take place within a few monolayers below the surface. The different models can be described as follows:

Photochemical surface models: valid for long pulses and/or higher irradiation fluences [69].

Thermal surface models: These models are mainly developed for metal ablation and do not consider the sharp ablation threshold, but can describe the occurrence of an *Arrhenius tail* [59,60,65,70].

Photochemical volume models: These models describe a sharp ablation threshold and a logarithmic increase of the ablation crater depth with the number of laser pulses, but the *Arrhenius tail* is not accounted for [3,5,47, 61,62]. A linear dependence can be described with models that consider the motion of the ablation front, but ignores the screening effects caused by the plasma plume.

Thermal volume models: These models are often oversimplified by neglecting the movement of the solid-gas interface and result therefore in very high temperatures [55,58].

Volume photothermal model: In this model by Arnold and Bityurin [71], a thermal surface model and a photochemical volume model have been combined. In this model it is assumed, that photothermal bond breaking takes place within the bulk polymer. When a density of broken bonds reaches a critical value, ablation begins. This model can account for sharp ablation thresholds and Arrhenius tails.

A new *coarse-grained chemical reaction model (CGCRM)* has been proposed by Garrison et al. [72,73]. In this model a kinetic Monte Carlos approach that includes a probabilistic element is used to predict when reactions occur. It is thereby possible to avoid the use of a chemically correct interaction potential. The CGCRM uses known chemical reactions along with their probabilities and exothermicities for a specific material to estimate the effect of chemical reactions on the ablation process.

The coarse grained molecular dynamics model was developed to study the role of thermal, mechanical and chemical reactions in the onset of the ablation process of PMMA [74–79]. In this model, the laser energy is absorbed in different ways, i.e., pure heating and Norrish type I and II reactions. Mechanical stresses and pressures are dominant for very short pulses in the stress confinement regime and can initiate ablation by a mechanical breakdown of the polymer in the case of pure heating. For longer pulse lengths, the ejection process is mainly thermally activated. This can be well described with thermal models based on thermally activated bond braking processes. The presence of small molecules and gaseous products cannot be accounted for by a purely thermal mechanism. A modeling of the photoablation channels requires a two-step ablation model that incorporates the effect of photolysis of the polymer and the creation of new species, which is then followed by a thermally activated removal step. The breathing sphere model was enhanced by Garrison et al. [80–83] to allow the photons to break a bond in the molecule and to describe subsequent abstraction and recombination reactions. The model was initially applied to chlorobenzene, where good correlation with experimental data was found.

The new concept that arises from these calculations is the difference in the temporal and spatial deposition of the available energy in photochemical and photothermal mechanisms. This concept provides the foundation to make specific comparisons with experiment and to explain experimental results as summarized below:

- It was found that photochemical reactions release additional energy into the irradiated sample and decrease the average cohesive energy and thereby decrease the value of the ablation threshold. The yield of emitted fragments becomes significant only above the ablation threshold.
- The presence of a shockwave with a high initial velocity, large clusters in the plume, and high velocities of particles in the plume are explained by the fast rise of energy deposition in time from 20 to 150 ps.
- The chemical reactions that take place above the surface after the laser pulse on longer timescales explain the higher background density in the plume with photochemical ablation than observed for photothermal ablation.
- The presence of a combination of a thermal mechanism below the ablation threshold and a volume ejection mechanism above the threshold explains why non-volatile products like HCl and the matrix are only observed below threshold and all products are observed above threshold.
- The absence of heat deposited below ~1.5 times the penetration depth may help to explain the cold etching process in far UV photoablation as is used commercially in the corrective eye surgery, LASIK.

The different models include many material parameters and several of these parameters are obtained from fitting of experimental data, and have to be adjusted to fit each polymer [9, 84].

In general it can be said, that polymers that show a photochemical ablation behavior at the irradiation wavelength would be preferable for structuring, as the damage of the surrounding material due to a thermal processes is minimized and less carbonization is observed. A conversion of the polymer into gaseous products is also of advantage, as no or only minor amounts of ablation products are redeposited on the structured surface and additional cleaning procedures may not be necessary.

7.2.5 Doped Polymers

Motivation

If we consider that the absorption of the laser photons is the basis for UV laser ablation with ns pulses and that many polymers absorb only at wavelengths <240 nm then it appears logical that methods were tested to extend the absorption of polymers by doping and therefore the number of applicable lasers. The ablation of doped polymers has been reviewed in 1997 by Lippert et al. [85] and the polymers and the ablation mechanism were classified according to the absorption properties of the absorber-polymer system. The properties changed from systems, where only the dopant is absorbing, to systems, where absorption occurs only in the polymer. It was suggested, that ablation results from a mixture of processes, that originate from the polymer and the dopant. The properties of the dopant result in processes that can dominate the ablation mechanisms.

An important factor is whether the dopant is decomposing or not. A photolabile dopant, that decomposes into gaseous products leads to pronounced surface swelling at low irradiation fluences, while this behavior is much less pronounced for "photostable" dopants. Thermoelastic stress can also be induced in the polymer below the ablation threshold fluence by localized heating and thermal expansion of the polymer. This stress is then released in acoustic waves and thermal conduction into the surrounding material. The resulting transient and quasi-static thermoelastic stresses can lead to material damage and even material ejection. At high fluences, very high ablation rates [33] can be achieved, but with the drawback of pronounced surface melting. In the case of photostable dopants, less surface swelling, lower ablation rates, and structures with higher quality are observed.

For all doped systems, it has to be considered that the amount of dopant is limited (typically ≤10 wt.%) and that polymer properties such as T_g may change (to lower values).

Different dopants were added to PMMA to investigate the ablation mechanism during UV irradiation. The dopants that were used ranged from polyaromatic compounds to compounds that contained photochemical active groups [85].

One group of dopants that was tested contains the triazene group (–N=N–N), as they are photochemically well studied [86–88] and also release a large amount of nitrogen when they are photochemically decomposed. Pronounced swelling has been observed by SEM analysis of the ablation craters at low irradiation fluences (see Fig. 7.14), which is caused by gaseous products produced by the decomposition of the photolabile dopants. It has been suggested that the released nitrogen and other gaseous ablation products act as carriers for larger ablation fragments.

With increasing fluence and dopant concentration, high ablation rates of up to 80 μm can be achieved, but pronounced signs of surface melting are

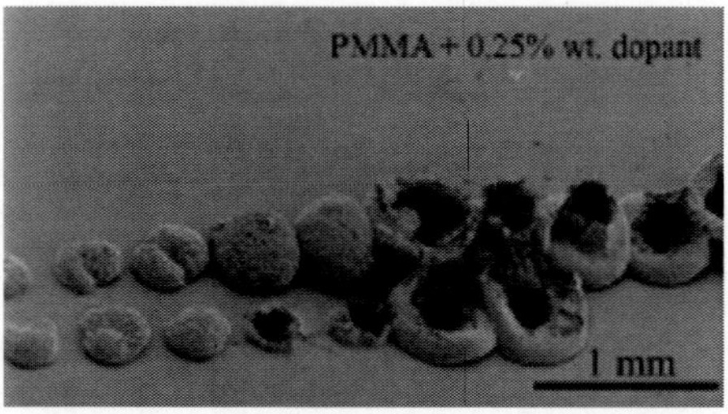

Fig. 7.14. Irradiated PMMA with different dopant concentration after irradiation with 308 nm. PMMA with 0.25 wt.% triazene was used for ablation with one laser pulse per position. The irradiation fluence increases from left to right. Pronounced swelling and bubble formation is visible

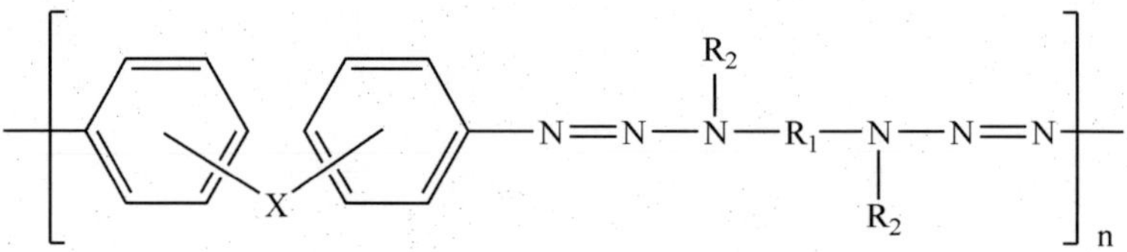

Fig. 7.15. Chemical structure of the triazene polymers

always visible [33] (see Fig. 7.7), which is an indication for the presence of a photothermal mechanism. A possible reason for these pronounced thermal effects could be that the maximum amount of dopant that can be added to the polymers is ≈10%, which limits achievable temperatures (energy/volume).

7.2.6 Designed Polymers: Triazene Polymers

New polymers have been developed to further improve the quality of the ablation process, i.e., to achieve higher ablation rates, lower threshold fluences, and better quality structures with no surface contamination and pronounced modification of the polymer. One approach was to incorporate the triazene unit into the polymer backbone. A unique feature of these triazene polymers (TP, chemical structure shown in Fig. 7.15) is the possibility to adjust the absorption maximum by varying the "X"-component in Fig. 7.15 [89]. The absorption maximum of such triazene polymers can be tuned from 290 to 360 nm and maximum linear absorption coefficients of up to 200, 000 cm^{-1} at 308 nm can be reached.

In the absorption spectra for two different triazene polymers with X = O, $R_1 = (CH_2)_6$, and $R_2 = CH_3$ or R_2 = t-butyl (shown in Fig. 7.16), two distinct absorption maxima can be distinguished. The R_1 and R_2 groups change the properties such as T_g, film forming and chromophore density. The absorption maximum around 200 nm can be assigned to the aromatic system, while the maximum around 330 nm corresponds to the triazene unit [90]. The absorption

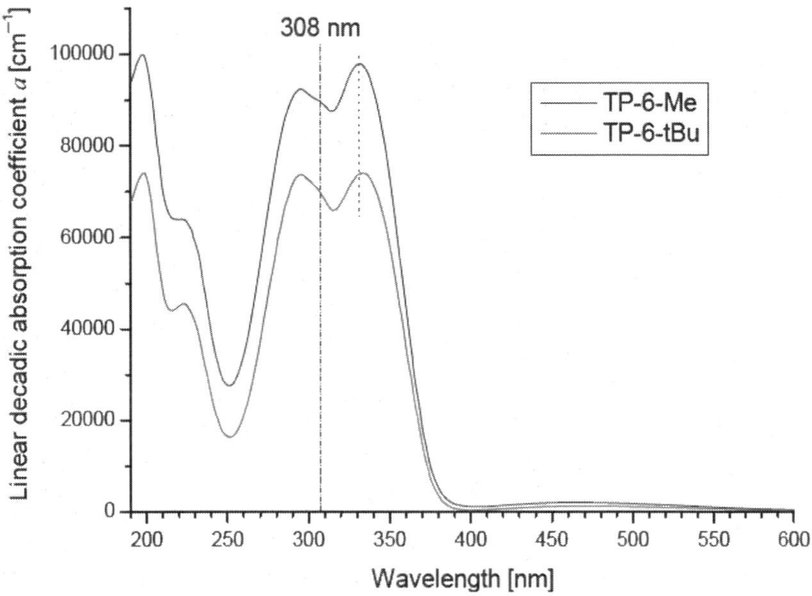

Fig. 7.16. Absorption spectra for two different triazene polymers with different chromophore density

coefficient of the triazene band can be influenced by R_2, i.e., with increasing bulkiness of the group the chromophore density decreases (as shown in Fig. 7.16 for Me vs. *t*-butyl). The two well-separated absorption maxima allow an excitation of different chromophores with different irradiation wavelengths such as 193, 248, and 308 nm, and thereby allow the study of their influence on the ablation behavior.

Higher ablation rates were measured for irradiation wavelengths that excite the triazene system (266, 308, and 351 nm) compared to the ablation rates for shorter wavelengths (248 and 193 nm) [6]. Also a clear and well defined ablation threshold fluence of 25 mJ cm^{-2} ($\pm$5 mJ cm^{-2}) is observed for a TP at an irradiation wavelength of 308 nm, while for irradiation with 248 nm a much broader range, 16–28 mJ cm^{-2} has been measured [90]. For irradiation at 248 nm, carbonization of the polymer was detected upon irradiation, whereas the surface of the polymer remained unchanged after several laser pulses for irradiation with 308 nm [91]. This is also an indication for the different ablation mechanisms at the irradiation wavelengths.

The triazene polymers are also well suited as probes for the ablation mechanism. Mass spectrometry was used to study the ablation products and to determine the different ablation mechanisms at the different irradiation wavelengths [92–94]. All decomposition products were identified with time resolved mass spectrometry for 248 nm and 308 nm irradiation. The proposed decomposition pathway for 308 nm irradiation is shown in Fig. 7.17, but similar

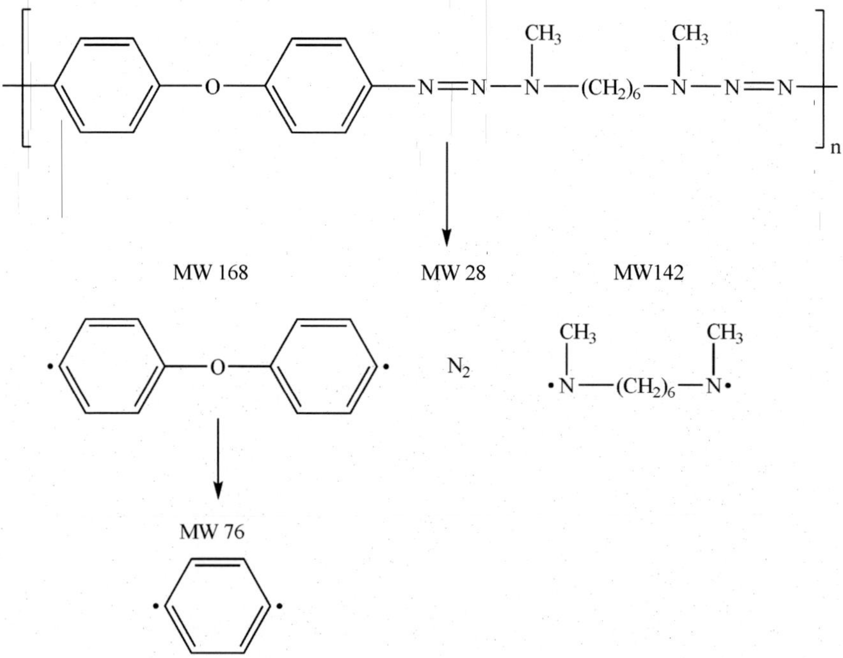

Fig. 7.17. Decomposition pathway for a TP measured by TOF-MS after irradiation with 308 nm

products were also observed for a thermal decomposition [87]. A clearer indication for the presence of a photochemical mechanism for 308 nm irradiation was given by the time-of-flight mass spectrometry (TOF-MS) data. Three different species of nitrogen were detected in the ablation plume (shown in Fig. 7.18): a very fast ground state neutral with up to 6 eV of kinetic energy, a slower ground state species with a broad energy distribution, which is most probable a thermal product, and a metastable (excited) neutral N_2 species that can only be created by an electronic excitation [95] because temperatures corresponding to this energy would be completely unreasonable.

It is interesting to compare the time of flight data and corresponding energies of the triazene polymer with data obtained for a polymer, i.e., Teflon, where we expect an unzipping and which also has a ceiling temperature (pronounced thermal decomposition pathway). In Fig. 7.19 the time of flight curve for the main product of Teflon decomposition, C_2F_4 with a mass of 100 amu, is shown for irradiation at 248 nm. A single curve is obtained which can be modeled by Maxwell Boltzmann distribution. The temperature, which is obtained from the analysis, is 987 K that is very reasonable compared to the temperature of 630 K for which the onset of unzipping is expected. The clear difference between the triazene polymer and Teflon confirm, again, the importance and influence of the material on the ablation mechanism.

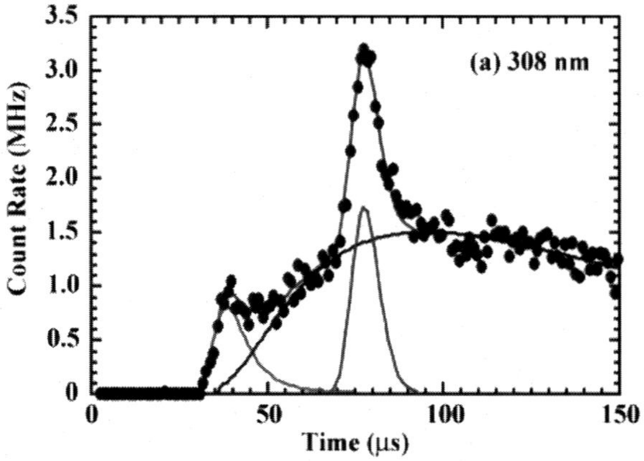

Fig. 7.18. Time of flight curves of the main product of decomposition of the triazene polymer, i.e., N_2. The irradiation wavelength was 308 nm with a fluence of 200 mJ cm^{-2}

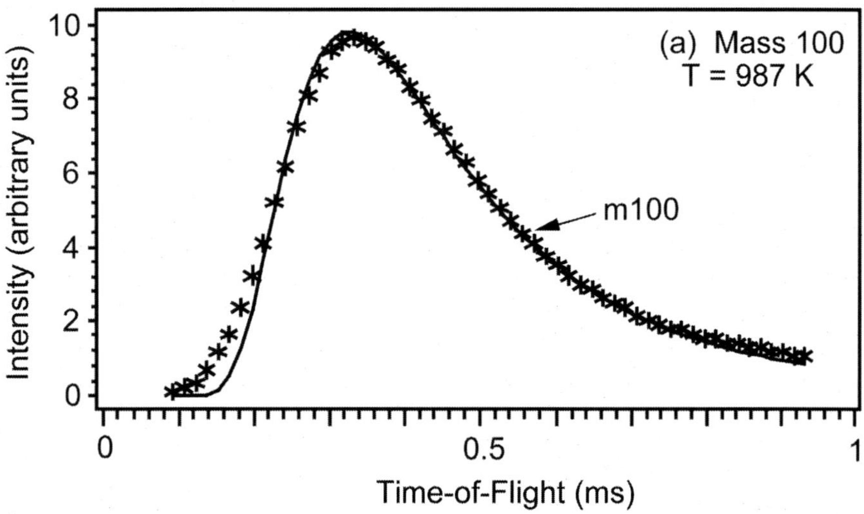

Fig. 7.19. Time of flight curve for the emission of the Teflon monomer for irradiation at 2 J cm^{-2} at 248 nm. The dark line shows a curve fit to a Maxwell Boltzmann distribution at a temperature of 987 K

Another method, which can be used to determine the ablation mechanism, is ns-interferometry. The ablation process could take place on a longer time scale (depending on the temperature) for a photothermal processes than for a photochemical reaction. First, swelling is observed and that is followed by etching [96,97], e.g., as discussed for a polyimide at 351 nm irradiation. This etching takes place on a microsecond time scale, which is much longer than the 30 ns excimer laser pulse. For the triazene polymer on the other hand the

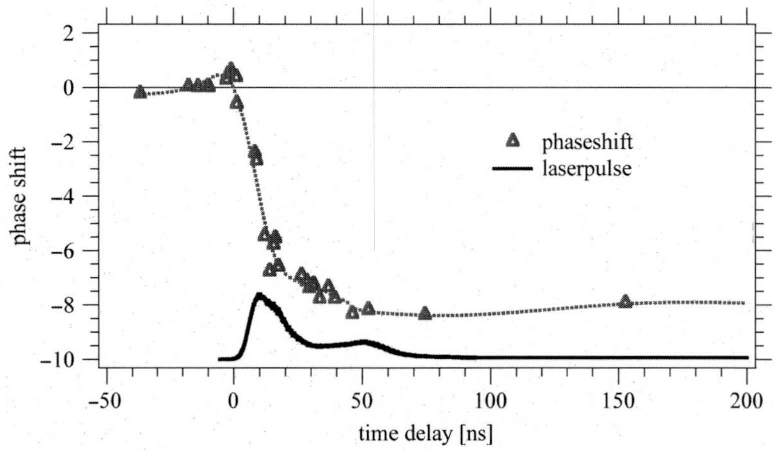

Fig. 7.20. Interference measurement for a triazene polymer during irradiation with 308 nm. The black curve represents the laser pulse, while the gray line corresponds to the phase shift, which is related to the ablation depth

etching starts and stops with the ablation laser pulse [98, 99] (see Fig. 7.20), which is again a clear indication for a photochemical process.

Irradiation experiments in the near-IR range at 800 nm with pulses in the pico- and femtosecond range were also performed. For femtosecond pulses, a lower ablation threshold fluence was found than for picosecond pulses, which indicates the presence of a thermal mechanism [100]. Also no complete removal of a thin triazene polymer film from a glass substrate was possible with 100 fs pulses. These short pulses in the near-IR do not yield much better results and are therefore no alternative to UV ablation [101].

The influence of the location of the predetermined "decomposition" site of in the polymer has been tested by incorporating the triazene unit into the side chains. The obtained ablation structures were less defined, and more pronounced thermal effects were observed [102]. Investigation of the polymer "between" the individual triazene units suggest that a higher triazene density results in better ablation results [7].

In Fig. 7.21 the ablation threshold fluences are plotted versus the polymer weight per triazene unit for TP1, two polyurethane polymers with the triazene unit in the polymer backbone (PUH-T1, PUH-T2) [103], two polyacrylates with the triazene unit in the polymer side chain (T-PAc1, T-PAc2) [102], two different triazene polymers with malonyl ester groups in the side chains (TM1 and TM2) [7] and a polyurethane polymer with the triazene unit in the side chain (PU-NO) [104]. A sudden increase of the ablation threshold fluence can be observed at about 285 amu/(triazene group) from ~25 to ~70 mJ cm^{-2}. Polymers above this jump have a higher ablation threshold fluence, as more bonds have to be broken to remove the larger remaining polymer fragments. Below or above this step, the ablation threshold fluence remains more or less

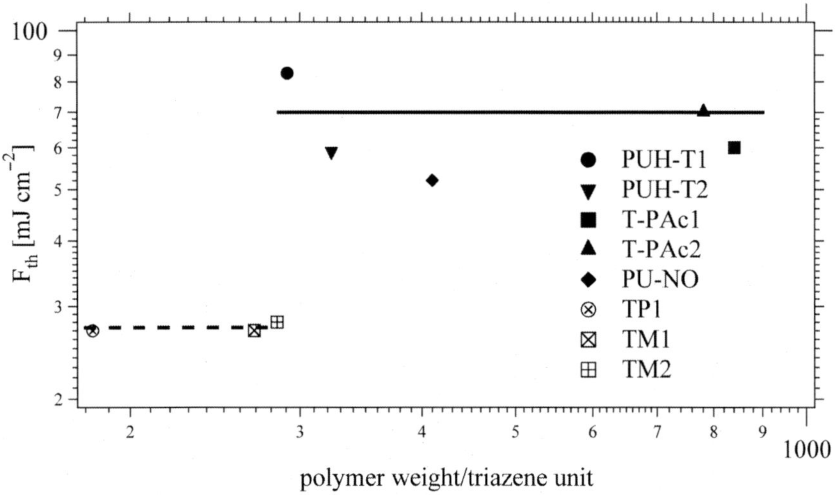

Fig. 7.21. The ablation threshold fluence versus polymer weight per triazene unit is shown for various triazene polymers. The two lines in the plot are shown as guidelines

constant, independent of the polymer weight per triazene unit. Why this sharp step is observed is not yet clear and must be studied in more detail.

7.2.7 Comparison of Designed and Commercially Available Polymers

Compared to commercially available polymers such as polyimides or other designed polymers, e.g., polyesters, the triazene polymers showed the highest ablation rates (up to 250 nm per pulse for 100 mJ cm^{-2} compared e.g., to 50 nm per pulse for Kapton) and the lowest ablation threshold fluence (20 mJ cm^{-2} for the triazene polymers compared to ~60 mJcm^{-2} for Kapton) for selected wavelengths. The structure produced in TP (Fig. 7.22 (*top*)) with 308 nm irradiation are much sharper than those in Kapton^{TM} (Fig. 7.22 (*bottom*)) and also no polymer debris is redeposited in and around the ablated structure in the case of the triazene polymer [98]. Kapton^{TM} was chosen as commercially available reference because it has a similar α_{lin} at 308 nm. The absence of redeposited material for TP corresponds well with ns-shadowgraphy measurements, where it was shown that no solid products are produced for 308 nm irradiation of TP [99].

All data obtained for TP strongly suggest that photochemical reactions play an important role during UV laser ablation, but also that photothermal processes are important. This is confirmed by the presence of the thermal N_2 products in the TOF curves and from an analysis of the threshold fluence for very thin films, where a clear influence of the thermal conductivity of the substrate was detected [105]. Photothermal processes will also always

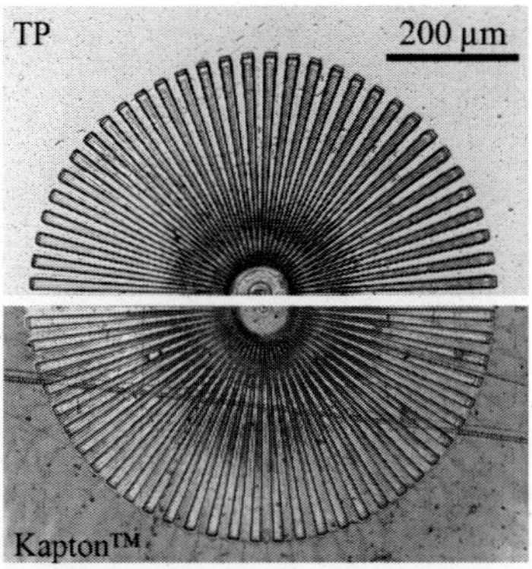

Fig. 7.22. SEM of Siemens Stars in TP (*top*) and Kapton^{TM} (*bottom*), both produced with five laser pulses at 308 nm

be present if the polymer decomposes exothermically during a photochemical decomposition and if the quantum yield of the photochemical reaction is not equal to one (which is most of time the case). The ablation of polymers will therefore always be a photophysical process (a mixture of photochemical and photothermal processes), where the ratio between the two mechanisms is a function of the irradiation wavelength and the polymer. In addition photomechanical process, such as pressure produced by trapped gaseous ablation products or shock and acoustic waves in the polymer, take place [106] and can lead to a damage of the polymer and are probably most important for picosecond pulses [107, 108].

A more pronounced photochemical part is preferable for material structuring, as it leads to a more uniform decomposition of the polymer and results in less debris. Additionally large quantities of gaseous products are produced and less material is redeposited in and around the ablated area. The designed polymers such as the TP show a clear advantage over commercially available polymers.

7.3 Deposition of Thin Films Using UV Lasers

Thin polymers films of polymers are normally prepared by solvent-based methods, such as spin coating and master blading. Other techniques have the disadvantages that only complete layers with no lateral resolution are formed and that the polymers must be soluble in certain solvents, which are appropriate for these techniques. Requirements are that the solvent does not evaporate too fast, but still fast enough, and that relative large concentrations of the polymer (may be up to 15 wt.%) should be soluble. More problems exist for these techniques in the case of multilayers, which can only be realized for certain solvent systems and limits the number of polymers that can be used.

One technique that can be used to deposit layers with lateral resolution is ink jet printing, but in this case, the solvent may also be a problem, i.e., a high concentration solution with high viscosity must be used, and multilayers may also not be possible (depending on the solvents). Therefore "solvent-free" techniques, such as laser-based techniques have been developed.

Laser-based direct-writing and printing operations are finding increasing applications for precise surface micromodification techniques by either controlled ablation processes or the tailored deposition of complex materials. Several methods have been developed for the targeted deposition of a broad range of various materials applying lasers [109]. Among them, PLD can be used to grow films of inorganic [18, 19] or organic materials [15, 16, 110–113] on a substrate. A typical setup for PLD is shown in Fig. 7.23.

This method appears to be very limited, i.e., only a few polymers (PMMA, Teflon) have been deposited in this way (with UV laser irradiation) successfully. This is not really surprising, if we consider that UV photons will induce reactions and that the above described decomposition classification is valid. The previously described decomposition mechanisms show also clearly that only polymers which depolymerize, i.e., form the monomer upon UV laser irradiation, may be used for this approach. The deposited polymer films will most probably have a different molecular weight and weight distribution than the starting material, and may also contain decomposition products. One possible approach to deposit thin polymer films by PLD is the application of mid-infrared radiation, which is tuned to certain absorption bands of the polymer. This approach is called RIR-PLD (resonance-infra-red) and is described in detail in Chap. 8 by R. Haglund. A modified approach to PLD that is more

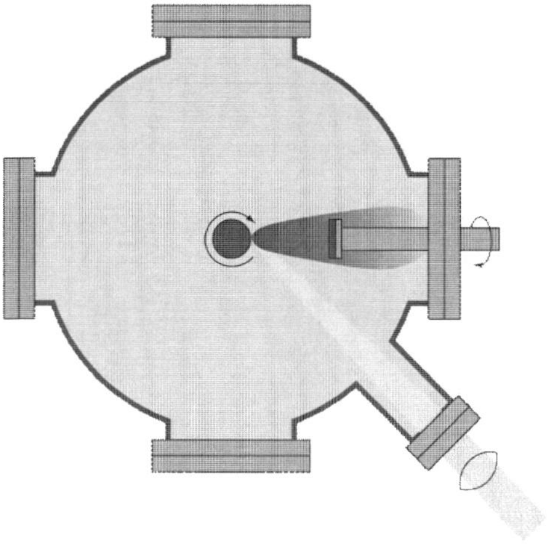

Fig. 7.23. Typical setup for PLD of organic or inorganic materials. The laser interacts with the moving target and vaporized the material to form a plasma (for inorganic materials) that is deposited on a substrate

gentle is MAPLE [110, 114], which uses in principle the same setup as PLD, with the main difference that the target consists of a frozen solution of the polymer. The laser is then used to vaporize the solvent that is removed by the pumps while the polymer chains are deposited intact on the substrate. This approach works of course only for polymer, which can be dissolved, and it should also work best when the laser is only/mainly absorbed by the solvent. The formation of high quality thin films is possible, although problems with the homogeneity of the films and trapped solvents exist. More details on MAPLE can be found in Chaps. 13 and 14 by A. Luches and I. Mihailescu.

Another versatile direct-writing method for the accurate microdeposition of a variety of different materials is based on LIFT techniques [21, 22, 115, 116], where the pressure increase from a vaporized material propels a transfer material onto a receiver. There are several different variations to this general approach:

1. Laser molecular implantation (LMI, scheme shown in Fig. 7.24).

For LMI a laser, which may be shaped by a mask, passes through a transparent substrate with a transparent polymer that is in contact with an absorbing polymer film containing the molecules that should be implanted, e.g., fluorescence probes, such as pyrene [117–119]. As absorbing polymer materials such as the triazene polymer can be utilized. It was possible with this approach to implant pyrene with a resolution given by the mask into the target polymer within a few 10s of nm. The disadvantage of this method is the limited number of polymers, which can be used (must be transparent with a quite low T_g) and that the implantation is only possible with a depth <100 nm.

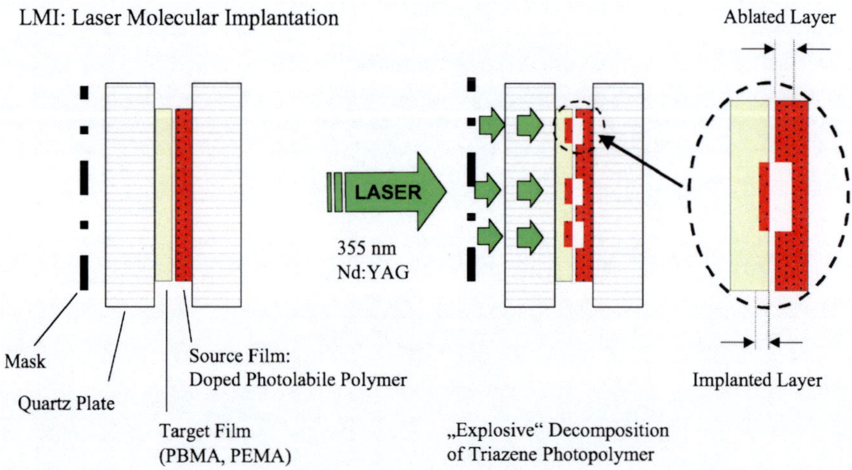

Fig. 7.24. Scheme of LMI

LITI: Laser-Induced Thermal Imaging

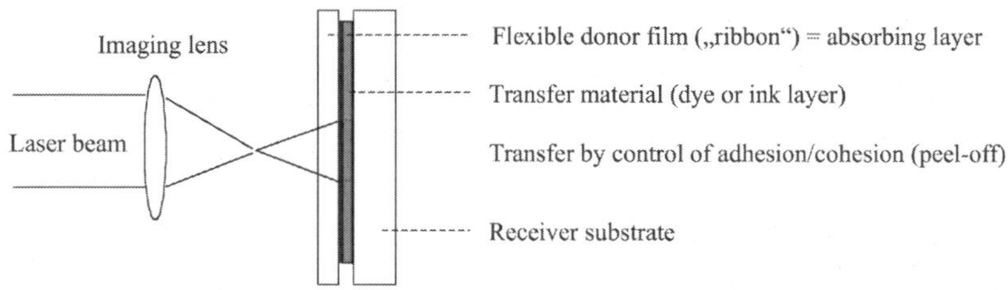

Fig. 7.25. Scheme of LITI

MAPLE-DW: Matrix-Assisted Pulsed Laser Evaporation Direct Write

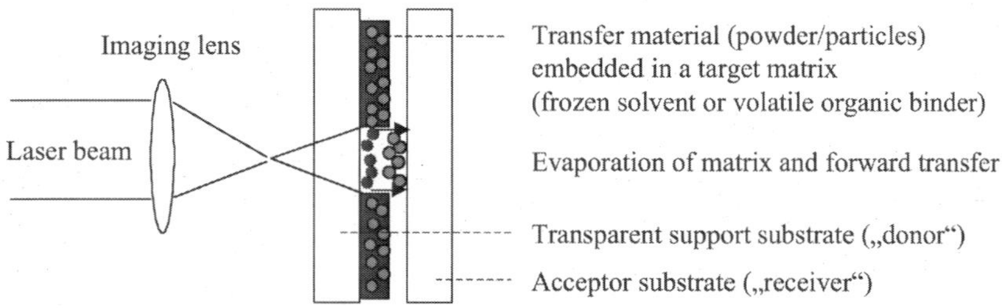

Fig. 7.26. Scheme of MAPLE-DW

2. Laser-induced thermal imaging (scheme shown in Fig. 7.25)

For LITI, typically IR lasers are used that are imaged through a flexible donor film onto an absorbing layer, which propels the material onto a receiver [120–122]. This may induce a thermal load to the transfer material. The thermal load may be detrimental to certain sensitive materials. Another key parameter to this technique is the control of adhesion and cohesion that can cause various problems.

3. Matrix-assisted pulsed laser evaporation direct write (MAPLE-DW, scheme shown in Fig. 7.26)

For MAPLE-DW the laser (mainly UV) is imaged through a transparent substrate onto the transfer material which is embedded in a matrix, e.g., frozen solvent or organic binder [110, 123–128]. The transfer material is then propelled over a gap to the receiver. There may be problems involved with the resolution of the transferred material (if it is liquid), a certain UV load, and that components of the matrix are also transferred.

4. Laser-induced forward transfer (LIFT, see Fig. 7.27).

For LIFT the UV laser is imaged through a transparent substrate on the transfer material and is transferred with or without a gap on the receiver [21, 22, 115, 116]. The main disadvantage of this method is the direct ablation

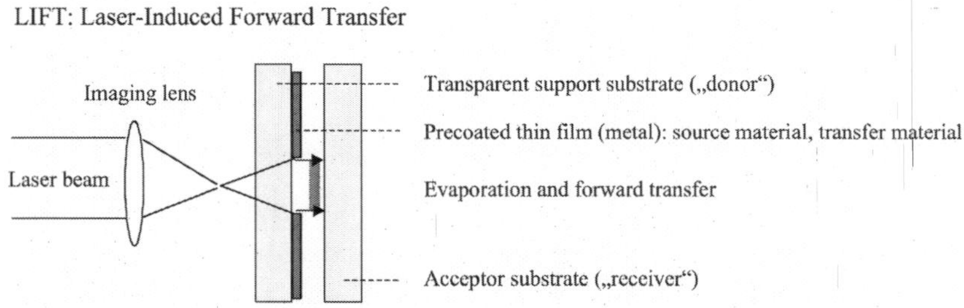

Fig. 7.27. Scheme of LIFT

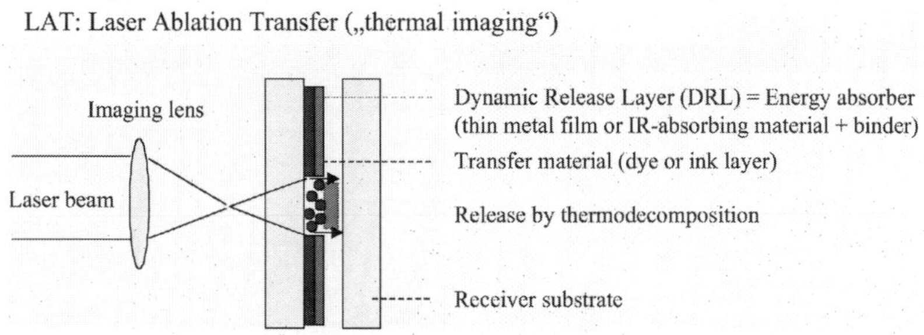

Fig. 7.28. Scheme of LAT

of the transfer material at the interface of the transfer. This may have a detrimental effect, i.e., decomposition, on the transfer material, especially if sensitive materials are used.

5. Laser ablation transfer (LAT, shown in Fig. 7.28).

In LAT an IR laser will be typically imaged onto an energy absorber which is also called dynamic release layer (DRL) or sacrificial layer [129–133]. The material is transferred over a gap onto the receiver, but the material may experience a thermal load and the fragments of the DRL that is often a metal may contaminate the transfer material. The use of thin intermediate films of metals (e.g., Ag, Au, Ti) or metal oxides (e.g., TiO_2) has been reported as absorbing layers for UV laser-based forward transfer applications of biomolecules [134–137] and cells [138]. This approach has then been called in the literature absorbing film assisted (AFA) LIFT [139–141] and Biological Laser Printing (BioLPTM) [142–146]. An alternative is the application of a thick polymer layer, which expands upon laser irradiation. This results in a "mechanical" transfer of the layer onto a receiver substrate [147].

In an attempt to overcome several of the above mentioned problems another variation of LIFT has been developed where UV lasers and a photosensitive, polymeric DRL is used. The advantage of this approach is that the control over the DRL thickness and laser fluence allows transferring sensitive material without any UV load as all photons are absorbed in the DRL layer. The complete composition of a polymeric DRL layer into gaseous fragments,

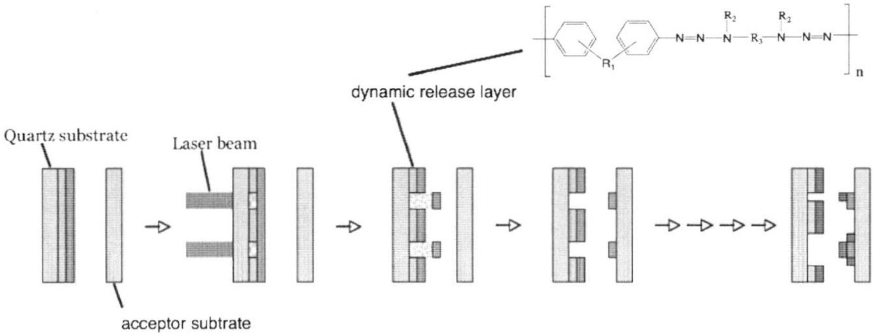

Fig. 7.29. Scheme of LIFT using a photosensitive DRL. The chemical structure of a DRL is included

as reported for the triazene polymers, allows then a contamination free transfer with ideally no thermal load of the material. A scheme of this approach is shown in Fig. 7.29, which also indicates the possibility to transfer multilayers.

From the chemical structure formula included in Fig. 7.29 it can be seen that two photocleavable aryltriazene (Ar–N=N–N–) chromophores per repeating unit are covalently incorporated into the polymer main chain. Exposure to UV irradiation causes a photolytic cleavage of the triazene chromophores, which leads to an irreversible evolution of elemental nitrogen and simultaneously to the fragmentation of the polymer into small molecules. Therefore, films of these photosensitive polymers proved to be excellently suitable for laser ablation applications since they can be cleanly ablated without carbonization or redeposition of debris already at fluences far below 100 mJ cm^{-2} [6, 10, 148]. The laser-triggered photofragmentation process results in an abrupt volume expansion, which has been used to transfer a range of materials, i.e., metals and ceramics [149], and also sensitive materials such as neuroblasts (biomaterial) [150], polymers (OLED, optical light emitting diode materials) and bilayers of a metal with an OLED material (MEH-PPV) [151,152]. With this approach (shown in Fig. 7.30 *left*) pixels of the OLED material were transferred with Al as electrode onto a transparent electrode (ITO).

The transferred OLED material was fully functional (see photo in Fig. 7.30 *right*) and the emission spectra revealed no indication of a thermal load (which would be indicative by a shoulder in the red), and gave the same slope efficiency as spin coated films [151–153]. This shows clearly that LIFT with a photosensitive polymeric DRL can be used to transfer sensitive material without thermal or UV load onto a receiver substrate with high lateral resolution.

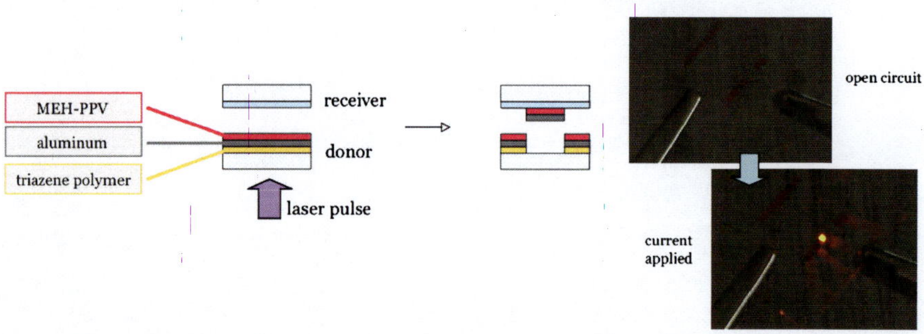

Fig. 7.30. Scheme of the bilayer transfer with DRL LIFT (*left*) and light emission of a transferred OLED (*right*)

7.4 Conclusion

UV laser ablation of polymers is a versatile method to structure polymers with high resolution. The mechanism of ablation is often discussed controversially, but it is necessary to keep in mind that polymers are complex systems with a wide variety of properties that can influence the ablation mechanism. Analyzing the data it appears that the ablation mechanism is a complex interrelated system, where photochemical and photothermal reactions are very important. The pressure jump, which is associated with the creation of small molecules and originates from both types of reaction, is also important for ablation. The importance of each effect is strongly dependent on the type of polymer, i.e., more photochemical features for specially designed polymers than for certain polyimides, the laser wavelengths (more photochemical features for shorter wavelengths), the pulse length and substrate.

UV laser ablation can also be utilized to deposit directly thin polymer films by PLD, but this is limited to certain polymers, i.e., polymers that decompose into the monomer upon decomposition. Alternative laser-based techniques, such as laser-induced forward transfer (LIFT), utilize the decomposition of a thin layer (either part of the material or in the form of a sacrificial layer) to transfer complete layers with high spatial resolution. This approach can be used to transfer pixels of sensitive materials to a substrate with a minimal thermal and UV load.

Acknowledgments

Financial support of the Paul Scherrer Institut and the Swiss National Science Foundation to support parts of this work is gratefully acknowledged. Contributions from R. Fardel, M. Nagel, F. Nüesch, and L.Urech are also gratefully acknowledged.

References

1. R. Srinivasan, V. Mayne-Banton, Appl. Phys. Lett. **41**, 576 (1982)
2. Y. Kawamura, K. Toyoda, S. Namba, Appl. Phys. Lett. **40**, 374 (1982)
3. R. Srinivasan, B. Braren, Chem. Rev. **89**, 1303 (1989)
4. D. Bäuerle, *Laser Processing and Chemistry*, 3rd edn. (Springer Verlag, Berlin, 2000)
5. S. Lazare, V. Granier, Laser Chem. **10**, 25 (1989)
6. T. Lippert, J.T. Dickinson, Chem. Rev. **103**, 453 (2003)
7. T. Lippert, in *Polymers and Light*, vol. 168, ed. by T. Lippert (Springer, Berlin, 2004) p. 51
8. P.E. Dyer, in *Photochemical Processing of Electronic Materials*, ed. by I.W. Boyd, R.B. Jackman (Academic, London, 1992) p. 360
9. N. Bityurin, B.S. Luk'yanchuk, M.H. Hong, T.C. Chong, Chem. Rev. **103**, 519 (2003)
10. T. Lippert, Plasma Process. Polym. **2**, 525 (2005)
11. N. Bityurin, Annu. Rep. Progr. Chem. C **101**, 216 (2005)
12. M. Karas, D. Bachmann, F. Hillenkamp, Anal. Chem. **57**, 2935 (1985)
13. R. Zenobi, R. Knochenmuss, Mass Spectrom. Rev. **17**, 337 (1998)
14. L.J. Radziemski, D.A. Cremers, *Laser-Induced Plasmas and Applications* (M. Dekker, New York, 1989)
15. G.B. Blanchet, Macromolecules **28**, 4603 (1995)
16. G.B. Blanchet, Chemtech **26**, 31 (1996)
17. W.B. Jiang, M.G. Norton, L. Tsung, J.T. Dickinson, J. Mater. Res. **10**, 1038 (1995)
18. R.W. Eason, *Pulsed Laser Deposition of Thin Films: Applications-Led Growth of Functional Materials* (John Wiley, Hoboken, NJ, 2007)
19. D.B. Chrisey, G.K. Hubler, *Pulsed Laser Deposition of Thin Films* (John Wiley, New York, 1994)
20. A. Pique, P. Wu, B.R. Ringeisen, D.M. Bubb, J.S. Melinger, R.A. McGill, D.B. Chrisey, Appl. Surf. Sci. **186**, 408 (2002)
21. J. Bohandy, B.F. Kim, F.J. Adrian, J. Appl. Phys. **60**, 1538 (1986)
22. K.D. Kyrkis, A.A. Andreadaki, D.G. Papazoglou, I. Zergioti, in *Recent Advances in Laser Processing of Materials*, vol. 241, ed. by J. Perrière, E. Millon, E. Fogarassy (Elsevier, Amsterdam, 2006) p. 213
23. H. Aoki (U.S. Patent 5736999, 1998)
24. R.S. Patel, T.A. Wassick, Proc. SPIE-Int. Soc. Opt. Eng. **2991**, 217 (1997)
25. G. Kopitkovas, L. Urech, T. Lippert, in *Recent Advances in Laser Processing of Materials,* ed. by E. Millon, J. Perriere, E. Fogarassy (Elsevier, Kidlington, 2006) p. 105
26. H. Klank, J.P. Kutter, O. Geschke, Lab Chip **2**, 242–246 (2002)
27. D. Snakenborg, H. Klank, J.P. Kutter, J. Micromech. Microeng. **14**, 182 (2004)
28. D. Gomez, F. Tekniker, I. Goenaga, I. Lizuain, M. Ozaita, Opt. Eng. **44**, 051105 (2005)
29. D.F. Farson, H.W. Choi, C.M. Lu, L.J. Lee, J. Laser Appl. **18**, 210 (2006)
30. S. Lazare, W.P. Guan, D. Drilhole, Appl. Surf. Sci. **96–98**, 605 (1996)
31. E.E. Ortelli, F. Geiger, T. Lippert, J. Wei, A. Wokaun, Macromolecules **33**, 5090 (2000)
32. P. Lemoine, W. Blau, A. Drury, C. Keely, Polymer **34**, 5020 (1993)

33. T. Lippert, A. Wokaun, J. Stebani, O. Nuyken, J. Ihlemann, Angew. Makromol. Chem. **213**, 127 (1993)
34. G. Bounos, A. Selimis, S. Georgiou, E. Rebollar, M. Castillejo, N. Bityurin, J. Appl. Phys. **100**, 114323 (2006)
35. E. Rebollar, G. Bounos, M. Oujja, S. Georgiou, M. Castillejo, J. Phys. Chem. B **110**, 16452 (2006)
36. R. Srinivasan, J. Appl. Phys. **70**, 7588 (1991)
37. R. Srinivasan, B. Braren, Appl. Phys. A **45**, 289 (1988)
38. P.E. Dyer, S.D. Jenkins, J. Sidhu, Appl. Phys. Lett. **52**, 1880 (1988)
39. P.E. Dyer, S.D. Jenkins, J. Sidhu, Appl. Phys. Lett. **49**, 453 (1986)
40. J. Wei, N. Hoogen, T. Lippert, O. Nuyken, A. Wokaun, J. Phys. Chem. B **105**, 1267 (2001)
41. F. Raimondi, S. Abolhassani, R. Brutsch, F. Geiger, T. Lippert, J. Wambach, J. Wei, A. Wokaun, J. Appl. Phys. **88**, 3659 (2000)
42. J. Krüger, W. Kautek, in *Polymers And Light*, vol. 168, ed. by T. Lippert (Springer, Berlin, 2004) p. 247
43. T. Dumont, R. Bischofberger, T. Lippert, A. Wokaun, Appl. Surf. Sci. **247**, 115 (2005)
44. T. Dumont, S. Lazare, T. Lippert, A. Wokaun, Appl. Phys. A **79**, 1271 (2004)
45. R. Srinivasan, B. Braren, D.E. Seeger, R.W. Dreyfus, Macromolecules **19**, 916 (1986)
46. R. Srinivasan, J. Appl. Phys. **73**, 2743 (1993)
47. J.E. Andrew, P.E. Dyer, D. Forster, P.H. Key, Appl. Phys. Lett. **43**, 717 (1983)
48. R. Srinivasan, B. Braren, J. Polym. Sci. A Polym. Chem. **22**, 2601 (1984)
49. R. Srinivasan, B. Braren, K.G. Casey, J. Appl. Phys. **68**, 1842 (1990)
50. S. Küper, M. Stuke, Appl. Phys. A **49**, 211 (1989)
51. S.V. Babu, G.C. D'Couto, F.D. Egitto, J. Appl. Phys. **72**, 692 (1992)
52. M. Himmelbauer, E. Arenholz, D. Bauerle, Appl. Phys. A **63**, 87 (1996)
53. G. Paraskevopoulos, D.L. Singleton, R.S. Irwin, R.S. Taylor, J. Appl.Phys. **70**, 1938 (1991)
54. R.S. Taylor, D.L. Singleton, G. Paraskevopoulos, Appl. Phys. Lett. **50**, 1779 (1987)
55. S. Küper, J. Brannon, K. Brannon, Appl. Phys. A **56**, 43 (1993)
56. S.R. Cain, J. Phys. Chem. **97**, 7572 (1993)
57. S.R. Cain, F.C. Burns, C.E. Otis, J. Appl. Phys. **71**, 4107 (1992)
58. G.C. D'Couto, S.V. Babu, J. Appl. Phys. **76**, 3052 (1994)
59. B. Lukyanchuk, N. Bityurin, M. Himmelbauer, N. Arnold, Nucl. Instrum. Methods Phys. Res. B **122**, 347 (1997)
60. N. Arnold, B. Luk'yanchuk, N. Bityurin, Appl. Surf. Sci. **129**, 184 (1998)
61. T.F. Deutsch, M.W. Geis, J. Appl. Phys. **54**, 7201 (1983)
62. E. Sutcliffe, R. Srinivasan, J. Appl. Phys. **60**, 3315 (1986)
63. G.D. Mahan, H.S. Cole, Y.S. Liu, H.R. Philipp, Appl. Phys. Lett. **53**, 2377 (1988)
64. V. Srinivasan, M.A. Smrtic, S.V. Babu, J. Appl. Phys. **59**, 3861 (1986)
65. H. Schmidt, J. Ihlemann, B. Wolff-Rottke, K. Luther, J. Troe, J. Appl. Phys. **83**, 5458 (1998)
66. B. Lukyanchuk, N. Bityurin, S. Anisimov, N. Arnold, D. Bauerle, Appl. Phys. A **62**, 397 (1996)
67. B. Lukyanchuk, N. Bityurin, S. Anisimov, D. Bauerle, Appl. Phys. A **57**, 367 (1993)

68. N. Bityurin, A. Malyshev, B. Luk'yanchuk, S. Anisimov, D. Bäuerle, Proc. SPIE-Int. Soc. Opt. Eng. **2802**, 103 (1996)
69. N. Bityurin, Appl. Surf. Sci. **139**, 354 (1999)
70. G.V. Treyz, R. Scarmozzino, R.M. Osgood, Appl. Phys. Lett. **55**, 346 (1989)
71. N. Arnold, N. Bityurin, Appl. Phys. A **68**, 615 (1999)
72. Y.G. Yingling, B.J. Garrison, J. Phys. Chem. B **109**, 16482 (2005)
73. Y.G. Yingling, B.J. Garrison, J. Phys. Chem. B **108**, 1815 (2004)
74. P.F. Conforti, M. Prasad, B.J. Garrison, Appl. Surf. Sci. **253**, 6386 (2007)
75. P.F. Conforti, M. Prasad, B.J. Garrison, J. Phys. Chem. C **111**, 12024 (2007)
76. M. Prasad, P.F. Conforti, B.J. Garrison, J. Appl. Phys. **101**, 103113 (2007)
77. M. Prasad, P.F. Conforti, B.J. Garrison, J. Chem. Phys. **127**, 084705 (2007)
78. M. Prasad, P.F. Conforti, B.J. Garrison, Y.G. Yingling, Appl. Surf. Sci. **253**, 6382 (2007)
79. Y.G. Yingling, B.J. Garrison, Appl. Surf. Sci. **253**, 6377 (2007)
80. Y.G. Yingling, L.V. Zhigilei, B.J. Garrison, J. Photochem. Photobiol. A Chem. **145**, 173 (2001)
81. L.V. Zhigilei, E. Leveugle, B.J. Garrison, Y.G. Yingling, M.I. Zeifman, Chem. Rev. **103**, 321 (2003)
82. Y.G. Yingling, B.J. Garrison, Chem. Phys. Lett. **364**, 237 (2002)
83. Y.G. Yingling, B.J. Garrison, Nucl. Instrum. Methods Phys. Res. B **202**, 188 (2003)
84. N. Bityurin, A. Malyshev, J. Appl. Phys. **92**, 605 (2002)
85. T. Lippert, A. Yabe, A. Wokaun, Adv. Mater. **9**, 105 (1997)
86. T. Lippert, J. Stebani, O. Nuyken, A. Stasko, A. Wokaun, J. Photochem. Photobiol. A Chem. **78**, 139 (1994)
87. O. Nuyken, J. Stebani, T. Lippert, A. Wokaun, A. Stasko, Macromol. Chem. Phys. **196**, 751 (1995)
88. A. Stasko, V. Adamcik, T. Lippert, A. Wokaun, J. Dauth, O. Nuyken, Makromol. Chem. Macromol. Chem. Phys. **194**, 3385 (1993)
89. O. Nuyken, J. Stebani, T. Lippert, A. Wokaun, A. Stasko, Macromol. Chem. Phys. **196**, 739 (1995)
90. T. Lippert, L.S. Bennett, T. Nakamura, H. Niino, A. Ouchi, A. Yabe, Appl. Phys. A **63**, 257 (1996)
91. T. Lippert, T. Nakamura, H. Niino, A. Yabe, Appl. Surf. Sci. **110**, 227 (1997)
92. T. Lippert, C. David, J.T. Dickinson, M. Hauer, U. Kogelschatz, S.C. Langford, O. Nuyken, C. Phipps, J. Robert, A. Wokaun, J. Photochem. Photobiol. A Chem. **145**, 145 (2001)
93. T. Lippert, A. Wokaun, S.C. Langford, J.T. Dickinson, Appl. Phys. A **69**, S655 (1999)
94. T. Lippert, S.C. Langford, A. Wokaun, G. Savas, J.T. Dickinson, J. Appl. Phys. **86**, 7116 (1999)
95. M. Hauer, J.T. Dickinson, S. Langford, T. Lippert, A. Wokaun, Appl. Surf. Sci. **197**, 791 (2002)
96. H. Furutani, H. Fukumura, H. Masuhara, J. Phys. Chem. **100**, 6871 (1996)
97. H. Furutani, H. Fukumura, H. Masuhara, S. Kambara, T. Kitaguchi, H. Tsukada, T. Ozawa, J. Phys. Chem. B **102**, 3395 (1998)
98. T. Lippert, J.T. Dickinson, M. Hauer, G. Kopitkovas, S.C. Langford, H. Masuhara, O. Nuyken, J. Robert, H. Salmio, T. Tada, K. Tomita, A. Wokaun, Appl. Surf. Sci. **197**, 746 (2002)

99. M. Hauer, D.J. Funk, T. Lippert, A. Wokaun, Appl. Surf. Sci. **208**, 107–112 (2003)
100. J. Bonse, S.M. Wiggins, J. Solis, T. Lippert, Appl. Surf. Sci. **247**, 440 (2005)
101. J. Bonse, S.M. Wiggins, J. Solis, T. Lippert, H. Sturm, Appl. Surf. Sci. **248**, 157 (2005)
102. E.C. Buruiana, T. Buruiana, H. Lenuta, T. Lippert, L. Urech, A. Wokaun, J. Polym. Sci. A Polym. Chem. **44**, 5271 (2006)
103. E.C. Buruiana, L. Hahui, T. Buruiana, L. Urech, T. Lippert, J. Photochem. Photobiol. A Chem. **195**, 337 (2008)
104. E.C. Buruiana, V. Melinte, T. Buruiana, T. Lippert, H. Yoshikawa, H. Mashuhara, J. Photochem. Photobiol. A Chemistry **171**, 261 (2005)
105. R. Fardel, M. Nagel, T. Lippert, F. Nuesch, A. Wokaun, B.S. Luk'yanchuk, Appl. Phys. A **90**, 661 (2008)
106. G. Paltauf, P.E. Dyer, Chem. Rev. **103**, 487 (2003)
107. D.E. Hare, D.D. Dlott, Appl. Phys. Lett. **64**, 715 (1994)
108. D.E. Hare, J. Franken, D.D. Dlott, J. Appl. Phys. **77**, 5950 (1995)
109. C.B. Arnold, P. Serra, A. Piqué, MRS Bull. **32**, 23 (2007)
110. D.B. Chrisey, A. Pique, R.A. McGill, J.S. Horwitz, B.R. Ringeisen, D.M. Bubb, P.K. Wu, Chem. Rev. **103**, 553 (2003)
111. X. Yang, Y. Tang, M. Yu, Q. Qin, Thin Solid Films **358**, 187 (2000)
112. B. Losekrug, A. Meschede, H.U. Krebs, Appl. Surf. Sci. **254**, 1312 (2007)
113. E. Suske, T. Scharf, H.U. Krebs, T. Junkers, M. Buback, J. Appl. Phys. **100**, 014906 (2006)
114. D.B. Chrisey, A. Pique, J. Fitz-Gerald, R.C.Y. Auyeung, R.A. McGill, H.D. Wu, M. Duignan, Appl. Surf. Sci. **154**, 593 (2000)
115. C. Germain, L. Charron, L. Lilge, Y.Y. Tsui, Appl. Surf. Sci. **253**, 8328 (2007)
116. A. Klini, A. Mourka, V. Dinca, C. Fotakis, F. Claeyssens, Appl. Phys. A **87**, 17 (2007)
117. H. Fukumura: J. Photochem. Photobiol. A Chem. **106**, 3 (1997)
118. H. Fukumura, Y. Kohji, K. Nagasawa, H. Masuhara, J. Am. Chem. Soc. **116**, 10304 (1994)
119. D.M. Karnakis, T. Lippert, N. Ichinose, S. Kawanishi, H. Fukumura, Appl. Surf. Sci. **129**, 781 (1998)
120. J.Y. Lee, S.T. Lee, Adv. Mater. **16**, 51 (2004)
121. H.P. Le, J. Imaging Sci. Technol. **42**, 49 (1998)
122. F. Pschenitzka, J.C. Sturm, Appl. Phys. Lett. **74**, 1913 (1999)
123. D.B. Chrisey, A. Piqué, J. Fitz-Gerald, R.C.Y. Auyeung, R.A. McGill, H.D. Wu, M. Duignan, Appl. Surf. Sci. **154–155**, 593 (2000)
124. J.M. Fitz-Gerald, H.D. Wu, A. Pique, J.S. Horwitz, R.C.Y. Auyeung, W. Chang, W.J. Kim, D.B. Chrisey, Integrated Ferroelectrics **29**, 13 (2000)
125. A. Pique, D.B. Chrisey, R.C.Y. Auyeung, J. Fitz-Gerald, H.D. Wu, R.A. McGill, S. Lakeou, P.K. Wu, V. Nguyen, M. Duignan, Appl. Phys. A **69**, S279 (1999)
126. A. Pique, R.A. McGill, D.B. Chrisey, D. Leonhardt, T.E. Mslna, B.J. Spargo, J.H. Callahan, R.W. Vachet, R. Chung, M.A. Bucaro, Thin Solid Films **356**, 536 (1999)
127. D.B. Chrisey, A. Pique, R. Modi, H.D. Wu, R.C.Y. Auyeung, H.D. Young, Appl. Surf. Sci. **168**, 345 (2000)
128. D. Young, R.C.Y. Auyeung, A. Pique, D.B. Chrisey, D.D. Dlott, Appl. Phys. Lett. **78**, 3169 (2001)

129. S.G. Koulikov, D.D. Dlott, J. Imaging Sci. Technol. **44**, 111 (2000)
130. I.Y.S. Lee, W.A. Tolbert, D.D. Dlott, M.M. Doxtader, D.M. Foley, D.R. Arnold, E.W. Ellis, J. Imaging Sci. Technol. **36**, 180 (1992)
131. W.A. Tolbert, I.Y.S. Lee, M.M. Doxtader, E.W. Ellis, D.D. Dlott, J. Imaging Sci. Technol. **37**, 411 (1993)
132. W.A. Tolbert, I.YS. Lee, X.N. Wen, D.D. Dlott, M.M. Doxtader, E.W. Ellis, J. Imaging Sci. Technol. **37**, 485 (1993)
133. G.R. Pinto, J. Imaging Sci. Technol. **38**, 565 (1994)
134. J.M. Fernandez-Pradas, M. Colina, P. Serra, J. Dominguez, J.L. Morenza, Thin Solid Films **453**, 27 (2004)
135. P. Serra, M. Colina, J.M. Fernandez-Pradas, L. Sevilla, J.L. Morenza, Appl. Phys. Lett. **85**, 1639 (2004)
136. P. Serra, J.M. Fernandez-Pradas, F.X. Berthet, M. Colina, J. Elvira, J.L. Morenza, Appl. Phys. A **79**, 949 (2004)
137. V. Dinca, E. Kasotakis, J. Catherine, A. Mourka, A. Mitraki, A. Popescu, M. Dinescu, M. Farsari, C. Fotakis, Appl. Surf. Sci. **254**, 1160 (2007)
138. B. Hopp, T. Smausz, N. Kresz, N. Barna, Z. Bor, L. Kolozsvari, D.B. Chrisey, A. Szabo, A. Nogradi, Tissue Eng. **11**, 1817 (2005)
139. B. Hopp, T. Smausz, Z. Antal, N. Kresz, Z. Bor, D.B. Chrisey, J. Appl. Phys. **96**, 3478 (2004)
140. B. Hopp, T. Smausz, N. Barna, C. Vass, Z. Antal, L. Kredics, D.B. Chrisey, J. Phys. D Appl. Phys. **38**, 833 (2005)
141. T. Smausz, B. Hopp, G. Kecskeméti, Z. Bor, Appl. Surf. Sci. **252**, 4738 (2006)
142. J.A. Barron, D.B. Krizman, B.R. Ringeisen, Ann. Biomed. Eng. **33**, 121 (2005)
143. J.A. Barron, R. Rosen, J. Jones-Meehan, B.J. Spargo, S. Belkin, B.R. Ringeisen, Biosens Bioelectron **20**, 246 (2004)
144. J.A. Barron, B.J. Spargo, B.R. Ringeisen, Appl. Phys. A **79**, 1027 (2004)
145. J.A. Barron, P. Wu, H.D. Ladouceur, B.R. Ringeisen, Biomed. Microdevices **6**, 139 (2004)
146. J.A. Barron, H.D. Young, D.D. Dlott, M.M. Darfler, D.B. Krizman, B.R. Ringeisen, Proteomics **5**, 4138 (2005)
147. N.T. Kattamis, P.E. Purnick, R. Weiss, C.B. Arnold, Appl. Phys. Lett. **91**, 171120 (2007)
148. T. Lippert, Adv. Polym. Sci. **168**, 51 (2004)
149. D.P. Banks, K. Kaur, R. Gazia, R. Fardel, M. Nagel, T. Lippert, R.W. Eason, Europhys. Lett. **83**, 38003 (2008)
150. A. Doraiswamy, R.J. Narayan, T. Lippert, L. Urech, A. Wokaun, M. Nagel, B. Hopp, M. Dinescu, R. Modi, R.C.Y. Auyeung, D.B. Chrisey, Appl. Surf. Sci. **252**, 4743 (2006)
151. R. Fardel, M. Nagel, F. Nusch, T. Lippert, A. Wokaun, Appl. Surf. Sci. **254**, 1322 (2007)
152. R. Fardel, M. Nagel, F. Nuesch, T. Lippert, A. Wokaun, Appl. Phys. Lett. **91**, 061103 (2007)
153. M. Nagel, R. Fardel, P. Feurer, M. Häberli, F. Nüesch, T. Lippert, A. Wokaun, Appl. Phys. A **92**, 781 (2008)

8

Deposition of Polymer and Organic Thin Films Using Tunable, Ultrashort-Pulse Mid-Infrared Lasers

Stephen L. Johnson, Michael R. Papantonakis, and Richard F. Haglund

Summary. Advanced processing techniques for organic and polymer materials are critical to the burgeoning use of these versatile materials in electronic and electro-optic thin-film devices. Tunable, ultrashort-pulse mid-infrared lasers offer a promising new tool for laser-mediated deposition of these thermally labile materials, because they can be operated in a regime that largely avoids pathways to electronic excitation and are thus less prone to inducing destructive photochemical processes. This paper outlines the fundamental issues in the use of tunable mid-infrared lasers in the processing of organic materials, from the perspective of thin-film deposition by laser ablation. Consideration of the effect on laser ablation of spatially and temporally dense vibrational excitation leads to a figure of merit for comparing the process yield of lasers with varying pulse durations, energies and intensities. The mechanisms of mid-infrared laser ablation are examined using recent studies of three model materials: polystyrene, poly(ethylene glycol) and poly(tetrafluoroethylene). Examples of thin-film deposition of technologically interesting polymers and nanoparticles by matrix-assisted pulsed laser evaporation (MAPLE) are presented. Finally, current developments in mid-infrared laser technology suitable for the processing of organic molecules, polymers and functional nanoparticles are presented.

8.1 Introduction and Motivation

From the earliest history of the laser, there has been an evident desire to be able to manipulate materials selectively by taking advantage of laser intensity, monochromaticity and coherence to achieve states of materials that could not be realized by conventional incoherent addition of thermal or electronic energy to the material. In accepting the Nobel Prize for Physics in 1964, Alexander Prochorov, for example, famously observed that:

> Construction of an oscillator for any given radiation frequency will greatly extend the region of application of lasers. It is clear that if we make a laser with a sweep frequency, we apparently shall be able to influence a molecule in such a way that definite bonds will be

> exited and, thus, chemical reactions will take place in certain directions. However, this problem will not be simple even after design of the appropriate lasers. But one thing is clear: the problem is extremely interesting and perhaps its solution will be able to make a revolution in a series of branches of chemical industry. [1]

The dream of this kind of selective photochemistry has been realized in only a few special cases. Yet, evidence for selective laser interactions with materials abounds; the only question is whether or not the selectivity is rooted in the excitation of specific quantum states, or in something of a more macroscopic character. This selectivity, for example, has long since been exploited in the use of ultraviolet lasers for micromachining of organic and polymer materials.

Just recently the significance of a breakthrough in laser processing of materials has become stunningly apparent with the first exhibition of televisions based on organic light-emitting diodes (OLEDs) rather than liquid-crystal or plasma-screen displays. The OLED televisions exhibit not only much higher brightness, color selection and contrast than the older technologies, but also require typically less than 10% of the power of conventional LCD or plasma displays. Impressive as these latter technologies are, they are burdened with a number of operationally and environmentally undesirable byproducts. In this intensely competitive and capital-intensive industry, the development of improved deposition methods could be an enormous step forward.

In this chapter, we describe recent studies of wavelength-selective, mid-infrared (MIR) laser processing of organic and polymeric materials. We begin by discussing the rationale for using a specific class of tunable lasers – those with high intensity, high pulse-repetition frequency and high average power – as processing tools for these thermally labile materials. We continue by presenting a model for evaluating the processing efficiency of lasers of varying operating characteristics, on the basis of the concept of *density of vibrational excitation* as the key to understanding how effective a given laser can be for inducing a specific material modification. We next consider MIR laser ablation of the neat model polymers polystyrene, poly(ethylene glycol) and poly(tetrafluoroethylene) (PTFE) using a continuously tunable, picosecond MIR free-electron laser. Further, we give technologically interesting examples of conducting and semiconducting polymers and nanoparticles deposited by matrix-assisted resonant infrared laser evaporation. We conclude by discussing tunable, solid-state MIR lasers that could become a viable process technology for organic and polymer thin films based on resonant MIR laser ablation.

8.1.1 Mechanism of Laser Ablation at High Vibrational Excitation Density

A distinguishing feature of laser-materials interactions is their *hierarchical* character. The initial photon interaction is with individual sites – atoms, molecular complexes, clusters, defect sites, grain boundaries – that require an atomic-scale description. The sum of these site-specific interactions produces a *mesoscale* effect that evolves into thermal and mechanical equilibrium. At the *macroscale*, materials modifications are characterized by their properties on the micro- to millimeter scale. From this "bottom-up" perspective, lasers that are currently used in material processing are relatively inefficient, because substantial fractions of the input laser energy are dissipated through delocalized degrees of freedom that do not produce the localized bond-breaking and nuclear motion needed to control material modification on the atomic scale.

Ultrashort-pulse, tunable, high repetition lasers operating in the MIR spectral region can potentially control the laser-materials' interaction at all of these levels. The initial atomic-scale interaction can be specified because the states addressed by vibrational excitation are usually identifiable, in contrast to the multiplicity of transient intermediate states that follow a multi- or multiple-photon electronic excitation. Mesoscale effects can be controlled by tuning the laser to set the local density of vibrational excitation and with it the mesoscale relaxation channels for the absorbed laser energy. The modification of macroscale materials will be ruled by the average power of the laser and by the pulse-repetition frequency. Such a mode-selective process can be initiated even for relatively weakly absorbing vibrational modes, and avoids collateral damage.

To achieve laser ablation by vibrational excitation, three conditions must hold: First, a threshold intensity is needed to initiate the process, generally a few $\mathrm{MW} \cdot \mathrm{cm}^{-2}$. Second, vibrational energy must be localized on a small group of atoms or a molecular-size cluster in the laser-irradiated solid for a time scale longer than a few vibrational periods. Third, the energy absorbed must initiate and sustain the breaking of bonds and the motion of atoms, ions, molecules and clusters. In laser ablation, the ejection of mesoscale volumes of material also creates a dense plume of ejecta in which the interactions of atoms, ions, molecules and clusters with each other and with laser light can play a significant role. Given these preconditions, we now turn to the role of excitation density in the mechanism of laser ablation using ultrashort-pulse MIR lasers. Specifically, we will be considering lasers with pulse structures similar to the free-electron laser (FEL), in which a macropulse some microseconds in duration comprises a train of thousands of picosecond micropulses with high intensities and hundreds of picoseconds separation from the next one.

8.1.2 The Role of Excitation Density in Materials Modification

The crucial role of temporal and spatial excitation density in material processing was first identified by Stoneham and Itoh [2], and it is central to the idea of wavelength-selective material processing in the "molecular fingerprint" region of the electromagnetic spectrum shown schematically in Fig. 8.1. In a typical laser ablation experiment, the macroscopic observable effects – such as total ablation yield – necessarily scale with energy deposited per unit volume (E/V):

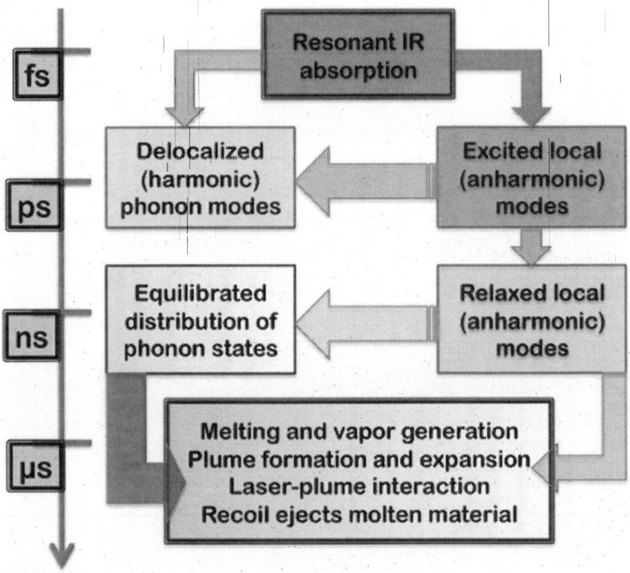

Fig. 8.1. Temporal sequence of processes that occur during resonant infrared laser ablation of polymers and organics using the macropulse of the free-electron laser

$$Yield \propto \left(\frac{E}{V}\right) \equiv F_{\mathrm{L}}\alpha\left(\omega, I\right) \cong I_0\tau_{\mathrm{L}}\left[\alpha_0\left(\omega\right) + \beta I\left(z, t\right)\right] \tag{8.1}$$

where F_{L} is the laser fluence, α is the absorption coefficient, ω and I are the laser frequency and intensity, respectively, and (z, t) are the relevant space (depth) and time coordinates.

When ultrashort pulses excite a resonant vibrational mode of a solid, the persistence of the deposited energy in that mode and the specific relaxation mechanisms that unload the deposited energy determine whether or not a non-thermal process results from the laser excitation. The time evolution of the laser-excited process follows from:

$$\frac{\mathrm{d}N_0}{\mathrm{d}t} = \eta N_0 \sigma_{(k)} \left(\frac{I}{\hbar\omega}\right)^k \tag{8.2}$$

where N_0 is the number of atoms or molecules per unit volume in the process volume, η is a quantum efficiency that includes loss channels that reduce the flux of modified atoms; $\sigma_{(k)}$ the kth order cross section for the laser interaction with the material; and $\Phi = (I \,/\, \hbar\omega)$ is the photon flux, the number of photons per unit time per unit area. The cross sections will of course be especially large for frequencies in the strong anharmonic absorption bands of organic and polymeric materials. A diagram of the most important of these bands in the molecular fingerprint region of the MIR is shown in Fig. 8.2.

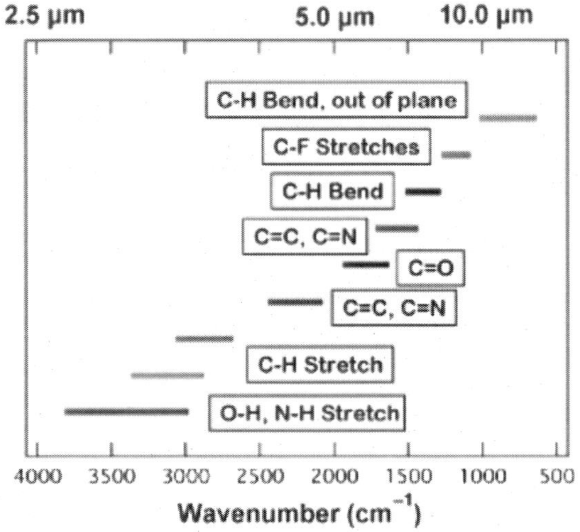

Fig. 8.2. Important absorption bands for organic and polymeric materials in the mid-infrared region of the vibrational spectrum, the so-called "molecular fingerprint" region

If one integrates (8.2) over the laser pulse for a process with a characteristic relaxation rate γ, one finds that the yield is proportional to the energy initially deposited in the mode,

$$\begin{aligned} \mathit{Yield} \propto E_\nu &\cong \int_0^{\tau_\mathrm{L}} \{I \cdot [\alpha_0(\omega) + \beta I] - \gamma E_\nu\} \mathrm{d}t = \left(1 - \mathrm{e}^{-\gamma\tau_\mathrm{L}}\right) \cdot \frac{I}{\gamma} [\alpha_0(\omega) + \beta I] \\ &\equiv \left(\frac{1 - \mathrm{e}^{-\gamma\tau_\mathrm{L}}}{\gamma\tau_\mathrm{L}}\right) F_\mathrm{L} \end{aligned} \tag{8.3}$$

The density of energy in the vibrational mode E_v has units of $J\,m^{-3}$, γ is the decay constant of the mode, α_0 is the linear optical absorption coefficient, β is the nonlinear absorption coefficient, I is the laser intensity and τ_L the laser pulse duration. The first factor in the last expression has a maximum value of 1 for small values of $\gamma\tau_\mathrm{L}$, corresponding to pulse durations shorter than $1/\gamma$; it falls roughly as $1/\gamma\tau_\mathrm{L}$ for τ_L much larger than $1/\gamma$. The probability of a multi-quantum vibrational transition depends on the spatial and temporal density of vibrational quanta produced by the laser irradiation. For the experiments presented here, the relevant quantity is the flux density in a 1-ps micropulse, with about 1 μJ of energy, in a few cubic nanometers volume.

Since the probability of initiating nonlinear processes scales with high powers of the flux density, rather than the fluence, picosecond and shorter pulse durations may well confer significant advantages for materials modification and processing. In this case, the use of ultrashort pulses also helps to insure that the desired, spatially localized electronic or vibrational excitation is not diluted by mixing with the delocalized, harmonic modes of the phonon bath during the laser pulse, consistent with experimental evidence. An additional result of the high excitation densities is probably the nonlinear absorption in the ablation phase, as the examples of dramatically reduced absorption depths in $NaNO_3$, $CaCO_3$ and amorphous SiO_2 show [3].

8.1.3 Laser Ablation at High Intensity and Pulse-Repetition Frequency

Gamaly et al. have proposed that thin-film deposition by laser ablation is optimized by using modest energy per individual laser pulse to ablate a small amount of material; relatively high intensity to enhance cross section; and high pulse-repetition frequency (PRF) to provide high average throughput [4]. Meeting the first criterion insures that collateral damage, particulate emission, and thermal loading are minimized, as suggested by the schematic in Fig. 8.3. The second follows from the consideration of reaction *rates*, which are proportional to intensities and cross sections, and not to fluence (8.2); in addition, at high intensities, nonlinear effects may add to the yield of desirable products. Third, high PRF yields both high throughput and optimal deposition from the nearly continuous vapor deposition achieved at high PRF.

Tunable, picosecond FELs are an ideal tool for testing these concepts [5]. With micropulse durations of 0.5–2 ps, these lasers deliver energy on a time scale comparable to or shorter than vibrational relaxation times, thus avoiding thermal equilibration during the laser pulse. Producing pulses with as much as tens of microjoules in these picosecond pulses, FELs are able to achieve intensities up to the 10^{10}–10^{11} W cm^{-2} range, easily enough to initiate multiphoton processes if desired and to sustain high reaction rates. And with micropulse repetition frequencies in the 50 MHz–3 GHz range, and macropulse repetition rates ranging from 30 Hz to kHz, they deliver extremely high average powers. Most importantly, because of their tunability, the material absorption, and hence the density of vibrational excitation, can be controlled.

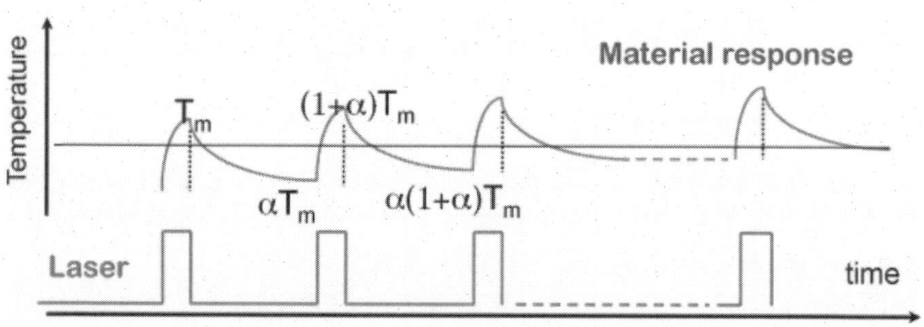

Fig. 8.3. Schematic of a multiple-pulse, high repetition-rate pulse sequence showing how relaxation in the interpulse interval helps to reduce the temperature rise in the irradiated material

8.1.4 Figures of Merit for Comparing Different Laser Processing Regimes

To compare the overall processing efficiency of different laser processing regimes, it is useful to develop a figure of merit (FOM) that is based on throughput or total yield for a particular process. As will become clearer from the experimental examples, the temperature reached by an individual micropulse as well as the number of pulses delivered by the laser are key parameters for calculating the effect of the laser-materials interaction. In the one-dimensional approximation to the heat-conduction equation, the target temperature during a single laser pulse reaches the value

$$\langle T\rangle = \sqrt{\frac{2}{\pi}\frac{I_{\mathrm{abs}}\sqrt{a\cdot\tau_{\mathrm{L}}}}{\kappa}}, a \equiv \frac{\kappa}{C_{\mathrm{p}}\rho_0} \tag{8.4}$$

If all the absorbed energy is converted into the desired materials modification (e.g., vaporization), then the specific energy input per unit volume is given by

$$\frac{E_{\mathrm{abs}}}{V} = C_{\mathrm{p}}\rho_0 \langle T\rangle = \rho_0\Omega \tag{8.5}$$

where Ω is the binding energy per atom, another material-dependent parameter. This makes it possible to express the temperature rise in the laser-irradiated volume as a function of laser and materials parameters as follows:

$$\Delta T = \frac{F_{\mathrm{abs}}(\omega)\cdot\alpha(\omega,I)}{C_{\mathrm{p}}\rho_0} \cong \frac{F_{\mathrm{abs}}\left[\alpha(\omega_0)+\beta I_{\mathrm{abs}}\right]}{C_{\mathrm{p}}\rho_0}, I_{\mathrm{abs}}\tau_{\mathrm{L}}^{1/2} = \sqrt{\frac{\pi a}{2}}\Omega \equiv f(a,\Omega) \tag{8.6}$$

The process rate – in this case, for laser ablation leading to the deposition of a thin film – for a laser producing pulses at a rate N_{pps} with an energy E_{L} per pulse is given by

$$Y = \eta(\omega,I)(E/V)N_{\mathrm{pps}} = \eta(\omega,I)\frac{E_{\mathrm{L}}f(\alpha,\Omega)}{L_{\mathrm{opt}}A(F_0,\omega)}N_{\mathrm{pps}} \tag{8.7}$$

Here L_{opt} is the optical absorption length, $A_0(\omega)$ is the laser spot size at the ablation threshold for the specified laser frequency. However, since absolute efficiencies are difficult to quantity, it is easier to calculate a *relative* FOM for two different lasers, as follows:

$$\begin{aligned} FOM^* &= \frac{Y_1}{Y_2} = \frac{\eta(\omega_1,I_1)(E_1/V_1)N_{\mathrm{pps}}(\omega_1)}{\eta(\omega_2,I_2)(E_2/V_2)N_{\mathrm{pps}}(\omega_2)} \\ &= \frac{\eta(\omega_1,I)E_{\mathrm{L}}N_{\mathrm{pps}}(\omega_1)}{\eta(\omega_2,I)E_{\mathrm{L}}N_{\mathrm{pps}}(\omega_2)} \times \frac{L_{\mathrm{opt}}(\omega_2)A(\omega_2)}{L_{\mathrm{opt}}(\omega_1)A(\omega_1)} \end{aligned} \tag{8.8}$$

Structurally, this equation comprises three factors: first, the ratio of the microscopic quantum efficiencies; second, the ratio of laser energy input to the material; and third, the frequency-dependent interaction volume, in essence, the mesoscopic volume that is being modified by the laser pulse – in this case, being ablated.

8.2 Resonant Infrared Pulsed Laser Ablation of Neat Targets

The experiments described below – desorption, ablation, film deposition – occur at high spatial and temporal density of *vibrational* excitation. Excitation of resonant infrared modes produces anharmonic vibrations whose wave functions couple on a time scale of 1–10 ps to the harmonic vibrations that constitute the phonon bath. Thus atomic motion and bond-breaking can begin before the energy leaks out of the excited anharmonic mode. Many vibrational excitations can be identified with specific molecular degrees of freedom, and in most cases do not generate electronic excitations that could produce photofragmentation, photochemistry and cross-linking. With ultrashort MIR pulses, non-thermal vibrational excitations leading to desorption, evaporation, ablation and ionization can in principle be initiated.

Resonant infrared laser ablation of neat polymer targets is characterized by complex effects that result from the unique thermal properties of polymers. In this section, we describe three examples of the *genre*. Ablation of poly(ethylene glycol) shows the effects of temperature-dependent absorption coefficients. Ablation of polystyrene shows the effects of absorption coefficients that vary with wavelength. Finally, ablation of PTFE shows the effects of irradiating the material at the fundamental (high) and overtone (low) densities of excitation.

8.2.1 Experimental Details

The ablation laser used in our experiments was an rf-linac driven FEL, with a wavelength that is continuously tunable from 2 to 10 μm [6]. The accelerator in the FEL is powered by an S-band klystron at 2.865 GHz that produces 4–6 μs macropulses at a repetition rate of 30 Hz. Typical macropulse energies are of the order 10–30 mJ. Within each macropulse, the small phase-space acceptance angle of the rf accelerator produces some 10^4 optical micropulses of approximately 1 ps duration. The optical bandwidth of the micropulses is typically 1% of the center frequency (FWHM) with a micropulse energy of several microjoules, yielding a peak unfocused irradiance of order $10^7\,\mathrm{W\,cm^{-2}}$. This bandwidth (about $16\,\mathrm{cm^{-1}}$) is large enough in many cases to overcome the anharmonicity of multiphoton vibrational excitation.

Films were deposited in a stainless-steel high-vacuum chamber in the pulsed laser deposition configuration shown in Fig. 8.4. Using a pyroelectric joulemeter, the transmission through the lens and entrance window was measured to be approximately 0.8. Focal spot sizes were estimated from knife-edge measurements and burns on thermal paper; total energy on target was extracted from transmission measurements. The deposited films were characterized by Fourier-transform infrared spectroscopy; X-ray diffractometry; X-ray photoelectron spectroscopy; scanning-probe and scanning-electron microscopy; and stylus profilometry. For measurements of plume dynamics, we used the layout in Fig. 8.5, in ambient environment.

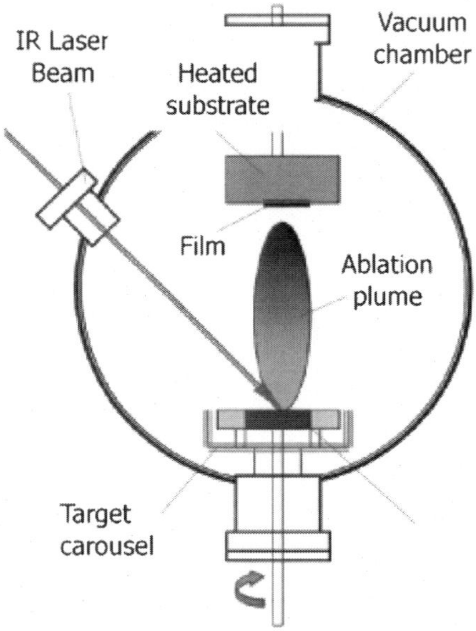

Fig. 8.4. Schematic of experimental layouts used in the experiments described in the text. Pulsed laser deposition in a vacuum chamber

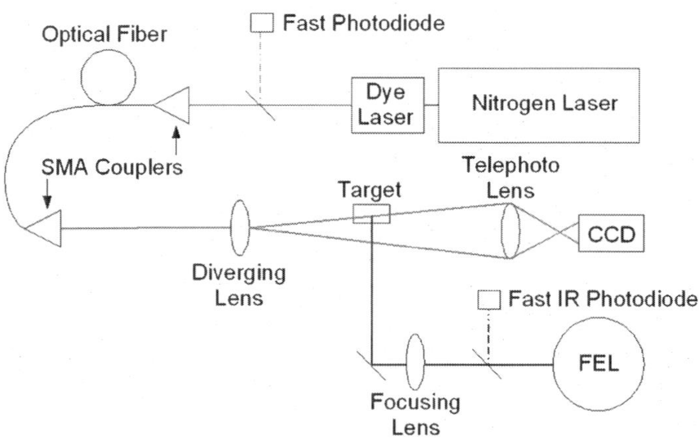

Fig. 8.5. Schematic of experimental layouts used in the experiments described in the text. Plume shadowgraphy measurements made in ambient air

8.2.2 Resonant Infrared Laser Ablation of Poly(Ethylene Glycol)

Polyethylene glycol is a useful model material for untangling thermal and wavelength-selective effects in resonant MIR laser ablation because of a temperature-dependent absorption coefficient in the O–H band that leads to narrowing and blue-shifting of the vibrational band with increasing

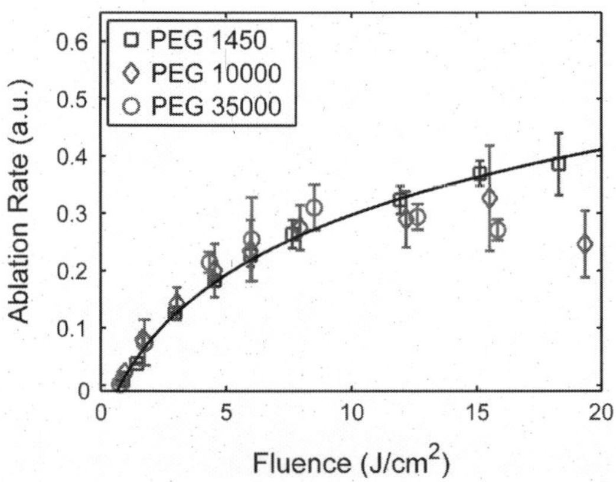

Fig. 8.6. Ablation rate of poly(ethylene glycol) of varying molecular weights as a function of macropulse fluence for an irradiation wavelength of 3.45 μm

temperature [7]. Moreover, the O–H band weakens as a function of molecular weight, leading to a change in the $1/e$ absorption depth from 100 μm for 1000 g mol^{-1} PEG to 1000 μm at 10,000 g mol^{-1}. In contrast, the C–H stretch (3.45 μm) does not exhibit this change; the band varies with laser intensity but does not shift with temperature. The difference in absorption properties leads to distinctive differences between ablation rates depending on FEL wavelength.

Figure 8.6 shows the ablation rate measured by a quartz crystal monitor in the vacuum chamber as a function of PEG molecular weight and fluence for irradiation of the C–H mode at 3.45 μm. The data fit reasonably well to a blowoff model [8] modified by a plume-shielding factor,

$$\delta \propto \frac{1}{\alpha\gamma} \ln\left[\gamma\frac{\Phi}{\Phi_{\text{thres}}} - \gamma + 1\right] \tag{8.9}$$

The plume-shielding factor γ ranges from 0, for no shielding, to 1 with complete shielding of the target by the plume. The shielding factor for the FEL ablation experiments turned out to be approximately 0.3 for a wide range of targets and experimental conditions.

This behavior contrasts sharply with the response of PEG 1450 when ablated at the 2.94 μm wavelength corresponding to the O–H mode, as shown in Fig. 8.7; the ablation rate as a function of fluence for ablation at 3.45 μm is shown for comparison, and follows the general trend of (8.4) as expected. The ablation rate for PEG 1450 at 2.94 μm, on the other hand, increases almost linearly, as one would expect for a steady-state model with no plume shielding, up to a fluence of 12 J cm^{-2}, where there is a dramatic increase in the ablation rate. Finite-element calculations indicate that at this fluence, the temperature rise is sufficient to fully liquefy the PEG surface. Since PEG decomposes at a temperature between 200 and 250°C into a monomer phase, the liquified

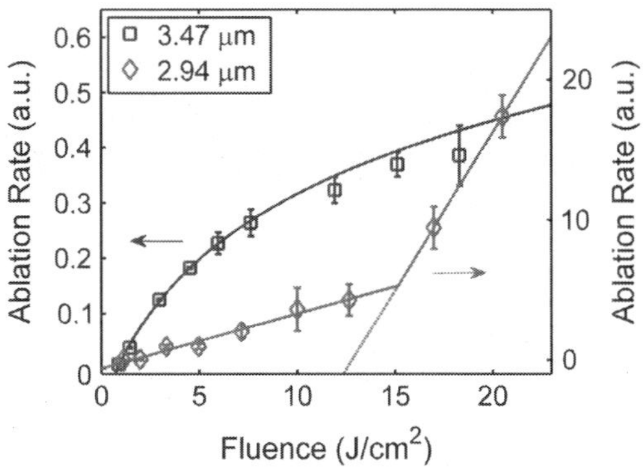

Fig. 8.7. Measured ablation rate for poly(ethylene glycol) (PEG 1450) at two wavelengths

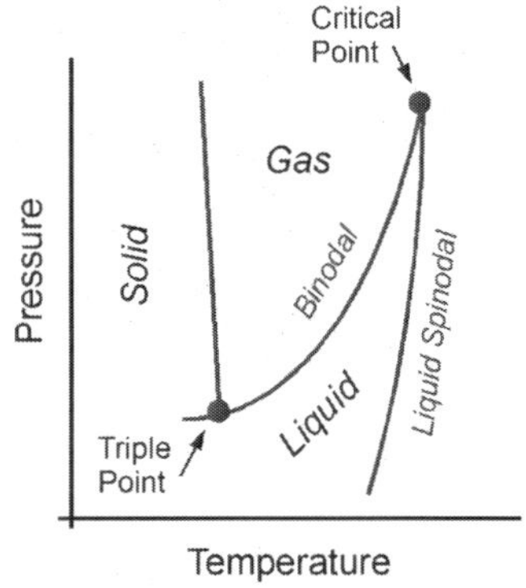

Fig. 8.8. Phase diagram applicable to ablation. Ablation of PEG 1450 occurs along the bimodal

ethylene glycol evolves along the binodal of the phase diagram (Fig. 8.8). The recoil force produced by the decomposed vapor plume then drives the ejection of liquid. This picture is corroborated by plume shadowgraphs taken during these experiments.

8.2.3 Resonant Infrared Laser Ablation of Polystyrene

The model of ablation suggested by the results on polyethylene glycol has been confirmed by extensive measurements and finite-element calculations of resonant MIR polystyrene ablation at several wavelengths [9]. Ablation was studied by pulsed nanosecond laser shadowgraphy, to understand both plume dynamics and the temporal onset of ablation. Commercially purchased polystyrene beads (molecular weight 224 kDa) were drop-cast into a polished aluminum target well, then extracted from the well and irradiated on the smooth side. Crater depths were measured by stylus profilometry as a

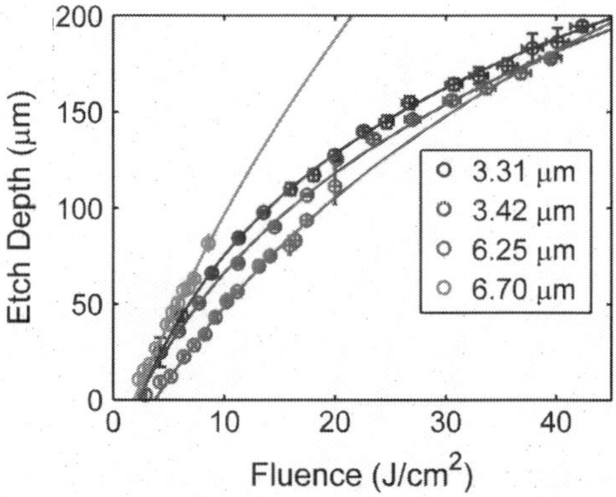

Fig. 8.9. Measured etch depth per laser macropulse (proportional to ablation rate) for polystyrene (MW 240 kDa) for the indicated laser wavelengths. Each wavelength corresponds to a different resonant vibrational mode

function of fluence for wavelengths of 3.31, 3.43, 6.25 and 6.70 μm with a stylus profilometer; the shape of the ablation crater was well-correlated with the Gaussian spatial profile of the laser pulse. Etch depths were then fit to (8.9) to extract values for the threshold fluence Φ_{th} and plume-shielding factor γ as a function of the experimentally determined absorption coefficients corresponding to these wavelengths (Fig. 8.9).

For fluences below the empirically determined ablation threshold (about 2 J cm^{-2}), the spatial distribution of temperatures around the ablation crater can then be correlated with the etch depth through the equation

$$T = T_{\mathrm{i}} + \frac{2\alpha\Phi}{\rho c_{\mathrm{p}}} \exp\left[-\frac{2\left(x^2+y^2\right)}{w^2} - \alpha z\right] \tag{8.10}$$

where T_{i} is the initial temperature, α is the absorption coefficient, Φ is the laser fluence, x, y and z are respectively the lateral and depth coordinates, w is the focal spot diameter, ρ is the density and c_{p} is the specific heat at constant pressure.

Given the large molecular weight of the polystyrene used in these experiments, it is likely that the PS undergoes decomposition before ejection into the gas phase. The plume shadowgraphs in Fig. 8.10 indicate that the laser heats the surface region of the film, driving it into thermodynamic instability. This superheated region then undergoes spinodal decomposition (in the vernacular, a "phase explosion") into a mixture of gas and liquid droplets. The recoil momentum of the expanding gas acts as a piston to compress the liquid polymer at the center of the ablation zone and drive liquid polymer from the periphery of the absorption volume at a non-normal angle, producing a hollow cylindrical column of ejected polystyrene (lowest row of shadowgraphs). At lower laser fluences, only the very center of the laser beam is capable of achieving spinodal decomposition, leading to the ejection of a thinner jet of

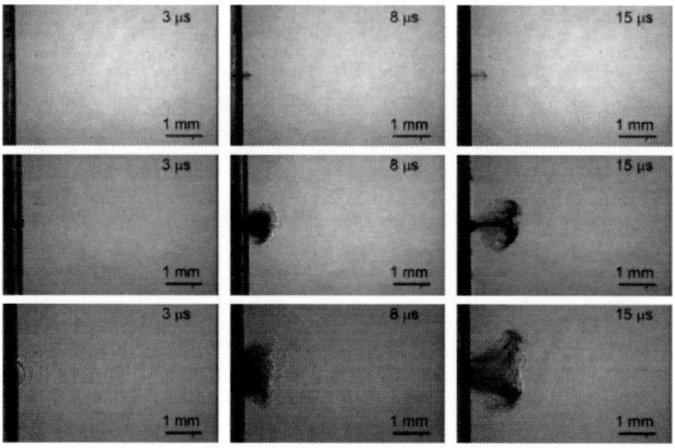

Fig. 8.10. Shadowgraphs of polystyrene ablation at a wavelength of 3.45 µm, at three different fluences, at intervals of 3 µs, 8 µs and 15 µs following the irradiation pulse. Top row, 2.3 J cm^{-2}, middle row, 5.4 J cm^{-2}, bottom row 9.2 J cm^{-2}

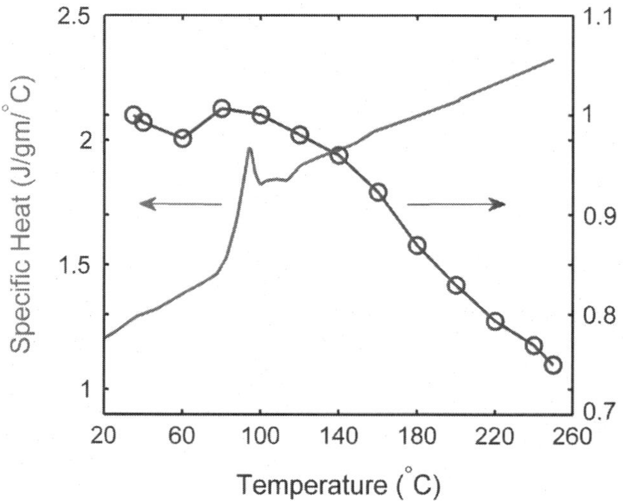

Fig. 8.11. Specific heat and normalized absorption of polystyrene as a function of temperature, for a laser fluence of 0.5 J cm^{-2} and a wavelength of 3.45 µm

material, as in the topmost row of Fig. 8.10. It will also be noticed that the first signs of a visible plume in the shadowgraphs are delayed as fluence drops to near threshold. Dynamical calculations show that this time to the appearance of the ablation plume varies inversely with fluence.

The detailed dynamics of the ablation process can be understood from the temperature dependence of the critical parameters of specific heat capacity and absorption coefficient. Finite-element calculations show that these quantities vary substantially as the temperature rises due to absorption of laser light, as seen in Fig. 8.11. The specific heat and the absorption coefficient vary in opposite directions as laser light is absorbed and the temperature of the polymer rises. One particularly noticeable feature in Fig. 8.11 is the spike in specific heat at 95°C, coincident with the glass-transition temperature. From that point on, the maximum surface temperature is lower than that predicted by a static heat-transfer calculation. In fact, the difference in temperature at a fluence of 0.5 J cm^{-2} (well below ablation threshold) is nearly a factor of two 4 μs into the ablation event. This illustrates the complex nature of the ablation process in materials that have such strongly varying thermophysical properties.

8.2.4 Resonant Infrared Laser Deposition of Poly(Tetrafluoroethylene)

While liquid-phase methods such as spin coating are the standard way of making thin polymer films, there are numerous technologically important polymers that are insoluble, such as PTFE. We have demonstrated that resonant infrared laser ablation can produce crystalline, smooth PTFE films at substantially lower temperatures than has been possible up to now using ultraviolet lasers [10].

For the studies described here, the FEL was tuned to wavelengths of 4.2 and 8.26 μm and a pulse duration of 6 μs. The surface roughness of the PTFE films were strongly dependent on the wavelength and fluence; however, at a fluence of 0.5 J cm^{-2} a 160-nm-thick film was deposited in only 7 min, with very few particles as large as 500 nm and most below ~30 nm.

One of the most attractive potential uses for thin films of PTFE is the coating of microstructures that need to take advantage of its unique properties. The film morphology resulting from irradiation of a bulk PTFE target was observed on a wire mesh, which served as a representative platform for a microstructure device. Figure 8.12 shows SEM images of wire mesh coated

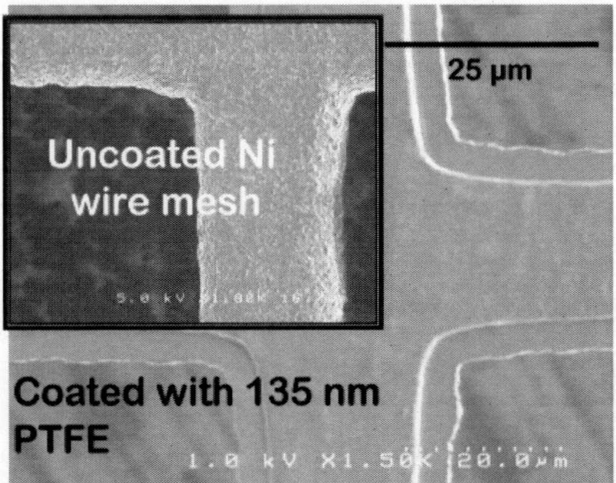

Fig. 8.12. Electron micrograph of a nickel wire mesh uncoated (inset) and coated with 135 nm of PTFE deposited by resonant IR ablation at a wavelength of 8.26 μm

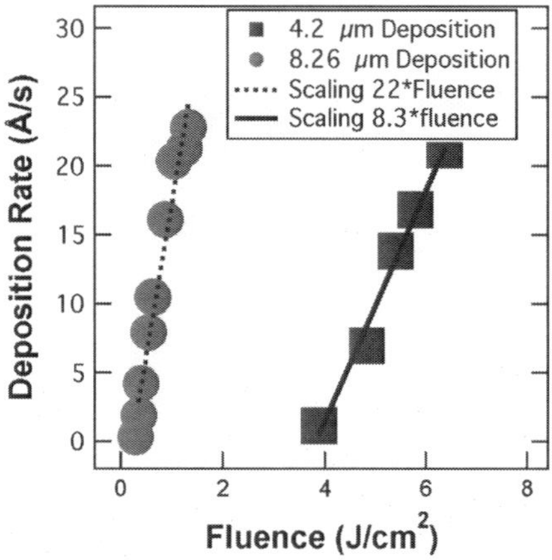

Fig. 8.13. Deposition rate of PTFE in Å/s for two different wavelengths: the CH_2 resonance at 8.26 μm, and the first overtone of that resonance at 4.2 μm

with PTFE deposited by 8.26 μm irradiation at a fluence of 0.5 J cm^{-2}. The film is shown to be particulate free and continuous across the 5 μm wide wire. Profilometry of a silicon wafer placed next to the wire during deposition showed the film to be 135 m thick for the 5-min deposition. XRD studies show that the deposited films are primarily crystalline, with a strong (100) peak, indicating orientation of the molecular chains parallel to the surface. The crystallinity of the films was improved by annealing in vacuo at temperatures well below the melting point before venting the system. The XPS spectra of films from both bulk and pressed pellet targets show excellent agreement with respect the reference bulk and pressed pellet targets.

Studies of deposition rate vs. fluence often reveal mechanisms about the operative ablation processes. Figure 8.13 shows the relative yield of PTFE for two FEL wavelengths with different absorption coefficients, and hence very different densities of vibrational excitation. Threshold fluences were estimated by extrapolating the data to a zero deposition rate, and found to be 0.26 and 3.5 J cm^{-2} for 8.26 and 4.2 μm irradiation, respectively. The weaker mode at 4.2 μm, with a $1/e$ optical penetration depth of ~30 μm, requires an order of magnitude higher intensity for the ablation onset than at 8.26 μm, where the penetration depth is only 1.3 μm. Thus at 8.26 μm, the density of excitation is much higher, with the energy confined to the first micron beneath the surface (calculate photon density). However, at 4.2 μm, absorption occurs primarily at defect sites, and ablation occurs by photomechanical processes.

8.3 Matrix-Assisted Resonant Infrared Pulsed Laser Deposition

A quite different mechanism of laser ablation is at work in the next category of experiments we describe here. Rather than ablating a neat target, the idea is to use a matrix as a carrier for the molecule to be deposited. The process of depositing films by dispersing the desired component in a matrix at low concentration in a matrix and then ablating at a resonant absorption of the matrix is often called matrix-assisted pulsed laser evaporation (MAPLE). The liquid solutions containing PEDOT:PSS or MEH-PPV were submerged in liquid nitrogen until frozen and the placed vertically upward inside a vacuum chamber. The FEL beam was directed through a BaF_2 window into the chamber and focused onto the target with a 500-mm focal length CaF_2 lens mounted outside the chamber. The substrate, at room temperature, was located ∼3 cm above the target to collect the vaporized plume. The chamber, evacuated by a turbomolecular pumping system, had a base pressure of order 10^{-5} Torr that rose to as much as 10^{-3} Torr during ablation. The beam was rastered across the surface of the target at a linear velocity of $3\,\mathrm{mm\,s^{-1}}$ while the target was rotated at 0.5 Hz to maintain a relatively even ablation track across the surface of the frozen target. The targets typically remained frozen for up to 20 min.

8.3.1 Deposition of the Conducting Polymer PEDOT:PSS

One of the most widely used conducting polymers is poly(3,4-ethylenedioxythiophene):poly(styrenesulfonate) (PEDOT:PSS). PEDOT, the conducting component of the blend, is polymerized with a counterion PSS to yield a highly conductive (∼$10\,\mathrm{S\,cm^{-1}}$), nearly transparent, water-soluble polymer thin film [11]. Combined with its ability to act as a buffer layer between ITO and an organic light emitter, PEDOT:PSS is an ideal hole-transport layer in a PLED [12].

For the experiments described here, PEDOT:PSS was obtained commercially from H.C. Starck as a 1% solution in H_2O (Baytron® P). For the experiment, three different PEDOT:PSS solutions were made into frozen targets: a native PEDOT:PSS solution containing only water as the solvent, a solution containing water and a co-matrix of *N*-methylpyrrolidinone (NMP) at varying concentrations, and a solution containing water and a co-matrix of isopropyl alcohol (IPA). Electroluminescent devices were fabricated both by RIR-PLD and by spin-coating for comparison purposes [13]. PEDOT:PSS films were deposited on glass cover slips for scanning electron microscopy (SEM) and four-point probe conductivity measurements, and on NaCl disks for FTIR spectroscopic analysis. In the conductivity measurements, the accuracy of the data was limited by the inhomogeneity of the film surface and thickness variations among samples.

For the various PEDOT:PSS solutions, the FEL was tuned to a vibrational resonance of one of the matrix components. In Fig. 8.14 the Fourier transform infrared (FTIR) spectra of the solvents used in these experiments are shown with the targeted vibrational modes for each solvent. Since the full width half maximum (FWHM) of the water O–H stretching resonance is reasonably well separated from the FWHM of the C–H stretching resonances of the IPA and NMP, it is possible to selectively tune the FEL to excite one matrix material

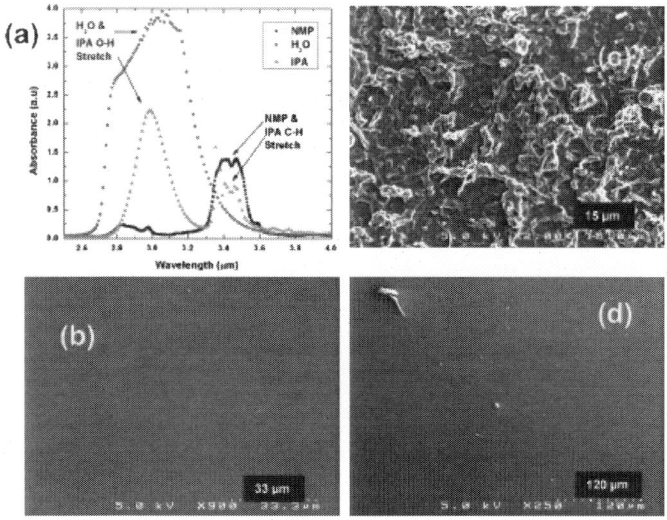

Fig. 8.14. Deposition of PEDOT:PSS films by matrix-assisted resonant IR pulsed laser deposition. (**a**) FTIR spectra of the frozen PEDOT:PSS solution, showing absorption bands of water, isopropyl alcohol (IPA) and *N*-methyl pyrrolidinone (NMP). (**b**) Electron micrograph of film grown by resonant ablation at the NMP wavelength. (**c**) Electron micrograph of film grown by resonant ablation at the water–ice wavelength, 2.94 μm. (**d**) Electron micrograph of film grown by resonant ablation at the IPA wavelength, 3.43 μm

even if two are present. The linewidth of the FEL in this region is typically 50 nm so that some overlap occurs, but most of the laser energy is deposited into one specific matrix absorption mode.

In an effort to produce electrically conductive films, *N*-methylpyrrolidinone (NMP), a known conductivity enhancer in spin-coated PEDOT:PSS films, was doped into the target material. The NMP concentration was varied between 5% and 50% by weight. As seen in Fig. 9.14a, the FWHM of the C–H stretch at 3.47 μm is well isolated from the O–H stretch of water at 3.05 μm. When either the H_2O or the NMP matrix is excited, films are conductive and exhibit an improved morphology compared to films deposited without a co-matrix. The FTIR spectra of the films also compared very well to a spectrum taken of a spin coated film. Conductivities of PEDOT:PSS films deposited from targets prepared with different concentrations of NMP are displayed in Table 8.1. When water is used as the excitation matrix, the film conductivities are dependent on NMP concentration, but independent of it when the NMP matrix is excited.

To further investigate the role of a co-matrix, isopropyl alcohol (IPA), also described by its manufacturer as a conductivity enhancer for PEDOT:PSS, was doped into the PEDOT:PSS solution prior to freezing. As seen from Fig. 8.14a, target irradiation at 3.05 μm deposits energy into both the water

Table 8.1. Comparison of resistivity of PEDOT:PSS films as a function of preparation method and NMP concentration, measured by four-point probe

Coating method	Percent NMP	Resistivity(Ω-cm)
Spin coated	0	14.7±2.9
IR-PLD 3.05 μm	0	N/A
IR-PLD 3.05 μm	10	39.5±22
IR-PLD 3.05 μm	50	11.7±8.9
IR-PLD 3.47 μm	10	5.65±0.5
IR-PLD 3.47 μm	50	5.20±1.7

and the IPA matrices since both contain an O–H bond, but 3.36–3.47 μm irradiation excites only the C–H bond in the IPA co-matrix. Since the O–H resonance in IPA has a lower cross section than that of H_2O, more of the photon energy at 3.05 μm is absorbed by the water matrix, but the absorbed laser light is shared between the two vibrational modes.

While the films deposited with the IPA co-matrix are not quite as smooth as those deposited with NMP, they are still superior to films deposited from unmixed PEDOT:PSS targets. The most interesting result of the four-point probe measurements on films deposited with IPA as a co-matrix is that films deposited using 3.05 μm irradiation (O–H stretch of water ice and IPA) are not conductive, whereas those deposited using 3.36 μm irradiation (the C–H stretch mode of the IPA) are. The PEDOT:PSS films deposited with an IPA co-matrix (not shown) are conductive at 3.36 μm, but not at 3.05 μm. This suggests that the role of the co-matrix is very complex, certainly more differentiated than the use of the matrix as a simple light absorber and carrier for the PEDOT:PSS that is typical of IR-MAPLE as it has been described in the literature [14].

8.3.2 Deposition of the Light-Emitting Polymer MEH-PPV

There is an extensive body of research on the properties of the light-emitting polymer poly[2-methoxy-5-(2-ethylhexyloxy)-1,4-phenylene vinylene] (MEH-PPV). There is a previous report of IR laser transfer of MEH-PPV where a comparison was made between UV and IR laser deposition and between neat polymer and frozen matrix targets [15]. However, a device structure (ITO/MEH-PPV/Al) was subsequently fabricated specifically to demonstrate electroluminescence [13] (see Figs. 8.15 and 8.16), whereas in reference [15] only photoluminescence is shown.

Deposition of MEH-PPV was rather more complicated than for the PEDOT:PSS, because the MEH-PPV powder is not easily soluble, even in DCB. For RIR-PLD, the FEL was tuned to either the C–H stretch of the DCB at 3.26 μm, or the C=C–C ring mode at 6.86 μm. FTIR spectroscopy was employed to analyze the bonding structure of the MEH-PPV after

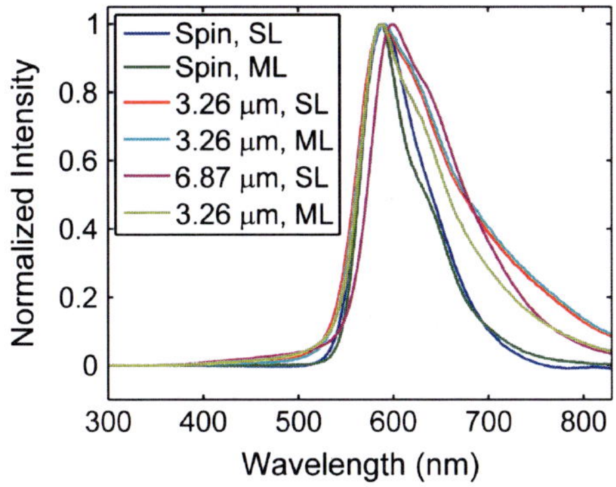

Fig. 8.15. Electroluminescence spectra of MEH-PPV in single-layer (SL) films and multi-layer (ML) devices with PEDOT:PSS, for various ablation wavelengths as indicated and for spin-coated samples

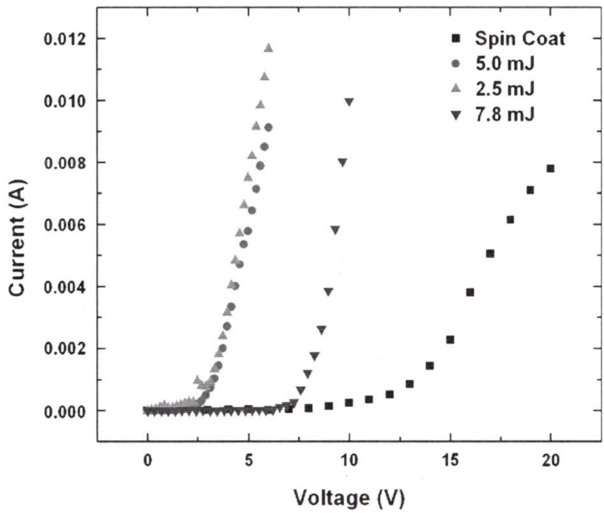

Fig. 8.16. I–V curves measured for spin-coated MEH-PPV compared to measurements for films deposited by resonant IR ablation at the indicated wavelengths

deposition. The similarity between the spin-coated spectrum and the spectra of laser-deposited MEH-PPV reveals that the local structure of the polymer is preserved throughout the laser deposition. It was also found that there was essentially no difference between the spectra of MEH-PPV deposited by laser ablation at different fluences. The fact that the spectra are independent of the laser fluence up to ∼5 times the ablation threshold) suggests that the matrix probably absorbs most of the laser energy and shields the polymer from interaction with the energetic matrix debris, allowing it to be deposited intact.

Since these results on MEH-PPV thin-film deposition were reported, other investigators using a fixed-frequency Er:YAG laser (2.94 μm) have shown that some control over the process can also be achieved by optimizing the matrix material [16].

To investigate the electrical behavior of the PLED devices, I–V curves were measured and are displayed in Fig. 8.16. It is immediately evident that the "turn-on voltage" of the FEL devices is much lower than that of the spin coated device. This is partly due to the difference in film thickness between the laser deposited devices and the spin coated device. The laser deposited films appeared by the eye to be slightly thinner than the spin coated devices, which would explain their lower turn-on voltage. Measurements with a stylus profilometer showed a very rough film, with an average thickness of order 35 ± 30 nm. It is doubtful, however, that the thicknesses of the various films differ by the same factor of $\sim$3 observed in the threshold voltage values.

The slopes of the I–V curves for the laser deposited devices are also steeper than that of the spin-coated device. In addition, it appears that there is a fluence dependence that suggests the possibility of laser-induced alteration of the MEH-PPV at high fluences. However, the fluence dependence may also simply be indicative of a thicker film and therefore a higher turn-on voltage. Whether or not the differences in the slope of the I–V curves is also a result of some laser-induced process is unknown at this point.

8.3.3 Deposition of Functionalized Nanoparticles

The ability to transfer layers of functionalized nanoparticles of controllable thickness to substrate platforms presents new opportunities for functional thin-film coatings. For example, TiO_2 nanoparticles deposited on light-emitting organic semiconductors have been shown to increase the efficiency of forward light emission by acting as miniature lenses, refocusing light emitted at high angles with respect to the surface normal. A stratified architecture could be produced consisting of alternating layers of nanoparticles with each other or another material such as a polymer. Alternating layers of nanoparticles would be difficult to obtain using solvent-based techniques, since the deposition process might disturb the previously deposited layers by hydrodynamic motion or, in the case of organic films, by dissolution of the layer beneath it. Moreover, conventional deposition techniques – for example, ink-jet deposition – exhibit poor surface distribution once the solvent in which the nanoparticles are suspended have dried.

Here we present an example of the successful deposition of fluorescent silica nanoparticles that were synthesized via the Stöber process [17]. The NPs were created with one dye in the core of the sphere (tetramethylrhodamine isothiocyanate; TRITC) and another dye (fluorescein isothiocyanate; FITC) covalently bound to surface of the NPs [18]. Several mL of a $\sim$1% wt. aqueous suspension of monodisperse, 160 nm diameter nanoparticles were placed into a hollowed disk and flash frozen with liquid nitrogen. The target holder was then introduced into a stainless-steel chamber under modest vacuum (10^{-4} torr). The FEL beam, tuned to a wavelength resonant with the solvent and rastered in one direction, was focused on the surface of the target. The laser fluence was set to a value just above the ablation threshold for the frozen target, typically just under 1 J cm^{-2}. During the plume expansion, the solvent was pumped away while the particles were collected on a substrate positioned several centimeters away. A dense single layer of nanoparticles was typically deposited in about 15 min.

The TRITC and FITC dyes have well separated peak absorption and emission wavelengths, making them attractive candidates for biological imaging. However, the reactivity and fragility of the dyes also made them a good test case of the potential damage that could result from the laser transfer process. Photoluminescence measurements were performed on the functionalized nanoparticles both before and after deposition, using an argon ion laser at 457 nm excitation wavelength. This wavelength is near the peak of the FITC absorption curve and in the tail of the TRITC absorption curve, which compensates for the higher concentration of TRITC in each nanoparticle. The spectra were normalized to the maximum intensity of the TRITC emission peak near 580 nm. Photoluminescence measurements performed on these doped silica nanoparticles show no significant alteration of their emission properties, suggesting that no damage occurred to the encapsulated or surface-bound dye either, during the ablation event (Fig. 8.17).

The laser ablation process results in line-of-sight particle transport, resulting in conformal coatings on the receiving platform; this is particularly useful for sensor bodies, such as cantilevers in atomic-force microscope structures. When used in conjunction with a shadow mask, resonant IR laser ablation can be used to pattern specific areas on a device platform. Figure 8.18 shows an optical image of patterned TiO_2 nanoparticles (50–100 nm in size) obtained by placing a TEM grid in contact with the silicon substrate. The lighter

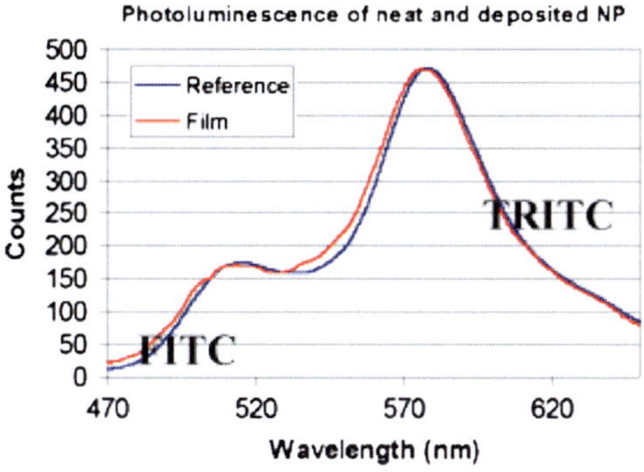

Fig. 8.17. Photoluminescence spectra of the TRITC and FITC organic molecules bound inside and onto 170 nm diameter fused silica nanoparticles. The spectra of the laser-deposited film and the reference are virtually identical

Fig. 8.18. Electron micrograph of TiO_2 nanoparticle layer deposited through a shadow mask. The square apertures in the mask are 100 μm in linear dimension, and the spacing between adjacent squares is 23 μm

areas represent the area blocked by the 23 μm wide wires. The minimum feature size is defined by the mask and its placement in contact with the substrate. Successive depositions of varied functionalized particles, in conjunction with appropriate masks, could allow one to impart different functionalities to specific areas of a device platform.

The principal criteria on the solvent or matrix for a successful nanoparticle transfer are (1) that it be compatible with material to be transferred and (2) have an appreciable optical absorption coefficient at an available laser wavelength. The kinetic energies of the plume expansion will limit the upper mass of the particles that can be transferred, but that limit has not yet been found. Polystyrene beads of order 600 nm have been successfully transferred using water as a matrix, for example.

8.4 Solid-State Lasers for Resonant MIR Ablation

The results achieved using the FEL are interesting from the standpoint of thin-film deposition research, but an FEL is far too expensive and complex to be used in the commercial production of micro- and nano-structured organic and polymeric films. Although the data shown in this paper represent experiments done with a FEL, other studies were performed with a compact, tabletop solid-state laser (Er:YAG), which was also found to produce similar coatings without chemically or physically altering functionalized nanoparticles. While a tunable laser allows greater flexibility in selecting an appropriate matrix material, table-top lasers are available at wavelengths convenient for common organic solvents. Not discussed in this work are favorable results using other organic solvents (e.g., benzene) which utilized laser wavelengths appropriate for those solvents (e.g., the aromatic C–H stretching mode of benzene at 3.27 μm).

Nevertheless, the experimental results achieved thus far point the way toward the development of laser systems more suited to commercial and industrial applications.

Why is the unique macropulse-micropulse time structure of the FEL so well suited to the ablation of intact polymers? The FEL irradiation has several critical properties that all seem necessary to provide efficient, but still gentle, ablation: (1) The O–H and C–H vibrational resonances are localized excitation modes of the monomer units of the polymer in question. (2) A micropulse delivering 1 μJ to one of these resonant excitations conveys of order one photon per nm^3 – a significant vibrational excitation density. (3) The picosecond micropulse has a duration shorter than the stress confinement time, given by

$$\tau_{\mathrm{micro}} << \tau_{\mathrm{stress}} \approx \frac{L_{\mathrm{opt}}}{C_{\mathrm{s}}} \sim 0.01 - 0.1\,\mu\mathrm{s} \tag{8.10}$$

where L_{opt} is the optical penetration depth and C_{s} is the speed of sound. This means that each micropulse delivers its quantum of energy in a time scale short compared to typical stress relaxation times – and that the absorbed laser energy will be fully equilibrated before the arrival of the next micropulse. (4) The FEL macropulse can be adjusted so that it has a duration less than the thermal confinement time, given by

$$\tau_{\mathrm{macro}} << \tau_{\mathrm{thermal}} \approx \frac{L_{\mathrm{opt}}^2}{\kappa} \sim 0.1 - 1\,\mu\mathrm{s} \tag{8.11}$$

where κ is the thermal diffusivity. This ensures that the total energy deposited goes into the absorption volume and does not diffuse into the cold surrounding material to create a heat-affected zone that "cooks" the polymers.

Thus the issue of replacing the FEL revolves around the question of whether or not it is possible to generate such a pulse structure in a solid-state laser, and to provide at least sufficient tunability to deposit energy in the wavelength ranges relevant to polymers. Most of these bands are shown in Fig. 8.2. The most important ones for the applications discussed here are the O–H, C–H and N–H stretching vibrations. However, excitation of bands in the 6–7 μm region produce efficient deposition of both MEH-PPV and the small light-emitting molecule $\mathrm{Alq_3}$. Moreover, one of the most interesting potential polymers – the insoluble polytetrafluoroethylene (trade name, Teflon®) – can be deposited by exciting the $\mathrm{C{-}F_2}$ stretching mode at 8.26 μm [10]. The challenge is how to reach these wavelengths using presently available nonlinear infrared materials.

Recent publications [19–21] suggest that the combination of a high repetition-rate pump laser such as Nd:YAG or Yb:YAG coupled to an optical parametric oscillator (OPO) [22] may well be capable of producing a high-repetition rate pulse structure with significant average power in the MIR range. However, it will clearly be a challenge to develop the macropulse–micropulse structure that appears to be so effective with the FEL. Moreover, the development of MIR materials to cover the wavelength bands of interest (see Fig. 8.7) and to find suitable sources of radiation to provide the idler beam for the OPO seems particularly difficult at this time. The only way at present to achieve wavelengths beyond 4 µm is to cascade successive OPO stages, with the attendant cost in efficiency at each stage. It may well be that the development of mode-locked Cr-doped ZnSe lasers, as well as the development of advanced IR frequency-conversion materials (e.g., glass-bonded GaAs laminates) will move us forward toward ablation of polymers with interesting bands in the 6–9 µm region of the spectrum.

8.5 Conclusion

In the seven years since the first demonstration of thin-film deposition by resonant infrared laser ablation [23], a wide variety of biocompatible [24], sensor [25], conducting [26], semiconducting [27] and thermoset [28] polymers have been successfully deposited by this method. While our understanding of the mechanism of resonant MIR laser ablation is by no means complete, there has been substantial progress, and with it increasing prospects of controlling the process at a level that would make it competitive with current solution-phase methods. Achieving that end would then also bring to bear on thin-film fabrication technologies all the advantages that accrue from the use of a solvent-free, conformal coating technique that does not damage fragile polymer and organic materials of interest. In the area of nanoparticle deposition, matrix-assisted MIR pulsed laser evaporation is already close to being commercially viable. All of this suggests both scientific challenges and bright technological prospects for laser materials modification with ultrashort-pulse, tunable, high repetition-rate lasers in the infrared. The most critical test for the near-term will be finding a reliable, table-top-scale replacement for the FEL, but even there the prospects are increasingly favorable.

Acknowledgments

Much of the research described in this paper, as well as the operations of the W.M. Keck Foundation Free-Electron Laser Center, was supported by the Medical FEL Program of the Department of Defense, administered by the Air Force Office of Scientific Research (Contract F49620–01–1–0429). Studies of nanoparticle deposition by matrix-assisted pulsed laser evaporation were supported by funds from the U.S. Department of Homeland Security, Science and Technology Directorate. While at the Naval Research Laboratory, M.R. Papantonakis was supported by a National Research Council Associateship. S.L. Johnson gratefully acknowledges financial support through a research assistantship from AppliFlex LLC.

References

1. B. Samuelsson, M. Sohlman, *Nobel Lectures Physics 1963–1970* (World Scientific, Singapore, 1998)
2. A.M. Stoneham, N. Itoh, Appl. Surf. Sci. **168**, 186 (2000)
3. N. Itoh, A.M. Stoneham, Radiat. Eff. Def. Sol. **155**, 277 (2001)
4. D.R. Ermer, M.R. Papantonakis, M. Baltz-Knorr, D. Nakazawa, R.F. Haglund, Appl. Phys. A **70**, 633 (2000)
5. E.G. Gamaly, A.V. Rode, B. Luther-Davies, J. Appl. Phys. **85**, 4213 (1999)
6. G.S. Edwards, S.J. Allen, R.F. Haglund, R.J. Nemanich, B. Redlich, J.D. Simon, W.C. Yang, Photochem. Photobiol. **81**, 711 (2005)
7. G.S. Edwards, D. Evertson, W. Gabella, R. Grant, T.L. King, J. Kozub, M. Mendenhall, J. Shen, R. Shores, S. Storms, R.H. Traeger, IEEE J. Sel. Top. Quant. Electron. **2**, 810 (1996)
8. S.L. Johnson, D.M. Bubb, K.E. Schriver, R.F. Haglund, J. Appl. Phys. **105**, 024901 (2009)
9. S.L. Johnson, Ph.D. Dissertation (unpublished), Vanderbilt University (2009)
10. M.R. Papantonakis, R.F. Haglund, Appl. Phys. A **79**, 1687 (2004)
11. B.L. Groenendaal, F. Jonas, D. Freitag, H. Pielartzik, J.R. Reynolds, Adv. Mater. **12**, 481 (2000)
12. M.P. de Jong, L.J. van Ijzendoorn, M.J.A. de Voigt, Appl. Phys. Lett. **77**, 2255 (2000)
13. S.A. Carter, M. Angelopoulos, S. Karg, P.J. Brock, J.C. Scott, Appl. Phys. Lett. **70**, 2067 (1997)
14. A. Berntsen, Y. Croonen, C. Liedenbaum, H. Schoo, R.J. Visser, J. Vleggaar, P. van de Weijer, Opt. Mater. **9**, 125 (1998)
15. D.B. Chrisey, A. Piqué, R.A. McGill, J.S. Horwitz, B.R. Ringeisen, D.M. Bubb, P.K. Wu, Chem. Rev. **103**, 553 (2003)
16. B. Toftmann, M.R. Papantonakis, R.C.Y. Auyeung, W. Kim, S.M. O'Malley, D.M. Bubb, J.S. Horwitz, J. Schou, P.M. Johansen, R.F. Haglund, Thin Solid Films **453–454**, 177 (2004)
17. R. Pate, K.R. Lantz, A.D. Stiff-Roberts, IEEE J. Sel. Top. Quant. Electron **14**, 1022 (2008)
18. H. Ow, D.R. Larson, M. Srivastava, B.A. Baird, W.W. Webb, U. Wiesner, Nano Lett. **5**, 113 (2005)
19. A. Burns, P. Sengupta, T. Zedayko, B. Baird, U. Wiesner, Small **2**, 723 (2006)
20. E. Innerhofer, T. Südmeyer, F. Brunner, R. Haring, A. Aschwanden, R. Paschotta, C. Honninger, M. Kumkar, U. Keller, Opt. Lett. **28**, 367 (2003)
21. U. Keller, Nature **424**, 831 (2003)
22. F. Brunner, E. Innerhofer, S.V. Marchese, T. Südmeyer, R. Paschotta, T. Usami, H. Ito, S. Kurimura, K. Kitamura, G. Arisholm, U. Keller, Opt. Lett. **29**, 1921 (2004)
23. E. Innerhofer, T. Südmeyer, F. Brunner, R. Paschotta, U. Keller, Laser Phys. Lett. **1**, 82 (2004)
24. D.M. Bubb, J.S. Horwitz, J.H. Callahan, R.A. McGill, E.J. Houser, D.B. Chrisey, M.R. Papantonakis, R.F. Haglund, M.C. Galicia, A. Vertes, J. Vac. Sci. Technol. **19**, 2698 (2001)

25. D.M. Bubb, B. Toftmann, R.F. Haglund, J.S. Horwitz, M.R. Papantonakis, R.A. McGill, P.W. Wu, D.B. Chrisey, Appl. Phys. A **74**, 123 (2002)
26. D.M. Bubb, J.S. Horwitz, R.A. McGill, D.B. Chrisey, M.R. Papantonakis, R.F. Haglund, B. Toftmann, Appl. Phys. Lett. **79**, 2847 (2001)
27. S.L. Johnson, H.K. Park, R.F. Haglund, Appl. Surf. Sci. **253**, 6430 (2007)
28. N.L. Dygert, A.M. Gies, K.E. Schriver and R.F. Haglund, Jr., Appl. Phys. A **89**, 481 (2007)

9

Fundamentals and Applications of MAPLE

Armando Luches and Anna Paola Caricato

Summary. Matrix-assisted pulsed laser evaporation (MAPLE) is an evolution of the pulsed laser deposition (PLD) technique. MAPLE preserves the advantages of the PLD technique (versatility, ease of use, high deposition rates) but in addition offers a gentle mechanism to transfer easy-to-decompose materials from the condensed phase into the vapor phase. The material of interest (polymers, biological cells, proteins, etc.) is diluted in a volatile, noninteracting (even under laser irradiation) solvent with a typical concentration of a few weight percent and frozen at the liquid nitrogen temperature. The frozen target is irradiated with a pulsed laser beam, whose energy is principally absorbed by the solvent and converted into thermal energy, allowing the solvent to vaporize. The molecules of the material of interest receive enough kinetic energy through collective collisions with the evaporating solvent to be transferred in the gas phase and finally deposited on a suitable substrate. Here, important results of MAPLE deposition of polymers are illustrated, and a novel application is presented: MAPLE deposition of nanoparticles and nanoparticle films. Finally, fundamentals of the MAPLE mechanism are discussed.

9.1 Introduction

Much effort is being expended to the development of new materials and material structures. One of the principal challenges is the development and integration of organic, inorganic, and biomaterials in devices to be employed in different areas such as biophysics, microelectronics, optoelectronics, and sensors. Efforts are also being made to deposit materials such as polymers, organic materials, and biomaterials (soft materials) in form of thin, uniform, and adherent films. The goal is the realization of controlled structures (or nanostructures) to be used, for instance, in drug delivery and tissue engineering [1], for gas and vapor detection [2], for light-emitting devices [3, 4], etc.

The choice of the deposition method depends above all on the particular physicochemical properties of the material of interest, the requirements for film quality, the type of substrate, and the production costs. There are techniques that can be used starting from the bulk material, such as pulsed laser deposition (PLD) [5–7] or vacuum evaporation [8]. Other simple methods involve liquid solutions of the material of interest in a volatile solvent, which include aerosol spraying, dip coating, and spin coating techniques [9]. For organic thin films, Langmuir–Blodgett dip coating using self-assembled monolayers is the common method to functionalize surfaces with a single biomolecular layer [10, 11].

Each of these deposition techniques has its own merits and drawbacks. In general, each one allows the treatment of a limited class of compounds. The ability to deposit a wide class of materials using a single technique would provide a great advantage for the development and implementation of devices based on soft materials. The deposition of this kind of materials in the form of thin or ultrathin films is not an easy task. There are several parameters that have to be controlled during and after the film deposition: preservation of the chemical integrity and the required physical–chemical properties of the materials, good thickness control, good adhesion to the substrate, and film uniformity over an extended area.

Laser-based deposition techniques present some advantages for thin film growth: monolayer thickness control, good film-to-substrate adhesion, minimum material consumption, low substrate temperature, and stoichiometry preservation among others. PLD is a well-known and powerful deposition method, but it is not well suited for the deposition of complex molecules such as polymers and biomaterials, with perhaps a few exceptions such as Teflon (polytetrefluoroethylene, PTFE) [12], polymethylmethacrylate (PMMA) [13], polypernaphthalene (PPN) [14] and maybe some others.

To avoid photochemical damage and decomposition caused by PLD, a new laser-based deposition technique has been introduced: the matrix-assisted pulsed laser evaporation (MAPLE) [15–18]. MAPLE is an evolution of the PLD technique. It is supposed to produce a "gentle" mechanism to transfer small- and large-molecular-weight species from the condensed phase into the vapor phase. Its main difference from PLD is the target structure: the material of interest (solute) is diluted in a volatile solvent matrix to form a homogeneous solution (solute concentrations typically up to several weight percent). The solution is frozen at liquid nitrogen temperature and then rapidly introduced into a vacuum chamber to act as a target for the laser-assisted deposition. The frozen target is irradiated with a pulsed laser beam, whose energy is principally absorbed by the solvent and converted into thermal energy, allowing the solvent to vaporize. By collective collisions with the evaporating solvent, the molecules of the material of interest receive enough kinetic energy to be transferred in the gas phase, thus covering a suitable substrate where they deposit as a thin film. The solvent is pumped away during the flight from the target to the substrate. Hence, the deposited film is composed only of the solute material. Since most of the laser energy is absorbed by the volatile matrix rather than by the solute molecules, their photochemical decomposition can be minimized. Moreover, the ablation onset in MAPLE is defined by the thermodynamic parameters of the highly volatile solvent, rather than the ones of the solute material, so deposition can proceed at much lower fluences $(0.05\text{–}0.5\,\mathrm{J\,cm^{-2}})$ as compared to conventional PLD (a few $\mathrm{J\,cm^{-2}}$). Using low fluences, thermal damage or decomposition of the solute molecules is of course greatly reduced. The films can be grown on substrate areas ranging from many square centimeters down to few square micrometers, and with thickness from a few nanometers to several micrometers. From a certain point of view, MAPLE could be considered the opposite of MALDI (matrix-assisted laser desorption/ionization) [19], where sometimes the solvent (e.g., glycerine) is almost transparent to the used laser light, which in contrast is strongly absorbed by dispersed particles (e.g., Co). The strong heating of the particles causes fast evaporation of the solvent and consequent evaporation/ionization of large biomolecules dispersed in the solvent. In other MALDI configurations, however, the solvent is the absorbing medium, like in MAPLE, but the application is very different: mass spectroscopy investigation, not film deposition.

In the past years, MAPLE has been extensively used to deposit films of polymers and biomaterials. Here, some examples will be given of polymer MAPLE deposition, with special attention to the work performed at the University of Salento; only some hints on MAPLE deposition of biomaterials will be given, as this is extensively dealt with in the chapter "Advanced biomimetic implants based on nanostructured coatings synthesized by pulsed laser technologies", by I.N. Mihailescu et al. Moreover, a new application, MAPLE deposition of nanoparticle films, will be introduced. Finally, fundamental features of the MAPLE technique will be discussed.

9.2 MAPLE Deposition Apparatus

The MAPLE deposition hardware does not substantially differ from the ones used in PLD. Excimer lasers (or Nd:YAG, third harmonic at 335 nm) are mostly used, since UV radiation couples with almost any target material, except in some particular cases where infrared laser sources are used to selectively dissociate solvent molecules. The main difference is the target holder, since it has to be kept at a low temperature during depositions. It means that a liquid nitrogen reservoir must be connected to the target holder. It is usually made of high-conductivity oxygen-free cooper, crossed by a stem of the same material supporting the target holder. The target must be rotated (3–10 Hz) to allow smooth erosion of the frozen solution. Feedthroughs and connectors have to be accurately designed, with properly chosen gaskets, to allow rotation at low temperature without seizing problems. A schematic diagram of a deposition system, very similar to the ones used for PLD, except for the target and target holder, is shown in Fig. 9.1. An image of the cooled target holder and substrate configuration is shown in Fig. 9.2.

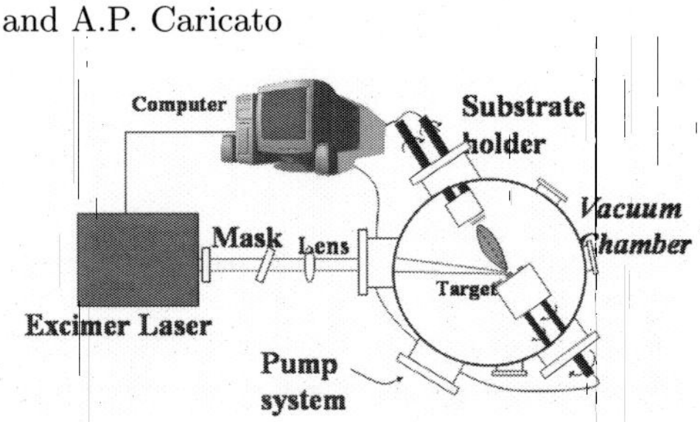

Fig. 9.1. Schematic diagram of a MAPLE deposition system, which is very similar to the ones used for PLD, except for the target and target holder

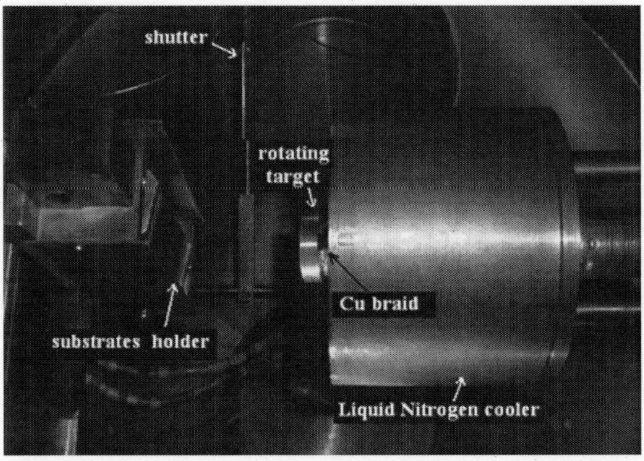

Fig. 9.2. Cooled target holder and substrate holder (MAPLE system at the University of Salento, Italy)

9.3 MAPLE Deposition of Polymers and Organic Materials

Attempts to deposit thin films of polymeric and organic materials were first made by PLD. Several types of polymers (polyethylene, polycarbonate, polyimide and PMMA) were ablated using UV lasers at energies near the ablation threshold [20]. A decrease in the molecular weight of the polymers forming the films was always noted. PTFE (Teflon) films were also deposited [21]. Film formation was supposed to occur via pyrolytic decomposition, followed by repolymerization. Since repolymerization can be incomplete, frequently, the properties of the PTFE films made byPLD are different from those of the target material [22]. Other polymers were deposited by PLD, with characteristics somewhat different from those of the target, at times claimed to be better than the original ones and useful for new applications. In any case, inhomogeneous films and inaccurate thickness control are common drawbacks. Moreover, since the deposition seems to proceed via a "depolymerization–monomer ablation–repolymerization" mechanism, PLD clearly cannot be used in general for complex polymers. Complex organic materials such as, for instance, polynaphthalene and polyacrylonitrile [23,24] were also deposited by PLD, generally with the same degradation problems presented by polymers.

MAPLE was developed to overcome the above difficulties presented by PLD of polymers and complex organic materials. Polymers are now by far the most used materials in MAPLE deposition, with excellent structural fidelity.

The first MAPLE studies were performed to deposit thin, homogeneous films of chemioselective polymers (a hydrogen bond acid functionalized polysiloxane known as SXFA) onto surface acoustic wave (SAW) devices. The MAPLE-deposited films (10–50 nm thick and highly uniform across the whole area) showed higher sensitivity and faster response times to various chemical vapors than films deposited by spray coating technique [25], due to the complete coverage of the substrate.

Recently, a deposition system explicitly designed for MAPLE technique, based on a Q-switched Ng:YAG laser ($\lambda = 1{,}064, 532$ and $355\,\text{nm}$, $\tau = 6\,\text{nm}$), was used to fabricate POOPT (poly [3-(4-octyloxyphenyl) thiophene]) thin films [26] with 20,000 consecutive laser pulses at the repetition rate of 10 Hz. The frozen target was composed of 0.56 wt.% of POOPT dissolved in chloroform. Laser fluences were $F \cong 4.7\,\text{J}\,\text{cm}^{-2}$ at 1,064 nm; $F \cong 136\,\text{mJ}\,\text{cm}^{-2}$ at 532 nm; and $F \cong 288\,\text{mJ}\,\text{cm}^{-2}$ at 355 nm). Results suggest a highly regular disposition of the polymer backbone. In the film deposited at 532 nm, the local chemical structure of the POOPT seems to have been preserved, while in the other two films, decomposition phenomena occurred. Furthermore, the film prepared at 532 nm appeared more homogeneous compared to the others, most probably due to the weaker energy per pulse of the laser radiation.

Tunno et al. and Caricato et al. [27, 28], used their customized MAPLE apparatus for the deposition of thin films of two compounds, poly(9,9-dioctylfluorene) (PF8) and a Ge-corrole derivative (Ge(TPC)OCH_3), both of great technological interest, using a KrF excimer laser.

As regards PF8, the influence of the laser fluence $(50\text{–}500\,\text{mJ}\,\text{cm}^{-2})$ and the nature of the solvent (chloroform, toluene, tetrahydrofuran) on the films properties have been studied. A detailed inquiry of the best deposition condition is of interest, since organic conjugated polymers are attracting considerable attention as a new material class to be used in photonics, for instance, as optically pumped distributed feedback lasers [29]. For polymer applications to organic lasers, a fine control of the active film thickness, in the scale of few nanometers, is extremely important in order to optimize the device performances. In fact, for laser devices exploiting active waveguides, the thickness affects the number of guided modes, their confinement in the active layer and the mode wavelength [30]. This means that a deposition technique such as MAPLE, allowing fine thickness control of polymeric films beyond the limits of spin coating and drop casting techniques, can be conveniently used. KrF excimer laser pulses ($\lambda = 248\,\text{nm}$; $\tau = 20\,\text{ns}$) at the repetition rate of 10 Hz were focused on the cooled target surface to have fluences in the range 50–500 $\text{mJ}\,\text{cm}^{-2}$. The target was prepared from different solutions of PF8 in chloroform ($CHCl_3$), toluene (C_7H_8) and tetrahydrofuran (THF – C_4H_8O), since they are among the most used solvents. A constant concentration of 0.5 wt.% was used in all the experiments. The cooled target holder was rotated at a frequency of 3 Hz, to allow uniform erosion. Before each deposition, 500 pulses were used to remove the surface layer of frozen aqueous vapor formed on the target, while a shutter screened the substrate. Five thousand laser pulses were employed for each deposition. Films were deposited on <100> Si substrate placed at 36 mm in front of the target.

The chemical composition of the deposited films was investigated by Fourier transform infrared (FTIR) spectroscopy in the mid-infrared (MIR) range $\left(700\text{–}3{,}600\,\mathrm{cm}^{-1}\right)$. To evaluate the influence of the three used solvents (THF, toluene and chloroform) on the structural properties of the deposited films, the infrared transmission spectra were acquired for films deposited at the same laser fluence $\left(200\,\mathrm{mJ\,cm}^{-2}\right)$. The spectra are shown in Fig. 9.3, together with the spectrum of a reference spin-coated film. A comparison of the spin-coated film spectra with those of the samples deposited by MAPLE suggests that chemical decomposition of PF8 takes place when chloroform is used as solvent. In fact, the vibrational bands are broader, the relative intensities are not preserved and some peaks are missing. The chemical degradation of PF8 deposited from a chloroform solution, which is a commonly used solvent in spin coating, is likely related to the presence of highly reactive radicals containing Cl released during excimer laser irradiation [31]. On the contrary, the FTIR spectrum of the MAPLE-deposited film using THF as solvent is very similar to that of the spin-coated film, since all the main peaks are present, preserving their relative intensities: the C–H stretching vibrations of the alkyl chain at 2,855 and 2,926 cm^{-1}, the C–H stretching mode of the aromatic ring at $2,953\,\mathrm{cm}^{-1}$, the aromatic ring-breathing vibration at 1,460 cm^{-1} and the alkyl C–H rocking mode at 813 cm^{-1}. Minor bands are also present, but it is important to notice that the peaks around 1,716 and 1,606 cm^{-1}, related to oxidized fluorene (fluorenone) [32], are not observed. This feature is particularly important, as PF8 is known to easily suffer photo-oxidation under light exposure, leading to yellow-light-emitting fluorenone defects. The increasing of the laser fluence has no effect on the peak presence or position, as can be inferred from Fig. 9.4, where FTIR spectra of samples deposited at laser fluences ranging from 200 to 500 $\mathrm{mJ\,cm}^{-2}$, using targets of PF8 diluted in toluene, are shown.

In order to investigate the role of the deposition conditions on the photoluminescence (PL) characteristics of the films, PL measurements were performed with a He–Cd laser ($\lambda = 325\,\mathrm{nm}$). The samples were excited at room temperature with a power density of about 40 $\mathrm{W\,cm}^{-2}$. The PL spectra of MAPLE-deposited PF8 films are shown in Fig. 9.5, together with that of a spin-coated film prepared using toluene as solvent. The PL spectrum of the spin-coated PF8 film shows a main peak at 442 nm, typical of the 0–0 line of the PF8 β-phase [33], followed by vibronic replicas at about 476, 503 and 540 nm. As regards the MAPLE-prepared samples, the PL spectrum of the film obtained from chloroform solutions is dominated by an extrinsic defect

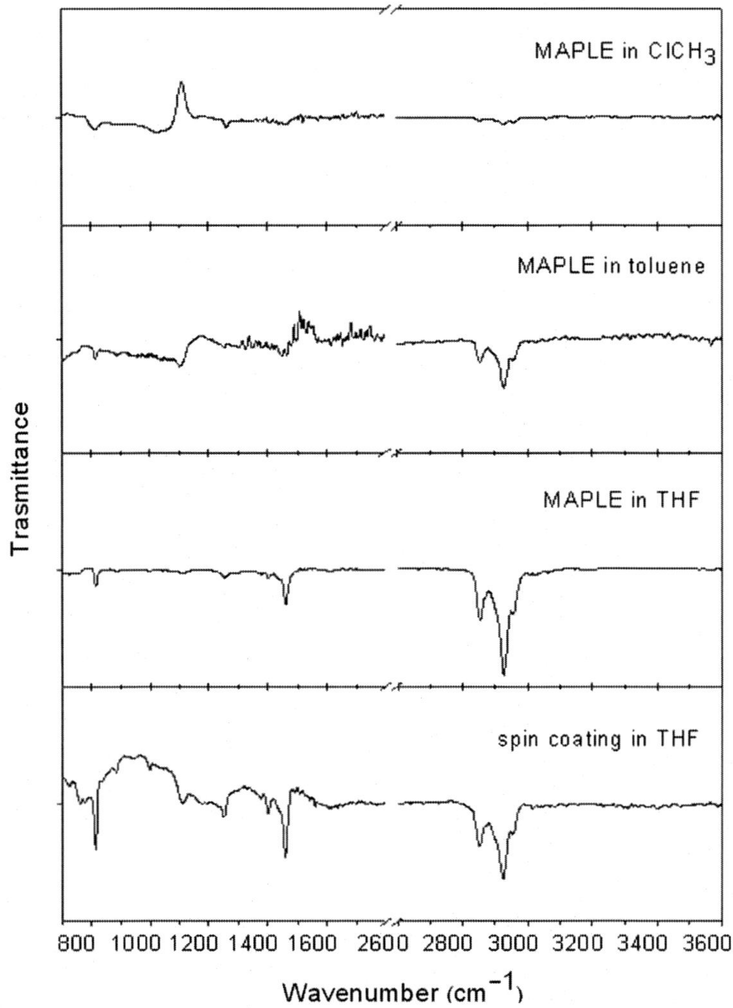

Fig. 9.3. FTIR transmittance spectra of MAPLE-deposited films ($F = 200\text{mJcm}^{-2}$) using different solvents (THF, chloroform and toluene). A spectrum of a spin-coated film is shown for reference

band peaked at about 570 nm and more than 150 nm broad, very similar to the fluorenone emission, with a further weak peak at about 440 nm, due to the PF8 β-phase 0–0 line. The strong defect emission is consistent with the FTIR results, confirming the chemical degradation of the polymer. The PL spectrum of the MAPLE film deposited from toluene solution, with a main peak at about 421 nm followed by a clear vibronic replica at 445 nm and two weaker higher order vibronic shoulders at about 480 and 515 nm, is similar to the PF8 glassy-phase spectrum. No evidence of defect emissions is present. It is interesting to observe that, despite the use of the same solvent, i.e., toluene,

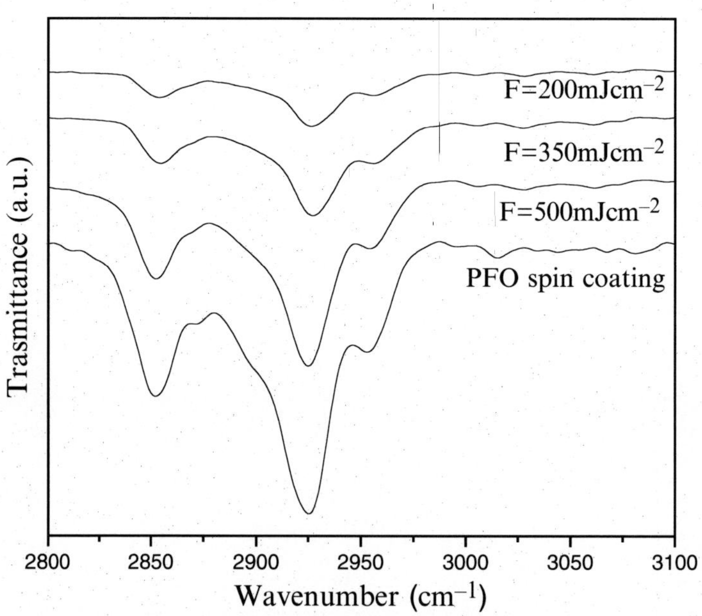

Fig. 9.4. FTIR transmission spectra of MAPLE-deposited PF8 films at different laser fluences, using toluene as solvent. A spectrum of a spin-coated film is shown for reference

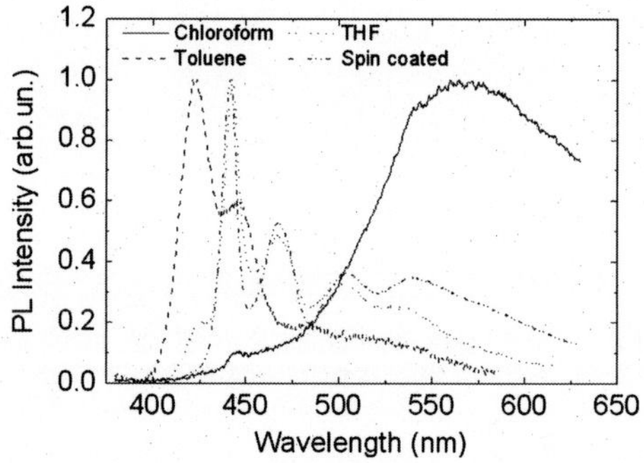

Fig. 9.5. PL spectra of PF8 films deposited with MAPLE and spin coating techniques. All the spectra are normalized to their peak value

the MAPLE film shows glassy-phase PF8 emission, while the spin-coated film shows β-phase emission only. As the β-phase is formed in spin-coated films from toluene due to an interplay between aggregation in solution and solvent-induced chain planarization in the solid phase [34], the absence of β-phase emission in the MAPLE-deposited film suggests that molecular aggregation does not occur and that negligible interaction between PF8 and toluene vapor takes place during the deposition process, thus avoiding the solvent-induced β-phase formation. This can be an indication that in the depositions from toluene solutions only the PF8 molecules reach the substrate, while the solvent molecules are pumped away during the process. This feature is actually different from the ones discussed by Mercado et al. [35], where chloroform and poly(lactide-*co*-glycolide (PLGA) were found to be co-deposited during the MAPLE deposition. One can suppose that the solvent can be co-deposited together with the solute if the solute–solvent interactions (usually quantified by the Hildebrand parameter δ [36]) are strong. Actually, the δ value for toluene is among the lowest for PF8 solvents. In any case, these and other problems related to MAPLE deposition of polymers will be discussed in a following section. The PL spectrum of the MAPLE film from THF shows a superposition of the emission features of the glassy and the β-phase, with a shoulder at about 420 nm (glassy phase 0–0 line) and a main peak at about 442 nm, due to the β-phase 0–0 line, followed by vibronic replicas up to the fourth order at 467, 502, 537 and 570 nm. Defect emissions are not observed. It must be underlined that, as efficient energy migration toward low energy emitting species usually takes place in polymeric films, the absence of defect-related emission in toluene and THF MAPLE films is a strong evidence that no relevant chemical modification of the PF8 emitting chromophors takes place during MAPLE deposition.

Other materials of interest are the porphyrin derivatives, since they can be used, for instance, as catalysis [37] and chemical sensors [38]. In the porphyrinoid area, the unique features shown by corroles, with a peculiar coordination chemistry and a photophysical behavior characterized by quantum yield values higher than those of the corresponding porphyrins [39], make these macrocycles a very promising material. Ge-corrole derivatives have been recently prepared, and their photophysical properties confirmed the high fluorescence quantum yield. The first evidence of phosphorescence emission from corrole complexes has also been recently reported [40]. Of course, these properties should be preserved when the material is used in thin film form. For this reason, the deposition method becomes decisive to realize devices based on corroles.

Caricato et al. [28] deposited Ge(TPC)OCH_3 films on silica substrates using the MAPLE technique. Ge(TPC)OCH_3 was prepared according to the method present in the literature [40] and diluted with a concentration of 0.01 wt.% in THF (C_4H_8O). The solution was gradually immersed in liquid nitrogen (77 K) and, after complete solidification, quickly mounted on the cooled (~113 K) target holder, which was rotated at a frequency of 3 Hz during laser irradiation. The vacuum chamber was evacuated down to 6×10^{-3} Pa before starting depositions. Target irradiations were performed with a KrF excimer laser ($\lambda = 248$ nm; $\tau = 20$ ns) at the repetition rate of 10 Hz and a fluence $F = 500\,\mathrm{mJ\,cm^{-2}}$. Before each deposition, 1,000 pulses were used to remove the surface layer of frozen aqueous vapor formed on the target, with a shutter screening the substrate. For each film deposition 10,000 laser pulses were employed. For comparison, Ge(TPC)OCH_3 thin films were deposited by the spin coating method starting from a THF solution with concentration of

0.1 wt.%, a ten times higher value with respect to the MAPLE solution, in order to obtain homogeneous coverage of the silica substrate.

The MAPLE-deposited Ge(TPC)OCH_3 film showed (Fig. 9.6) the typical morphology of organic films [35,41,42] obtained with this deposition method. The substrate was completely covered by the film, but a really uniform deposition was not obtained. In fact, the surface presented circular and irregular regions and gave a high roughness value (48 ± 9 nm). The presence of these characteristic features was recently investigated using molecular dynamic

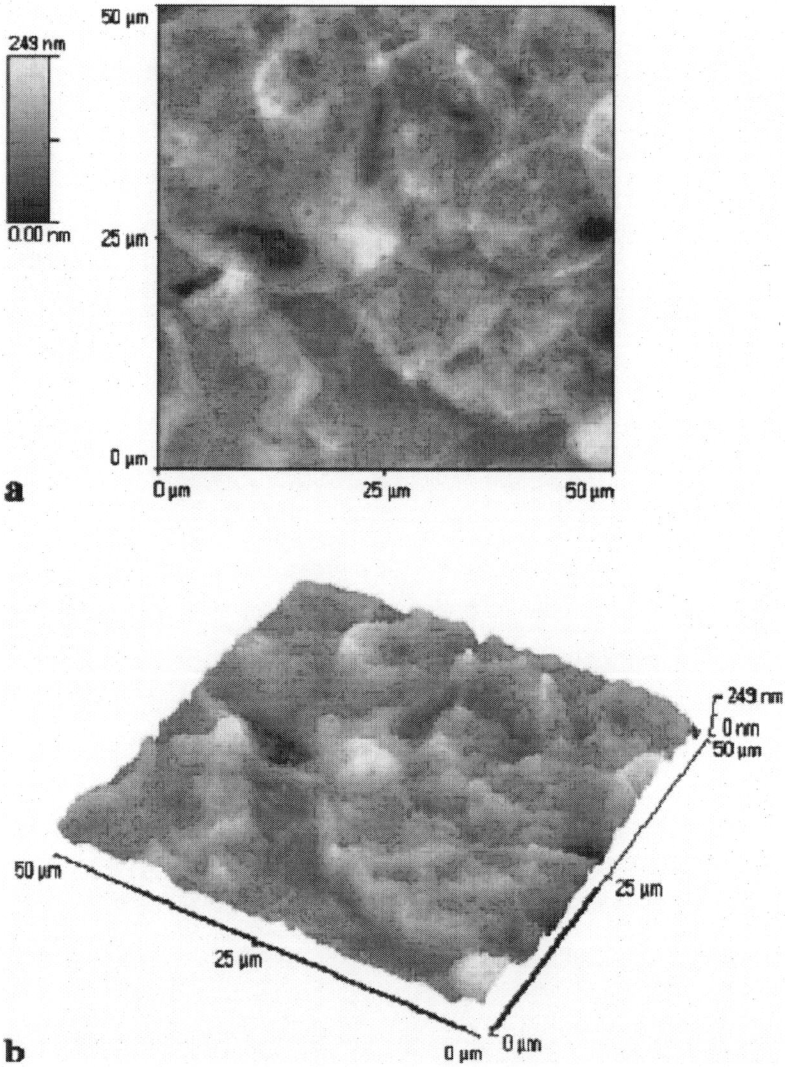

Fig. 9.6. (**a**) 2D and (**b**) 3D AFM micrographs of a typical MAPLE-deposited Ge(TPC)OCH_3 thin film

simulations by Leveugle and Zhigilei [43], which will be discussed later. On the other hand, the measurements performed on spin-coated films showed a thinner layer with the presence of scattered agglomerates and a nonuniform and incomplete coverage of the substrate.

The UV–vis absorption spectra of both films and solution at a concentration of 0.01 wt.% are shown in Fig. 9.7. The absorption spectrum of the MAPLE-deposited film presents the characteristic Soret band at 415 nm and the Q bands in the range 550–600 nm. The spectra of the films deposited by MAPLE and spin coating techniques present a slight red shift in the Soret and Q band positions with respect to the solution, probably due to the formation of corrole aggregates during the film growth. The Soret band in the spectrum of the solution has a full width at half-maximum (FWHM) of ∼20 nm. The same band is red-shifted by 5 nm with an FWHM of ∼20 nm in the spectrum of the MAPLE-deposited film, and red-shifted by 9 nm with an FWHM of ∼29 nm in the spectrum of the spin-coated film. The higher red shift and FWHM of the spectrum of the spin-coated film suggest the higher presence of corrole aggregates in the spin-coated film compared to the MAPLE-deposited one. Moreover, it can be noticed that the spectrum relative to the film obtained by MAPLE deposition starting from a concentration of 0.01 wt.% is comparable with that of the spin-coated film deposited using a concentration of 0.1 wt.%.

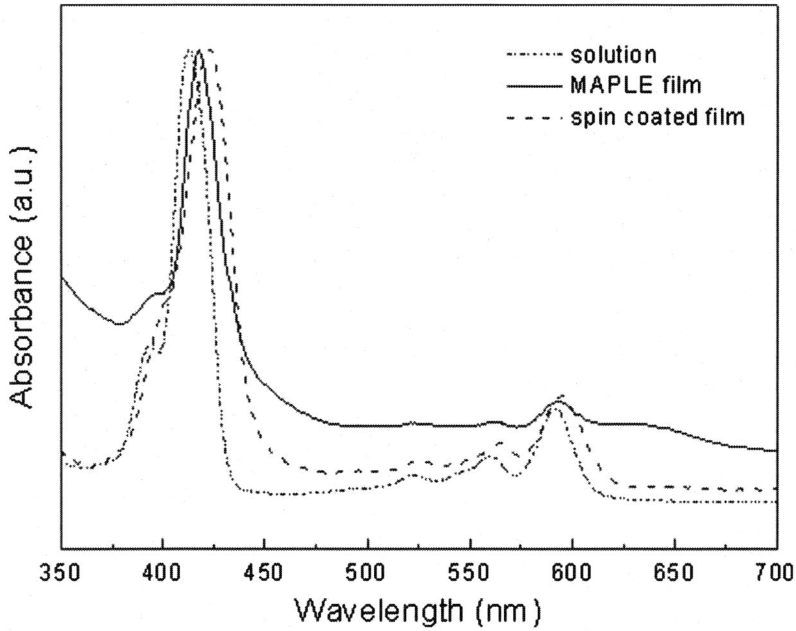

Fig. 9.7. Absorbance spectra of the spin-coated films, of the MAPLE-deposited films and of the solution (0.01 wt.%). All spectra are normalized to the main peak centered at 417 nm

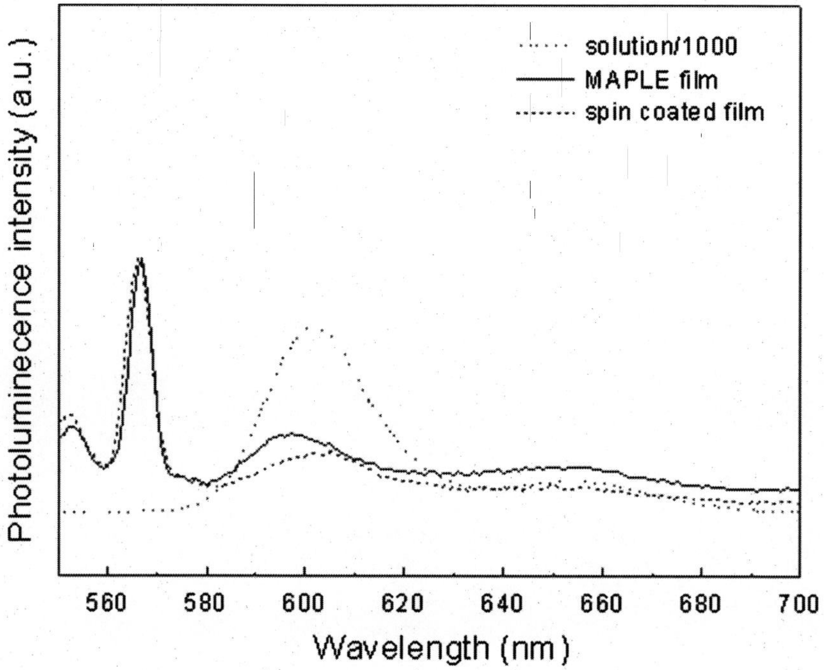

Fig. 9.8. PL spectra of MAPLE-deposited and spin-coated films. Spectra are normalized to the main peak at the wavelength of 566 nm. The PL spectrum of the solution is reduced by a factor 1,000. The excitation wavelength is 417 nm

The PL measurements were performed by using excitation radiation at 417 nm wavelength since Ge(TPC)OCH_3 has its maximum absorption at this wavelength. A comparison of the PL spectra of the solution, of the MAPLE-deposited film and of the spin-coated film is shown in Fig. 9.8. The MAPLE-deposited film presented well-evident luminescence peaks located around 600 and 650 nm, attributed to emission from Q bands, clearly indicating, together with the results in the UV–vis absorption spectra, the successful transfer of the organic material to the substrate. It can be noticed that the shift of the PL peak around 600 nm with respect to the absorption peak is ∼4 nm for the MAPLE film and ∼9 nm for the spin-coated film. Even this feature can be attributed to the possible presence of aggregates, in different concentrations.

The luminescence spectrum of the MAPLE-deposited film was acquired periodically to evaluate the aging effects. No detectable variation was recorded over a period of 1 month. This feature, together with the complete substrate coverage and the high roughness value, makes the MAPLE-deposited films appropriate for sensor applications.

9.4 MAPLE Deposition of Biomaterials

The MAPLE process has also been successfully used in the growth of biomaterials, and in particular of active protein thin films. Since this application is extensively illustrated in Chap. 10 of this book, here only two examples will be described briefly: the first on early deposition of horseradish peroxidase (HRP) and insulin and the second on recent deposition of bovine serum albumin (BSA).

Thin and uniform films of HRP and insulin have been deposited on a variety of substrates such as Si, NaCl as well as Au- and Pt-coated Si [44]. MALDI time-of-flight and electro-spray spectra demonstrated near-intact transfer of HRP and insulin with little or no decomposition. Infrared spectra of the HRP films showed that the chemical and physical structure of the protein was maintained after the MAPLE deposition. Moreover, activity tests indicated that most of the transferred protein retains its chemical and physical structure as well as its biological activity. These results represented the first demonstration that pure films of intact and active biomolecules could be deposited using the MAPLE technique.

Additional studies have since been performed on many other biomaterials, which, after deposition under well-tailored MAPLE parameters, also maintained their function and chemical structure in the form of thin films. One of the most recent results is related to MAPLE deposition of BSA. BSA is the main component of blood proteins for all animals and plays a great role in the body. In detail, BSA is a midsized protein with a molecular weight of 66 kDa. BSA protein solutions were prepared [45] using two different solvents: deionized water and phosphate buffered saline (PBS), which are typical solvents for diluting proteins. Besides solvents, the BSA concentration (1 and 2 wt.%) and the ArF ($\lambda = 193\,\text{nm}$, $\tau = 20\,\text{ns}$) laser fluence $\left(75\text{–}500\,\text{mJ}\,\text{cm}^{-2}\right)$ were varied to check the quality of the deposited films and the onset of threshold effects in the deposition rate.

FTIR was used to check whether the molecular structure was preserved after the laser–target interaction. The major BSA absorption bands at 1,653 and 1,550 cm^{-1} are present in the MAPLE-deposited BSA films. They are usually called amide I (C–O stretching) and amide II (N–H bending) bands, respectively. A minor band at 1,250 cm^{-1}, called amide III, is less evident (Fig. 9.9). It is interesting to note that no red shift of these signals is present in the deposited films. It means that the secondary protein structure is preserved in the α-helix state and no transformation in β-sheet configuration is formed after the laser irradiation [46]. This result confirms the absence of interaction of the protein molecules with the laser, thus avoiding their denaturation. In fact, the β-sheet protein configuration should result in the presence in the FTIR spectrum of the shift of amide I band to $1,628\,\text{cm}^{-1}$ [47].

Atomic force microscopy (AFM) was used to characterize BSA film surfaces. Figure 9.10a shows a typical AFM image of a BSA film deposited with $F = 150\,\text{mJ}\,\text{cm}^{-2}$ from a solution with a 1 wt.% BSA concentration in PBS.

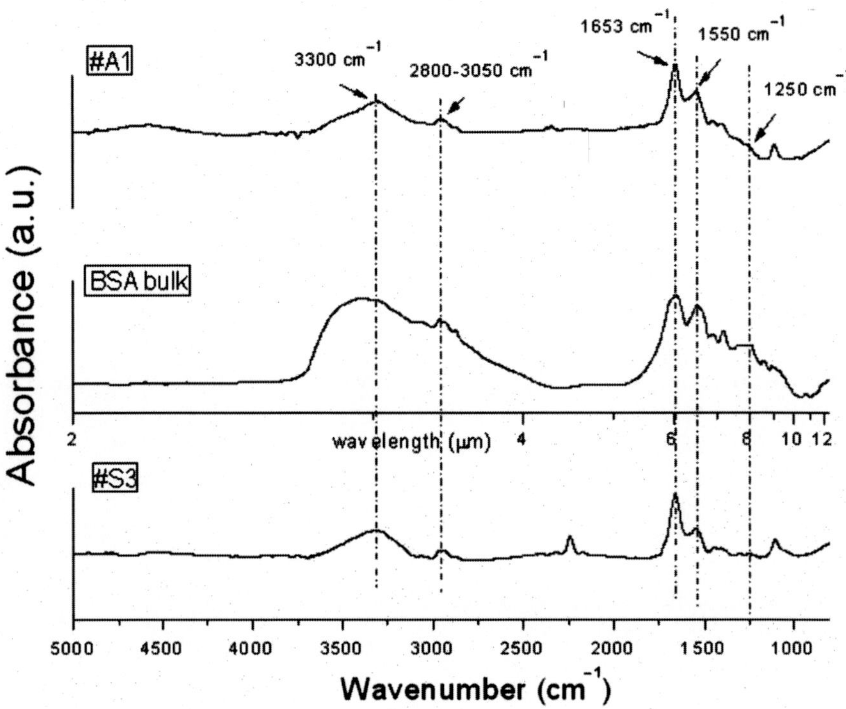

Fig. 9.9. FTIR absorbance spectra of sample # A1 (BSA, 2 wt.%, in deionized water; $F = 250\,\mathrm{mJ\,cm^{-2}}$), # S1 (BSA, 2 wt.%, in deionized water; $F = 150\,\mathrm{mJ\,cm^{-2}}$) and BSA bulk. BSA bulk sample was obtained from a pressed tablet employing 3 mg of protein dispersed in 250 mg of potassium bromide (KBr), a salt transparent to infrared wavelengths

In Fig. 9.10b, the three-dimensional reconstruction is shown. The film is uniformly covered by circular shaped structures with dimensions from hundreds of nanometers to a few microns. The well-known tendency of BSA to aggregate into macromolecular assemblies could explain the presence of objects of these sizes.

Biological tests were performed to heck the protein integrity after MAPLE deposition. It is possible to determine the molecular weight of a macromolecule by using sodium dodecyl sulfate polyacrylamide gel electrophoresis (SDS-PAGE) [48]. Results are shown in Fig. 9.11. The molecular weights of the BSA protein at different depth are listed on the left side of Fig. 9.11. At the top, the value 66 kDa corresponds to the entire protein, while at the bottom the smallest protein fragment (14.2 kDa) can be seen. In this figure, R indicates the column relative to the BSA protein used as reference, the second and third columns show two MAPLE-deposited samples, at $F = 150\,\mathrm{mJ\,cm^{-2}}$ and $F = 500\,\mathrm{mJ\,cm^{-2}}$, respectively. For this kind of analysis, the samples were

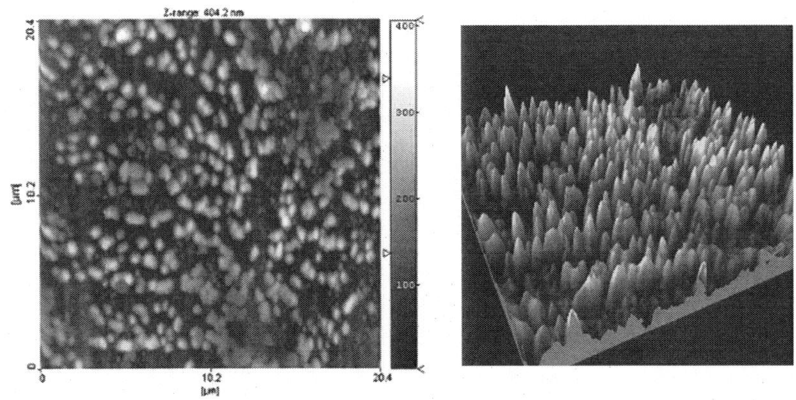

Fig. 9.10. (**a**) AFM and (**b**) 3D extrapolation images of a BSA thin film (BSA, 1 wt.% in PBS; $F = 150\,\mathrm{mJ\,cm^{-2}}$)

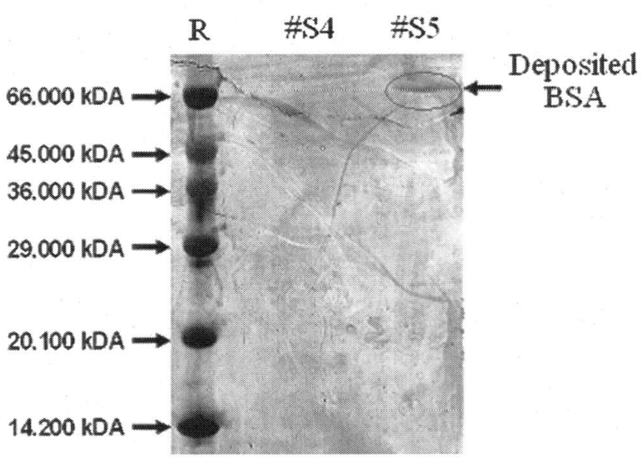

Fig. 9.11. SDS-PAGE image of pure BSA (R) and BSA films (# S4: BSA 2 wt.% in PBS, $F = 150\,\mathrm{mJ\,cm^{-2}}$, 30,000 pulses; # S5: BSA 2 wt.% in PBS, $F = 500\,\mathrm{mJ\,cm^{-2}}$, 20,000 pulses)

deposited on KBr substrates, which were then dissolved in a fixed water quantity. This solution was inserted into a gel, by using loading bores, to carry out SDS-PAGE. Proteins are not detected in the second column. It means that the film contained an insufficient quantity of protein to be immobilized and detected, due to the low deposition rate produced by the very low fluence $(150\,\mathrm{mJ\,cm^{-2}})$ used for this sample. In contrast, it can be seen that in the third column only the band relative to the entire BSA protein is present, indicating that the fluence of $500\,\mathrm{mJ\,cm^{-2}}$ is high enough to produce a reasonable deposition rate, without damaging the solute.

9.5 MAPLE Deposition of Nanoparticle Films

At present, the synthesis as well as the study of the properties and possible applications of nanoparticles of different materials is one of the most active research areas. Nanostructured films are expected to offer flexible tools for applications in different fields such as catalysis, sensors, information storage, nonlinear optics and many others. The functional properties of the nanostructured films are largely determined by the size, composition, morphology and surface properties of the nanoparticles forming the film. Nanoparticles can have very different properties with respect to the relative bulk material due, for example, to quantum confinement effects [49].

High-quality nanoparticle films can be obtained by different methods such as molecular beam epitaxy (MBE) [50] and metal-organic chemical vapor deposition (MOCVD) [51]. However, deposition processes are long and delicate; nanoparticles have a size distribution of 10–20% and more and are usually embedded in semiconductor or dielectric layers, thus preventing their growth on nonplanar substrates and their mixing with other materials. PLD is a much faster method to fabricate nanoparticles and nanoparticle films directly from bulk targets. Target ablation is performed in a noble gas or nitrogen atmosphere at pressures of the order of 100 Pa with nanosecond pulses [52] or in vacuum with femtosecond pulses [53]. Even though the results are of interest, the problem of the large size distribution poses strong limits to the extensive use of PLD for nanoparticle film fabrication. Recently, nanoparticles with well-tailored size and low size dispersion have been obtained by chemical growth techniques which are relatively easy and cheap. These techniques can produce spherical colloidal nanoparticles of different materials [54, 55]. Colloidal nanoparticles offer great versatility, as they are grown in solution, and they can be incorporated into polymer and glass matrices and into different photonic structures, including microcavities and photonic crystals. The nanocrystals can also be deposited as thin films by spin coating and drop casting. These deposition methods are quite simple and cheap, but they do not ensure a good control of the deposited film thickness and uniform coverage of the substrate, particularly on large areas.

The difficulties in the realization of uniform close-packed nanoparticle thin films set a very strong limit to their possible applications. This is why a versatile, fast deposition technique, such as MAPLE, was considered as a very attractive alternative. The colloidal nanosized particles, prepared with the required uniform dimensions, can be diluted in a volatile solvent and frozen at the liquid nitrogen temperature, thus forming the target to be irradiated. Films of carbon nanotubes [56] were deposited by this method. Recently, TiO_2 and SnO_2 colloidal nanoparticle films were prepared by MAPLE deposition, preserving the size and crystalline phase of the starting particles [57, 58]. The interest was in testing nanoparticle films as gas sensors. In fact, the electrical properties of semiconducting oxides such as TiO_2 and SnO_2 are influenced by gaseous ambients [59]. The use of nanoparticles increases the sensitivity of the sensor. This improvement has been attributed to the higher density of surface sites available for gas adsorption in nanocrystalline materials, as compared to that provided by the corresponding bulk material [60].

A summary of the methodology and of the obtained results is given here.

9.5.1 MAPLE Deposition of TiO_2 Nanoparticle Films

TiO_2 colloidal nanoparticles (size 10 nm) in the anatase phase were prepared by using standard procedures [61].The nanoparticles were diluted in deionized water with a concentration of 0.2 wt.% and kept in an ultrasonic bath for 10 min to prevent aggregation. Afterwards, the solution was gradually immersed in liquid nitrogen (77 K) and solidified to form the MAPLE target. Then it was quickly mounted inside a vacuum chamber on a rotating target holder, and cooled with liquid nitrogen to guarantee a low and constant temperature (~113 K). The vacuum chamber was evacuated down to 5×10^{-4} Pa, and the frozen target was irradiated with an ArF ($\lambda = 193$ nm) excimer laser with fluence $F = 550\,\mathrm{mJ\,cm^{-2}}$, pulse duration of 20 ns and repetition rate of 10 Hz. During the irradiation process, the pressure in the chamber rose to 4×10^{-3} Pa, due to the laser-induced evaporation process of the solvent. The nanoparticles were deposited on substrates placed at 36 mm in front of the target, which was rotated at the frequency of 3 Hz to allow uniform erosion. The number of subsequent laser pulses applied to deposit a single film was 6,500. Before starting depositions, 200 pulses were used in order to remove the surface layer of frozen water vapor formed on the target, while a shutter screened the substrate. The films were deposited on different substrates: silica, <100>Si and interdigitated alumina (Al_2O_3) slabs, for the different characterizations. The interdigitated alumina substrates, 2 mm × 2 mm used for the gas sensing tests, were equipped on the back face with a 50 nm Ti/500 nm Pt meander as heater, and on the front face with Pt interdigitated contacts in order to polarize the sensors and read the electrical current values during the sensing tests.

High-resolution SEM images of the TiO_2 nanoparticle films deposited on silicon substrates showed that the nanoparticles preserved the starting dimensions, although there was a tendency to form aggregates. By a comparison with TiO_2 nanoparticle films deposited by the spin coating technique, starting from a solution with the same TiO_2 nanoparticle concentration (i.e., 0.2 wt.%) used for the MAPLE-deposited films, a much more uniform coverage for the MAPLE-deposited film can be observed (Fig. 9.12). A very interesting result is that a uniform film of nanoparticles following the substrate morphology was obtained also on rough Al_2O_3 substrates used for gas sensing measurements (Fig. 9.13). This feature is essential to get a good electrical response. In fact, rough substrates improve the performance of the sensors by increasing the active area of the sensing films, and metal conductors formed on insulating substrates are necessary for electrical signal transfer. In Fig. 9.13, the darker

a

b

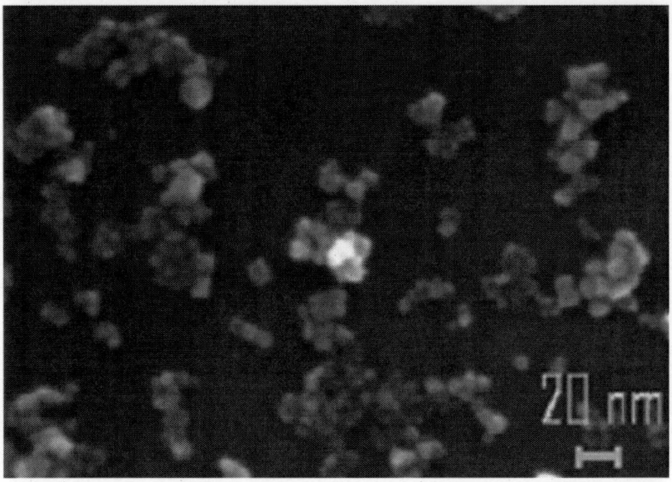

Fig. 9.12. High-resolution SEM micrographs of a TiO_2 nanoparticle film deposited on a silicon substrate (**a**) by MAPLE and (**b**) by spin coating

area corresponds to part of the substrate not covered by the TiO_2 nanoparticles and shows the morphology of the alumina grains. The brighter area shows the morphology of the TiO_2 film deposited onto the substrate. The deposition of TiO_2 spherical nanoparticles and agglomerates of nanoparticles is well evidenced in the inset of Fig. 9.13, showing the area of the TiO_2-covered alumina at higher magnification. However, as already noted, a trend toward the formation of spherical clusters of TiO_2 nanoparticles was observed both on Si and alumina substrates.

The preservation of the anatase crystal phase was evidenced by X-ray diffraction (XRD) spectra, where the characteristic peaks of the anatase phase at $2\theta = 25°$, corresponding to the reflection by the <101> crystallographic plane, are well evident. In Fig. 9.14, the XRD spectra of the starting solution and of the film (inset) are shown. Obviously, because of the small thickness of the film (≈30 nm, measured by forming a scratch on the film and making

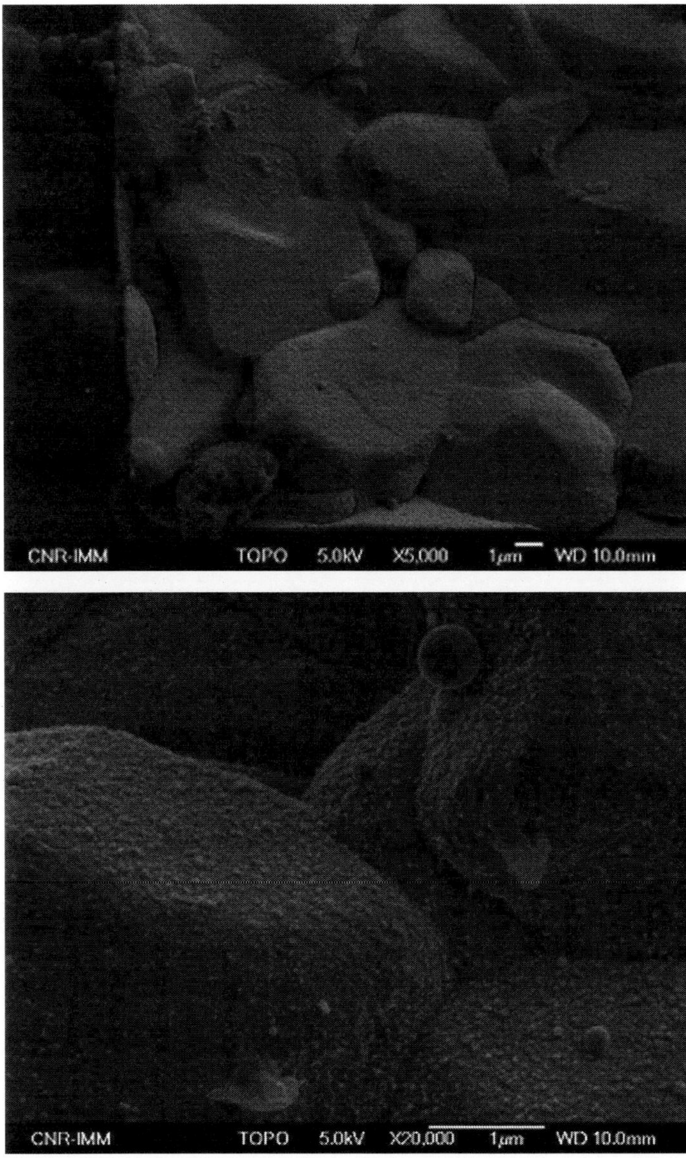

Fig. 9.13. High-resolution SEM micrograph (**a**) of TiO_2 nanoparticle thin film MAPLE-deposited onto a rough alumina substrate. The darker and brighter zones define the regions of the substrate uncovered and covered by the thin sensing layer, respectively; (**b**) zoom of the image remarking the presence of TiO_2 nanoparticles

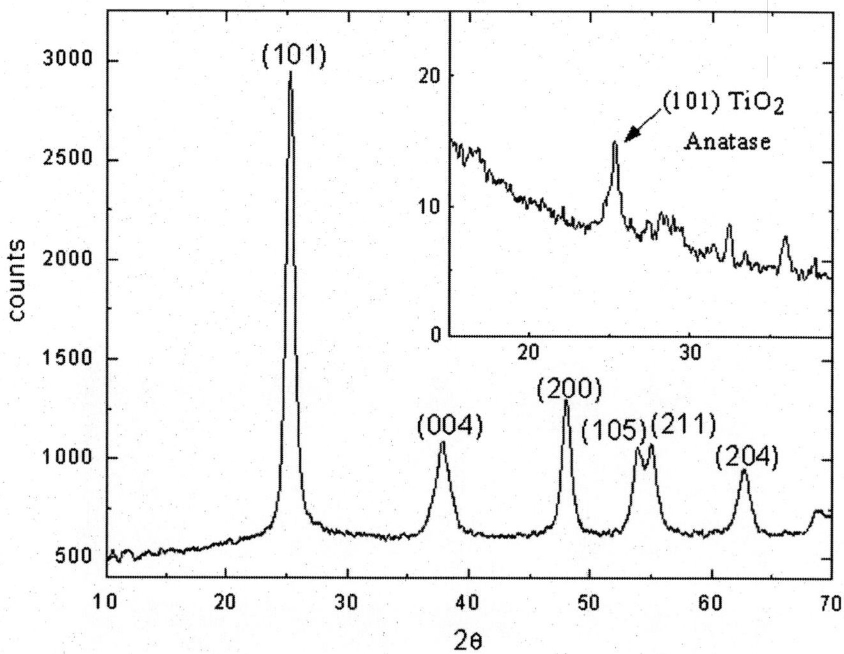

Fig. 9.14. XRD spectra of the TiO_2 nanocrystals in the starting solution and in the MAPLE-deposited thin film (inset)

an AFM measurement transversal to the scratch), the signal coming from the film is lower and noisier with respect to that from the solution.

The total transmittance and the reflectance UV–vis spectra of a TiO_2 nanoparticle film deposited by MAPLE technique on a silica substrate were acquired in the 200–800 nm spectral range. In general, the transmittance spectrum is characterized by a sharp fall at wavelengths shorter than ~360 nm, corresponding to the energy threshold for band-edge absorption of TiO_2. In order to determine the nature of the band gap of the nanostructured material, the spectral behavior near the fundamental absorption edge can be calculated by considering the expression [62]:

$$\alpha h\nu = B\ (h\nu - E_{\mathrm{g}})^p\ , \tag{9.1}$$

where E_{g} is the optical energy gap corresponding to the transition indicated by the value of p. In particular, p is 1/2, 3/2, 2 and 3 for direct allowed, direct forbidden, indirect allowed and indirect forbidden transitions, respectively. The factor B depends on the transition probability and can be assumed to be constant within the investigated optical frequency range. By plotting $(\alpha h\nu)^2$ against the photon energy, a linear trend, up to photon energies of about 5 eV, corresponding to allowed direct transition ($p = 1/2$) was obtained. The optical energy gap turned out to be nearly 3.6 eV, in agreement with the value reported in the literature for TiO_2 thin films [63].

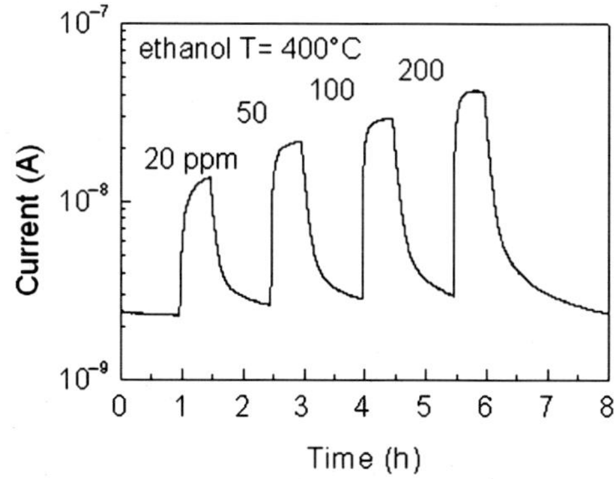

Fig. 9.15. Typical dynamic response of the MAPLE-deposited TiO_2 sensing layer for different concentrations of ethanol vapors at the working temperature of 673 K

Gas test measurements were carried out in the constant temperature mode by recording the dynamic changes of the electrical resistance caused by the exposure to different concentrations of ethanol and acetone vapors (20–200 ppm in dry air). The sensor working temperature was varied from 523 to 773 K to find the best operating temperature. As an example, Fig. 9.15 shows the dynamic changes in electrical resistance of a MAPLE-deposited TiO_2 nanoparticle film at the working temperature of 673 K for ethanol vapors at different concentrations. Surface reactions with reducing chemical species such as ethanol and acetone cause an increase in electrical conductance. The relative variation of signal in electrical current is very high (up to about one order of magnitude) even at very low concentrations of both the considered vapors. These very good gas-sensing properties toward ethanol and acetone may be attributed to the nanoscale dimensions of the TiO_2 particles. The measurements highlight also that the signal in electrical current of the MAPLE-deposited TiO_2 nanoparticle thin film is stable. Moreover, also the signal recovery is complete when the air flux is restored after the gas test. It can be also noticed that the response time is very fast ($\sim$4 min) and is not influenced by the working temperature. It has been shown that the working temperature has a strong influence on the gas response and on the dynamic behavior of the responses. The best sensitivities are measured for temperatures of 623 and 673 K for ethanol and acetone vapors, respectively. The calibration curves showed a higher sensitivity of the TiO_2 nanoparticle film to ethanol compared to acetone.

9.5.2 MAPLE Deposition of SnO_2 Nanoparticle Films

SnO_2 colloidal nanoparticles in the cassiterite phase of 3.6 ± 0.6 nm diameter, with one monolayer of trioctylphosphine capping layer, were prepared [54] and then diluted in toluene with concentration of 0.2 wt.%. Before the freezing procedure at the liquid nitrogen temperature, the nanoparticle solution was kept in an ultrasonic bath for 10 min to prevent aggregation. The solution was then frozen at liquid nitrogen temperature. The frozen target was irradiated with a KrF (248 nm, $\tau = 20$ ns) excimer laser at the fluence of 350 mJ cm^{-2}. Each film was deposited with 6,000 laser pulses at the repetition rate of 10 Hz. The target, which was rotated at the frequency of 3 Hz to allow uniform erosion, was placed in front of the substrate at a distance of 36 mm. Films were deposited on <100> Si single crystals and silica substrates in order to perform the different characterizations.

The as-deposited film consists of uniformly distributed elongated structures (Fig. 9.16). It was not possible to get higher magnifications than that one reported in this picture because of charging effects. The average film thickness was evaluated to be 150 $\pm$ 50 nm, which was done by making a scratch on it and followed by a scan with a mechanical profilometer. The FTIR spectrum shows different absorption bands, which are ascribed predominantly to the nanoparticle capping layer. The presence of the capping layer after MAPLE deposition was confirmed also by the UV–vis absorption spectrum, recorded in the spectral range of 200–800 nm, which was very similar to the absorption spectrum of the capping material diluted in hexane (the hexane absorption in the investigated spectral range is negligible).

The use of trioctylphosphine is necessary to avoid nanoparticle precipitation. However, the use of this capping layer determines, as a consequence, the presence of this agent also in the toluene solution with an estimated concentration of 10% in volume. Trioctylphosphine has a vapor pressure of 120 Pa at 293 K, which is much lower than that of toluene (2,900 Pa). The consequence

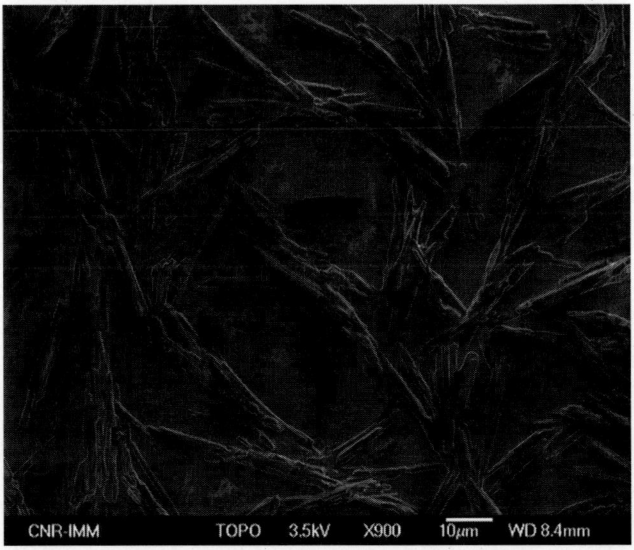

Fig. 9.16. SEM micrograph of the as-deposited SnO_2 sample on <100> Si substrate

is that it is not effectively pumped out during the MAPLE process and consequently it reaches the substrate, contributing to the composition of the deposited film.

To remove the capping layer, the MAPLE- deposited films were heated at 673 K, either in vacuum ($\sim$4–5 $\times$ 10^{-3} Pa) or in dry air. Similar results were obtained in both cases. No trace of the capping layer is present on the film after annealing. It is constituted by a uniform distribution of nanoparticles (Fig. 9.17). Occasionally, circular shaped zones were present on the film, formed by a less compact nanoparticle distribution. One of these areas is shown in Fig. 9.17b. From this figure, it was possible to measure the dimensions of a single nanoparticle. The average nanoparticle dimension was 4 ± 1 nm, in accordance with the nanoparticle dimension of the starting solution. Some bigger nanoparticles with dimensions of 10–20 nm were also noted, due either to nanoparticle coalescence caused by the postdeposition annealing or to the presence of these agglomerates in the starting solution itself.

The FTIR analysis was repeated on the annealed films. All the bands ascribed to trioctylphosphine or toluene are not present now. Only one peak was clearly visible at 667 cm^{-1}. This peak is due to the vibration of the antisymmetric O–Sn–O bridging bond [64]. FTIR spectra confirm the solvent and capping elimination at moderate temperature and the formation of a uniform SnO_2 nanoparticle film.

UV–vis absorption spectra were recorded in order to characterize the optical absorbance of the nanocrystalline SnO_2 film. The optical energy gap E_g was then determined [62] to be 4.24 eV (direct allowed transition). This value is higher than that 3.6 eV, reported in the literature [64] for bulk SnO_2. The observed blue shift of the absorption edge is attributed to the small dimensions of the nanocrystalline particles. The optical band gap of the nanocrystalline particles depends on the particle radius, due to quantum confinement of electrons and holes, as reported by different authors [65, 66]. The dependence of absorption onset on the particle size is based on the effective mass approximation (EMA), and the increase in the optical band gap of a nanocrystalline semiconductor may be represented by [66]:

$$E^* = E_g + \frac{h^2}{8R^2}\left(\frac{1}{m_e} + \frac{1}{m_h}\right) - \frac{1.8e^2}{\varepsilon R}, \tag{9.2}$$

where E^* and E_g are the cluster and bulk-state band gap energies, respectively; m_e and m_h are the effective mass of electrons and holes, respectively; ε is the dielectric constant of the semiconductor and R is the average particle size, while h is the Plank constant and e is the electron charge. Generally, for SnO_2 $E_g = 3.6\,\text{eV}, \varepsilon = 12$, $m_e = 0.3m_o$ and $m_h = 0.8m_o$, where m_o is the free electron mass [67, 68]. By this relation, for an energy gap of 4.24 eV as deduced from the absorption spectrum, an average particle size of 3.2 nm can be calculated. This value is in agreement with that evaluated form SEM analysis for the single nanoparticle and with the starting particle dimension.

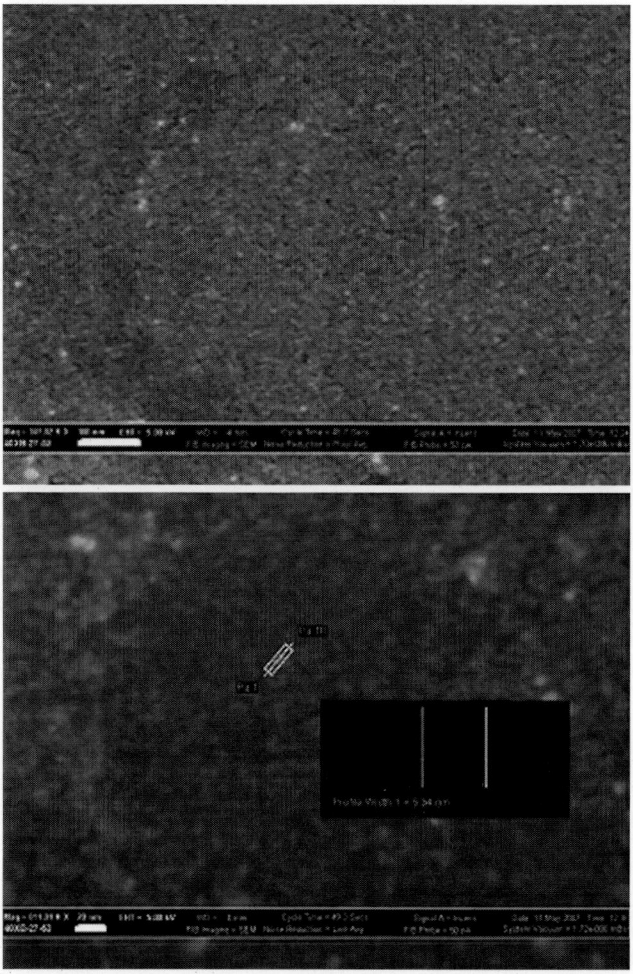

Fig. 9.17. (**a**) general overview of the annealed SnO_2 nanoparticle film morphology (the white marker is 100 nm); (**b**) higher magnification of a zone of the film (the white marker is 20 nm)

SnO_2 nanoparticle films were deposited on interdigitated alumina substrates, following the same procedure. After deposition, the devices were soldered onto a commercial TO-8 socket and hosted in a suitable test chamber. First, gas sensing tests were carried out on the as-deposited MAPLE SnO_2 films. As expected, no response to low concentration of ethanol (200 ppm in dry air) was observed, due to the presence of the nanoparticle capping layer of trioctylphosphine, until the sensor working temperature reached 673–693 K. At these temperatures, a very low and unstable variation of the sensor signal was detected. The sensor showed also a high resistance $\left(\sim 10^{10}\text{–}10^{11}\,\Omega\right)$. This behavior can be linked to the very small SnO_2 nanoparticle size (∼3.2 nm), to the small thickness of the film (about two monolayers) and to the electrical transport properties through the network of nanoparticles. It is generally agreed that the gas sensing properties of the nanoparticle film enhances with decreasing grain size in metal oxide films. But, a critical limit of a few nanometers should obviously exist. It must be noted that the understanding of the gas detection mechanism and the transduction function of the gas sensing layer is complex, and no comprehensive theory related to the electrical and gas sensing properties of such nanostructured materials is reported in literature. An improvement of the electrical properties of the MAPLE-deposited SnO_2 nanoparticle gas sensors was obtained after annealing at 873 K in air. This can probably be due to better electrical connection among the nanoparticles.

9.6 Discussion

The MAPLE technique was introduced to deposit thin and ultrathin organic, bioorganic and composite films with minimum chemical and structural modification of the target material, which occurs during traditional PLD. The goal can be obtained by dissolving the material of interest in a volatile solvent and freezing the solution, which is then used as the target for pulsed laser ablation. In the last years, many, many papers have demonstrated that with an appropriate choice of the laser wavelength, fluence, pulse duration and type of solvent, polymers, organic molecules and biomolecules can be deposited as thin films without significant modification of their chemical structure and functionality. However, despite the good results reported and the successful expansion of the technique to new applications, as for instance nanoparticle film deposition, the simple picture of the MAPLE process (the laser interaction vaporizes part of the solvent and the guest molecules receive enough kinetic energy by mechanical collisions to pass in the vapor phase) is being questioned. Recently, Leveugle and Zhigilei [43] observed that the initial picture of the ejection and transport of individual polymer molecules in MAPLE introduced by Chrisey et al. [18] cannot explain the results of high-resolution SEM and AFM imaging of many MAPLE-deposited films, where significant surface roughness and well-defined aggregates with characteristic sizes ranging from tens of nanometers to tens of micrometers are evidenced [see,e.g., 2, 69–72]. Moreover, Leveugle and Zhigilei observed that the frequent formation of large polymer features produces unexpected results when the original polymer concentration in the target is low and the polymer molecules are dissolved in the matrix down to the molecular level. Starting from these experimental evidences, Leveugle and Zhigilei formulated a computational model to get a better understanding of the relation between the basic mechanisms of laser interaction with the target material, nonequilibrium processes caused by the fast deposition of laser energy, parameters of the ejected plasma plume and the resulting morphological characteristics of the growing film. They observe that so far simulations have been focused on the analysis of matrix-assisted ejection of individual macromolecules or of several solute molecules, but at a concentration too low to allow any interaction among the molecules during the ejection process [73–75]. Since the concentrations of polymer molecules in MAPLE experiments are relatively high (typically 0.1–5 wt.%), the collective behavior of multiple polymer molecules may play an important role in defining the mechanisms of molecular ejection and the morphological characteristics of the deposited films. To take into account this collective behavior, the laser-induced molecular ejection from a MAPLE target is described by a coarse-grained molecular dynamics (MD) model combining the breathing sphere model for simulation of the molecular matrix and the bead-and-spring model for polymer molecules. The model adapts a coarse-grained representation of the molecules by spherical particles with real translational degrees of freedom, but approximate representation of the internal vibrational modes by a single internal degree of freedom. The internal degree of freedom, or breathing mode, is implemented by allowing the particles to change their sizes, and is used for simulation of molecular excitation by photon absorption and vibrational relaxation of the excited molecules. In the bead-and-spring model used to describe the polymer molecules in a MAPLE target, the "beads" representing functional groups of a polymer molecule (monomers) are connected by anharmonic springs with strengths appropriate for chemical bonding. Simulations were performed for MAPLE targets with concentrations of polymer molecules of 1, 3 and 6 wt.%, as well as for pure matrix. Computational cells with initial dimensions of 40–40–60 nm^3 (676,961 molecules in the case of pure matrix) were used in the simulations, with the polymer chains randomly and uniformly distributed in the sample. Each chain contains 100 monomer units and has the total molecular weight of 10 kDa. Time steps of 5 and 2 fs are used in the integration of the MD equations of motion in simulations performed for pure matrix targets and for polymer–solvent targets, respectively. Irradiation at a wavelength of 337 nm (3.68 eV), with pulse duration of 50 ps, is simulated. The probability of a molecule to be excited is modulated by Lambert–Beer's law to reproduce the exponential attenuation of the laser light with depth (optical penetration depth in pure matrix of 50 nm). The laser fluences $(3\text{–}9\,\mathrm{mJ\,cm^{-2}})$ are chosen to cover the range from below the ablation threshold (3.5 $\mathrm{mJ\,cm^{-2}}$ for pure matrix) up to more than twice the ablation threshold. In this irradiation regime, heat conduction does not contribute to the energy redistribution during the laser pulse and the thermal energy is largely confined within the absorbing region. The conditions of thermal confinement are also characteristic for most MAPLE experiments performed with nanosecond laser pulses. Thus, although the length and time scales of the simulations are very different from the ones in a typical MAPLE experiment, the fact that in the simulations and experiments the MAPLE process takes place under the same physical regime of thermal confinement suggests that the ejection mechanisms are similar, even though at larger time and length scales. Nevertheless, the authors emphasize that the goal of their computational study is not to reproduce quantitatively the behavior of a particular molecular system but to investigate the general characteristics of the ablation process for low-concentration polymer solutions in the thermal confinement regime.

In the ablation of MAPLE target, the presence of polymer molecules does not radically alter the general picture of the ablation process: explosive disintegration and expansion of the overheated matrix driving the ejection process and entraining the polymer molecules along. However, the presence of the polymer molecules has significant implications on the dynamics of the ablation process and quantitative parameters of the ejected plume. From computations, it turns out that the polymer molecules have a clear tendency to extend along the flow in the ablation plume. While in the simulation for pure matrix the liquid emerging from the "phase explosion" quickly transforms into spherical droplets as soon as they separate from the target, the presence of polymer chains determines the formation of complex matrix–polymer liquid structures elongated in the direction of the ablation plume expansion. As the polymer concentration increases, the chains become more entangled with the formation of intricate elongated structures, which can reach the substrate, resulting in the formation of complex surface morphology. Since in all simulations the polymer molecules are ejected only as parts of large matrix–polymer clusters that do not have enough thermal energy to evaporate during the flight to the substrate, one can expect that the growth of polymer films in MAPLE proceeds mainly through the deposition of matrix–polymer clusters. These results go against the model of the ejection and transport of individual polymer molecules in MAPLE. In all the simulations performed above the ablation threshold, the ejected plume consists of a mixture of individual matrix molecules, small matrix clusters, and larger clusters/droplets composed of both matrix molecules and polymer chains. Evaporation of the volatile matrix in flight and after the deposition on the substrate can also be responsible for the formation of the surface polymer features observed in the SEM images of MAPLE-deposited films.

Moreover, Leveugle and Zhigilei observe that MAPLE film depositions are always performed in a multipulse laser irradiation regime, and significant structural, morphological and compositional changes may accumulate in the surface region of a target irradiated by multiple laser pulses. Snapshots of the MAPLE target surfaces taken at the ends of the simulations reveal general characteristics of the new surface regions left behind by the ablation process. In addition to the formation of rough target surface morphology, simulations predict that the composition of the surface region of the target can be significantly altered by the ablation process. One can expect that the effect of the increasing polymer concentration in the target may accumulate during multipulse irradiation, especially at low laser fluences close to the ablation threshold, and for targets with low initial polymer loading. In addition to the direct effect on the mechanisms of molecular ejection, the compositional and morphological changes in the surface region can have implications on optical properties of the surface, thermodynamic properties of the laser-modified target material and even heat transfer mechanisms in the heat-affected region of the target.

It can be observed that the simulations performed by Leveugle and Zhigilei support the experimental evidence of minimal chemical modification of polymer molecules in MAPLE film deposition. Indeed, no photothermal bond scission events were detected in any of the simulations performed for polymer concentrations up to 6 wt.% and laser fluences up to more than twice the ablation threshold. But simulations also show that polymer molecules are always ejected as parts of matrix–polymer clusters with a broad cluster size distribution. The ejection of molecular clusters and droplets seems to be inherently connected to the basic mechanism of laser ablation – explosive decomposition of a surface region of the target overheated up to the limit of its thermodynamic stability. Cluster formation can decrease with decreasing of polymer concentration in the target. But, the consequence is the decrease of the efficiency of the MAPLE technique for polymer film fabrication. Also, a proper choice of the substrate temperature as a function of the polymer glass transition and polymer melting temperature can reduce the surface features [71, 76]. The choice of the solvent is also of great importance, not only to avoid chemical and photochemical reactions with the solute: solvent easily evacuated by the pumping system are not detected in the deposited films.

Possibly, the general picture polymer film deposition can be extended to biomolecular and other complex material deposition by the MAPLE technique.

9.7 Conclusions

The MAPLE technique allows deposition of thin films of organic and biological materials with minimum modification of the chemical structure and functionality of the deposited molecules. It is quite easy, with a proper choice of the deposition parameters, to avoid photochemical and photothermal molecular fragmentation and to deposit thin films with thickness control and surface coverage that cannot be achieved by solvent-based coating methods. In contrast, the fabrication of very smooth films appears much more difficult to be obtained. High-resolution imaging and computer simulations seem to contest the original simple model of molecule-by-molecule deposition. Aggregation of polymers and presence of residual matrix molecules result in local corrugations of the deposits. The roughness of the growing films can be, at least partially, controlled by limiting the laser fluence and the solute concentration and controlling the substrate temperature. In any case, MAPLE has opened very perspective roads for the deposition of performing films of complex organic and biological molecules, and the area of MAPLE applications is rapidly expanding. For instance, it has been demonstrated that the MAPLE technique is very promising for the deposition of metal oxide nanoparticles for gas sensing applications. TiO_2 and SnO_2 nanoparticle films were deposited on flat and rough substrates, starting from colloidal solutions. A uniform coverage on both kinds of substrates was obtained, and the crystalline phase and nanoparticle dimensions were preserved.

Acknowledgments

The authors acknowledge the NATO CLG 982748 grant support for the sensor investigations.

References

1. P.K. Wu, B.R. Ringeisen, D.B. Krizman, C.G. Frondoza, M. Brooks, D.M. Bubb, R.C.Y. Auyeung, A. Piqué, B. Spargo, R.A. McGill, D.B. Chrisey, Rev. Sci. Instrum. **74**, 2546 (2003)
2. R. Frycek, M. Jelınek, T. Kocourek, P. Fitl, M. Vrnata, V. Myslık, M. Vrbova, Thin Solid Films **495**, 308 (2006)
3. A.W. Grice, D.C. Bradley, M.T. Bernius, M. Inbasekaran, P.K. Wu, E.P. Woo, Appl. Phys. Lett. **73**, 629 (1998)
4. M.T. Bernius, M. Inbasekaran, J. O'Brien, W.S. Wu, Adv. Mater. **12**, 737 (2000)
5. D. Bäuerle, *Laser Processing and Chemistry* (Springer, Berlin, 2000)
6. D.B. Chrisey, G.K. Hubler (eds) *Pulsed Laser Deposition of Thin Films* (Wiley, New York, 1994)
7. R. Eason, *Pulsed Laser Deposition of Thin Films: Applications-Led Growth of Functional Materials* (Wiley, New York, 2006)
8. S.C.K. Misra, M.K. Ram, S.S. Pandey, B.D. Malhotra, A. Chandra, Appl. Phys. Lett. **60**, 1219 (1992)
9. L.E. Scriven, Mater. Res. Soc. **121**, 717 (1988)
10. S. Sun, P. Ho-Si, D.J. Harrison, Langmuir **7**, 727 (1991)
11. I. Fujiwara, M. Ohnishi, J. Seto, Langmuir **8**, 2219 (1992)
12. G.B. Blanchet, S.I. Shah, Appl. Phys. Lett. **62**, 1026 (1993)
13. G.B. Blanchet, Macromolecules **28**, 4603 (1995)
14. M. Yudasaka, Y. Tasaka, M. Tanaka, H. Kamo, Y. Ohki, S. Usami, S. Yoshimura, Appl. Phys. Lett. **64**, 3237 (1994)
15. R.A. McGill, D.B. Chrisey, Patent No. 6025036 (2000)
16. A. Piqué, R.A. McGill, D.B. Chrisey, D. Leonhardt, T.E. Mslna, B.J. Spargo, J.H. Callahan, R.W. Vachet, R. Chung, M.A. Bucaro, Thin Solid Films **355–356**, 536 (1999)
17. A. Piqué, P. Wu, B.R. Ringeisen, D.M. Bubb, J.S. Melinger, R.A. McGill, D.B. Chrisey, Appl. Surf. Sci. **186**, 408 (2002)
18. D.B. Chrisey, A. Piqué, R.A. McGill, J.S. Horwitz, B.R. Ringeisen, D.M. Bubb, P.K. Wu, Chem. Rev. **103**, 553 (2003)
19. M. Karas, D. Bachman, U. Bahr, F. Hillenkamp. Int. J. Mass Spectrom. Ion Process. **78**, 53 (1987)
20. S.G. Hansen, T.E. Robitaille, J. Appl. Phys. **64**, 2122 (1988)

21. G.B. Blanchet, C.R. Fincher, C.L. Jackson, S.L. Shah, K.H. Gardner, Science **262**, 719 (1993)
22. R. Schwodiauer, S.B. Gogonea, S. Bauer, J. Heitz, E. Arenholz, D. Bauerle, Appl. Phys. Lett. **73**, 2941 (1998)
23. M. Yudasaka, Y. Tasaka, M. Tanaka, H. Kamo, Y. Ohki, S. Usami, S. Yoshimura, Appl. Phys. Lett. **64**, 3237 (1994)
24. S. Nishio, T. Chiba, A. Matsuzaki, H. Sato, J. Appl. Phys. **79**, 7198 (1996)
25. R.A. McGill, R. Chung, D.B. Chrisey, P.C. Dorsey, P. Matthews, A. Piqué, T.E. Mlsna, J.I. Stepnowski, IEEE Trans. Ultrason. Ferroelectr. **45**, 1370 (1998)
26. F. Bloisi, A. Cassinese, R. Papa, L. Vicari, V. Califano, Thin Solid Films **516**, 1594 (2008)
27. T. Tunno, A.P. Caricato, M.E. Caruso, A. Luches, M. Martino, F. Romano, D. Valerini, Appl. Surf. Sci. **253**, 6461 (2007)
28. A.P. Caricato, M. Lomascolo, A. Luches, F. Mandoj, M.G. Manera, M. Mastroianni, M. Martino, R. Paolesse, R. Rella, F. Romano, T. Tunno, D. Valerini, Appl. Phys. A **93**, 651 (2008)
29. G. Heliotis, R. Xia, G.A. Turnbull, A. Piers, W.L. Barnes, I.D.W. Samuel, D.D.C. Bradley, Adv. Funct. Mater. **14**, 91 (2004)
30. E.M. Calzado, J.M. Villalvilla, P.G. Boj, J.A. Quintana, M.A. Díaz-García, J. Appl. Phys. **97**, 093103 (2005)
31. D.M. Bubb, P.K. Wub, J.S. Horwitz, J.H. Callahan, M. Galicia, A. Vertes, R.A. McGill, E.J. Houser, B.R. Ringeisen, D.B. Chrisey, J. Appl. Phys. **91**, 2055 (2002)
32. M. Ariu, M. Sims, M.D. Rahn, J. Hill, A.M. Fox, D.G. Lidzey, M. Oda, J. Cabanillas-Gonzales, D.C. Bradley, Phys. Rev. B **67**, 195333 (2003)
33. M. Sims, D.D.C. Bradley, M. Ariu, M. Koeberg, A. Asimakis, M. Grell, D.G. Lidzey, Adv. Funct. Mater. **14**, 765 (2004)
34. M.E. Caruso, S. Lattante, R. Cingolani, M. Anni, Appl. Phys. Lett. **88**, 181906 (2006)
35. A.L. Mercado, C.E. Almond, J.G. Hoekstra, J.M. Fitz-Gerald, Appl. Phys. A **81**, 591 (2006)
36. T.T. Khan, P. Sreerunothai, L.M. Hert, M.J. Banach, A. Kohler, Phys. Rev. B **69**, 085201 (2004)
37. A. Mahammed, Z. Gross, J. Am. Chem. Soc. **127**, 2883 (2005)
38. J. Radecki, I. Stenka, E. Dolusic, W. Dehaen, Electrochim. Acta **51**, 2282 (2006)
39. T. Ding, E.A. Alemán, D.A. Mordarelle, C.J. Ziegler, J. Phys. Chem. A **109**, 7411 (2005)
40. S. Nardis, F. Mandoj, R. Paolesse, F.R. Fronczek, K.M. Smith, L. Prodi, M. Montalti, G. Battistini, Eur. J. Inorg. Chem. **16**, 2345 (2007)
41. E.J. Houser, D.B. Chrisey, M. Bercu, N.D. Scarisoreanu, A. Purice, D. Colceag, C. Constantinescu, A. Moldovan, M. Dinescu, Appl. Surf. Sci. **252**, 4871 (2006)
42. B. Toftmann, K. Rodrigo, J. Schou, R. Pedrys, Appl. Surf. Sci. **247**, 211 (2005)
43. E. Leveugle, L.V. Zhigilei, J. Appl. Phys. **102**, 074914 (2007)
44. B.R. Ringeisen, J. Callahan, P.K. Wu, A. Piqué, B. Spargo, R.A. McGill, M. Bucaro, H. Kim, D.M. Bubb, D.B. Chrisey, Langmuir **17**, 3472 (2001)
45. T. Tunno, PhD Thesis (2007)
46. J. Sagawa, S. Nagare, M. Senna, Appl. Surf. Sci. **244**, 611 (2005)
47. G. Shanmugam, P. Polavarapu, Biophys. Chem. **111**, 73 (2004)
48. U.K. Laemmli, Nature **227**, 680 (1970)

49. K.W. Kolasinski, *Surface Science: Foundations of Catalysis and Nanoscience* (Wiley-Blackwell, New York, 2008)
50. Y. Wu, M. Takeguchi, K. Furuya, Jpn. J. Appl. Phys. **38**, 7241 (1999)
51. Y. Sun, T. Egawa, C. Shao, L. Zhang, X. Yao, Jpn. J. Appl. Phys. **43**, 3544 (2004)
52. H.M. Lam, M.H. Hong, S. Yuan, T.C. Chong, Appl. Phys. A **79**, 2099 (2004)
53. S. Amoruso, R. Bruzzese, N. Spinelli, R. Velotta, M. Vitiello, X. Wang, G. Ausanio, V. Iannotti, L. Lanotte, Appl. Phys. Lett. **84**, 4502 (2004)
54. M. Epifani, J. Arbiol, R. Díaz, M.J. Perálvarez, P. Siciliano, R. Rella, Chem. Mater. **17**, 6468 (2005)
55. Y.J. Kim, Y.S. Kim, S.Y. Chang, D.H. Cha, Y.S. Choi, W.I. Lee, New J. Chem. **31**, 260 (2007)
56. P.K. Wu, J. Fitz-Gerald, A. Piqué, D.B. Chrisey, R.A. McGill, Mater. Res. Soc. Symp. Proc. **617**, J2.3 (2000)
57. A.P. Caricato, M. Catalano, G. Ciccarella, M. Martino, R. Rella, F. Romano, J. Spadavecchia, A. Taurino, T. Tunno, D. Valerini, Dig J. Nanomater. Bios. **1**, 43 (2006)
58. A.P. Caricato, S. Capone, M. Epifani, M. Lomascolo, A. Luches, M. Martino, F. Romano, R. Rella, P. Siciliano, J. Spadavecchia, T. Tunno, D. Valerini, Proc. SPIE – The International Society for Optical Engineering **6985**, 69850H (2008)
59. M.H. Madhushudana Reddy, A.N. Chandorkar, Sens. Actuators B **9**, 1 (1992)
60. D.S. Vlachos, A.C. Xenoulis, NanoStruct. Mater. **10**, 1355 (1998)
61. N. Pinna, G. Neri, M. Antonietti, M. Niederberger, Angew. Chem. Int. Ed. **43**, 4345 (2004)
62. G. Burns, *Solid State Physics* (Academic, NewYork, 1985)
63. J.P.G. Coutinho, M.T.C.M. Barbosa, J. Fluoresc. **16**, 387 (2006)
64. F. Gu, S.F. Wang, C.F. Song, M.K. Lu, Y.X. Qi, G.J. Zhou, D. Xu, D.R. Yuan, Chem. Phys. Lett. **372**, 451 (2003)
65. Y. Suda, H. Kawasaki, J. Namba, K. Iwatsuji, K. Doi, K. Wada, Surf. Coat. Technol. **174–175**, 1293 (2003)
66. E.J.H. Lee, C. Ribeiro, T.R. Giraldi, E. Longo, E.R. Leite, J.A. Varela, Appl. Phys. Lett. **84**, 1745 (2004)
67. S. Das, S. Kar, S. Chaudhuri, J. Appl. Phys. **99**, 114303 (2006)
68. L.E. Brus, J. Chem. Phys. **80**, 4403 (1984)
69. R. Cristescu, D. Mihaiescu, G. Socol, I. Stamatin, I.N. Mihailescu, D.B. Chrisey, Appl. Phys. Mater. Sci. Process. **79**, 1023 (2004)
70. A.T. Sellinger, E.M. Leveugle, K. Gogick, L.V. Zhigilei, J.M. Fitz-Gerald, J. Vac. Sci. Technol. A **24**, 1618 (2006)
71. K. Rodrigo, P. Czuba, B. Toftmann, J. Schou, R. Pedrys, Appl. Surf. Sci. **252**, 4824 (2006)
72. A.T. Sellinger, E. Leveugle, K. Gogick, G. Peman, L.V. Zhigilei, J.M. Fitz-Gerald, J. Phys. Conf. Ser. **59**, 314 (2007)
73. Y. Dou, N. Winograd, B.J. Garrison, L.V. Zhigilei, J. Phys. Chem. B **107**, 2362 (2003)
74. T.E. Itina, L.V. Zhigilei, B.J. Garrison, Nucl. Instrum. Methods Phys. Res. B **180**, 238 (2001)
75. T.E. Itina, L.V. Zhigilei, B.J. Garrison, J. Phys. Chem. B **106**, 303 (2002)
76. A. Gutiérrez-Llorente, G. Horowitz, R. Pérez-Casero, J. Perrière, J.L. Fave, A. Yassar, C. Sant, Org. Electron. **5**, 29 (2004)

10

Advanced Biomimetic Implants Based on Nanostructured Coatings Synthesized by Pulsed Laser Technologies

Ion N. Mihailescu, Carmen Ristoscu, Adriana Bigi, and Isaac Mayer

Summary. Calcium phosphates (CaPs) are alternative substitutes for human bones and so primary candidates for the manufacture of medical implants. Unfortunately, they do not withstand stress in bulk. To overcome this obstacle, a solution was developed to cover metallic implants with functional biomimetic layers.

Pulsed laser deposition (PLD) has proved to be a competitive method to grow high-quality biomaterial thin films. Nevertheless, in case of very complex, delicate biomolecules (such as organic and biopolymeric materials), PLD provokes an irreversible damage of the chemical bonds and thus the compositional change in the deposited film. This disadvantage is eliminated by matrix-assisted pulsed laser evaporation (MAPLE), capable of transferring large-molecular-mass compounds at very low temperatures.

In this chapter, we review the potential of PLD applied to different types of simple and/or doped CaP coatings. We also scrutinize the extension of MAPLE to the synthesis of hybrid organic–inorganic bionanocomposites for advanced biomimetic implants. Expected development and progress of the new in vitro and in vivo studies are discussed.

10.1 Introduction

In history, biomaterials date back thousands of years. Bones of skeletons or mummies from ancient cultures (Egypt, Babylon, Greece, Italy, Central and South America) prove that different materials such as gold, silver, copper, lead, wood, nacre, ivory, dog bones and teeth served to replace parts of the human bone system. They were apparently used without much knowledge of, or concern for, biocompatibility, a term which was defined as late as only 50 years ago. The current definition widely accepted by the international scientific community is that biomaterials are those materials, different from drugs, that can be used to treat, improve or replace whatever tissue, organ or body function.

Bone substitutions at high load-bearing sites, as in the case of hip and femoral components, bone plates and dental prosthesis, require biomaterials with reliable strength, high toughness and resistance to wear and corrosion [1]. Metals such as titanium and its alloys or cobalt alloys offer both biocompatibility and suitable mechanical characteristics. In particular, titanium and its alloys are extensively employed thanks to their high tensile strength associated with ductility, toughness and lightness, resulting in a high strength-to-weight ratio. Nonetheless, bone implant failures, which are mainly ascribed to low and incomplete osteointegration, and to stress shielding, occur frequently [2]. The integration with bone tissue can be improved and accelerated by the presence of a calcium phosphate (CaP) coating on the metal implant surface [3]. CaP materials are bioactive and osteoconductive and promote direct attachment to bone [4–6]. Calcium hydroxyapatite (HA) and other CaP coatings have been extensively applied with the aim to improve adhesion between hard tissue and metal implants, and to combine the mechanical advantages of the metal with the excellent bioactivity of CaP [6]. To this aim, a variety of physical and chemical methods, including plasma spray, pulsed laser technologies, sputtering, electrodeposition, anodic deposition and anodic spark deposition, sol–gel dipping and biomimetic deposition, have been developed [7–12].

Pulsed laser deposition (PLD) has proved to be a competitive method to grow high-quality biomaterial thin films, as it preserves the complex stoichiometry of deposited compounds [13]. Nevertheless, in case of the very complex delicate biomolecules, PLD provokes an irreversible damage of the chemical bonds and thus compositional changes in the deposited film [14]. This is eliminated by the deposition technique called matrix-assisted pulsed laser evaporation (MAPLE), which is capable of transferring large-molecular-mass compounds at very low temperatures [15].

We review in this chapter the capabilities of PLD as applied to different types of simple CaP coatings, carbonated and/or doped with divalent metallic ions or drugs. We also examine the recent extension of MAPLE to the synthesis of hybrid organic–inorganic bionanocomposites for advanced biomimetic implants.

10.1.1 Pulsed Laser Deposition Technologies

PLD has demonstrated to be a versatile technique for thin film processing with a high diversity of structural and morphological characteristics. Many independent parameters can be changed under control in order to select the optimum deposition regimes of some specific structures and thin films [14,16,18].

Growing thin films by PLD has numerous advantages, some of which are given below:

1. The laser source is placed outside the deposition chamber offering increased flexibility in handling the material, laying out the geometrical setup and adjusting deposition parameters.
2. Most of the solid and liquid materials can be ablated.
3. Laser pulses enable the control of the growth rate of the coatings very accurately (down to a few fractions of an angstrom).

4. Most of the ablated material is located inside the plasma generated under laser pulse action.
5. The stoichiometry of coating materials generally coincides with that of the target even for complex, highly unstable compounds.
6. The coatings adhere well because of the high plasma energy incident on the deposition substrate.
7. Species with metastable or nonequilibrium states and new phases can be synthesized.

The general PLD layout (Fig. 10.1) used in our experiments of CaP thin film synthesis is as follows. The laser beam generated by a pulsed UV excimer laser source enters the reaction chamber through a quartz window. In the studies presented in this chapter, we used KrF* excimer laser sources generating pulses of 7.4 or 25 ns duration at 248 nm. The beam is focused onto target surface by means of an (Anti-reflective) AR cylindrical lens situated outside the deposition chamber. It hits the target surface at oblique incidence. The target is rotated and translated during multipulse irradiation to avoid piercing. A temperature controller is used to monitor the heating and cooling of the substrate.

The reaction chamber is initially evacuated down to a residual pressure of 10^{-4} Pa. The dynamic pressure during experiments is kept constant using a flow controller.

The basic physical processes involved in PLD and other examples of nanosecond PLD applications are presented in Chap. 5 of this book.

MAPLE was developed as a complementary method to PLD and is used for delicate (organic or biologic) material deposition. MAPLE essentially differs from PLD by the target preparation, laser–material interaction and transfer mechanisms (Fig. 10.2). It provides a more gentle mechanism for transferring many different compounds, including large-molecular-weight species, such as

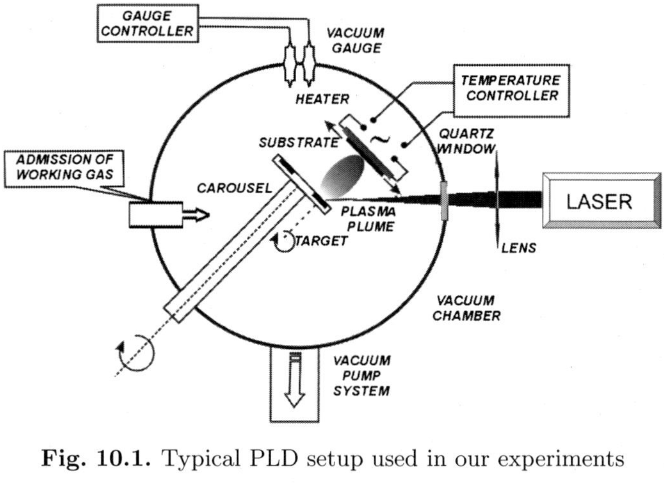

Fig. 10.1. Typical PLD setup used in our experiments

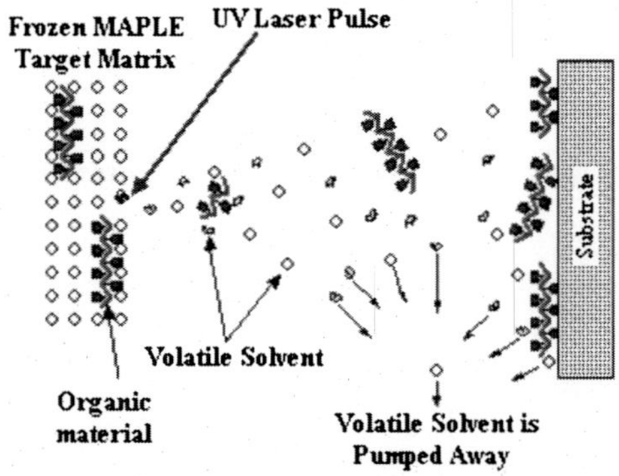

Fig. 10.2. Schematic of the MAPLE evaporation process

polymeric molecules. Specific to MAPLE is the use of a cryogenic composite target, a dilute mixture of the organic material to be deposited and a light-absorbent, high vapor-pressure solvent matrix.

Ideally, the incident laser pulse used for MAPLE initiates two photothermal processes in the matrix: evaporation of the frozen composite target and release of the organic material into the chamber. Solvent molecules are evaporated, and evacuated by the pumping system. The organic molecules attain sufficient kinetic energy through collective collisions with the evaporating solvent molecules to be transferred in gas phase to the substrate. By optimization of the MAPLE deposition conditions (laser wavelength, repetition rate, fluence, solvent type, concentration, temperature, background gas and pressure), the process can proceed without significant organic material decomposition [14, 15, 19]. Most of the laser energy is absorbed by the solvent molecules and not by the fragile solute.

The collective action of the rapidly evaporating volatile solvent acts to softly desorb the fragile solute through soft collisions and deposits the solute as a uniform thin film whose material properties, such as chemical structure and functionality, are preserved. The sticking coefficient of the solvent to the substrate is nearly zero, and the evaporated solvent is rapidly pumped away [15].

The solvent and concentration should be selected so that the following conditions are met:

1. The solute, in case of organic material, can dissolve to form a dilute, particulate-free solution.
2. Most of the laser energy has to be absorbed by the solvent molecules rather than the solute molecules.
3. There is no photochemical reaction between the solvent and the solute.

A more detailed presentation of MAPLE mechanisms and more results are reviewed in Chap. 9 of this book.

10.1.2 Calcium Phosphates

The successful performance of biomaterials for bone tissue repair is closely related to their ability to create a bond with the host living tissue. This is achieved by the formation of a biologically active bone-like apatite layer on the surface of the material in contact with physiological fluids [20, 21]. CaPs (Table 10.1) are vital biological compounds [22,23]. They offer a suitable interface, have unlimited availability and display excellent biocompatibility and osteoconductivity [24, 25]. CaPs display an almost ubiquitous presence in the body of vertebrates, and are the main inorganic components of physiologically and pathologically mineralized tissues, such as bone and teeth.

From a crystallographic point of view, HA ($Ca_{10}(PO_4)_6(OH)_2$) is more similar to natural bone tissue apatite than other CaPs, and it has long been considered as the best structural material for bone growth. However, biological apatite differs from synthetic stoichiometric HA in several aspects. The inorganic phase of hard tissues, such as bone and teeth, is a carbonated apatite with small crystal dimensions, low degree of crystallinity and non-stoichiometry, due to the presence of a number of foreign

Table 10.1. Biologically relevant calcium phosphates

Compound	Abbreviation	Chemical formula	Ca/P	Solubility at 37°C $-\log(K_{ps})$ [25]
Monocalcium phosphate monohydrate	MCPM	$Ca(H_2PO_4)_2 \cdot H_2O$	0.5	–
Monocalcium phosphate anhydrous	MCPA	$Ca(H_2PO_4)_2$	0.5	–
Dicalcium phosphate dihydrate	DCPD	$CaHPO_4 \cdot 2H_2O$	1.0	6.63
Dicalcium phosphate anhydrous	DCPA	$CaHPO_4$	1.0	7.02
Octacalcium phosphate	OCP	$Ca_8H_2(PO_4)_6 \cdot 5\,H_2O$	1.33	95.9
α-Tricalcium phosphate	α-TCP	α-$Ca_3(PO_4)_2$	1.5	25.5
β-Tricalcium phosphate	β-TCP	β-$Ca_3(PO_4)_2$	1.5	29.5
Hydroxyapatite	HA	$Ca_{10}(PO_4)_6(OH)_2$	1.67	117.2
Tetracalcium phosphate	TTCP	$Ca_4(PO_4)_2O$	2.0	37–42

ions. In fact, the deposition of the inorganic crystals in mineralized tissues occurs in an environment rich in a variety of ions and in presence of acidic macromolecules, which control nucleation, polymorphism, growth, chemical composition, shape, dimensions, orientation and texture of the crystals [26]. A further possible disadvantage of stoichiometric HA is its extremely slow resorption rate [27]. According to the biomimetic approach, a material aimed at repairing the skeletal system must resemble the biological one in composition, structure, morphology and functionality. This explains the increasing interest in nanocrystalline apatite, apatites enriched with biologically active ions and molecules, as well as in more resorbable CaPs, such as α- and β-tricalcium phosphate (α-TCP; β-TCP), octacalcium phosphate (OCP) and dicalcium phosphate dihydrate (DCDP), which can easily transform into nanocrystalline apatite [25, 28–30].

10.2 HA Coatings

HA is widely used in dentistry and orthopedics to repair bone defects and as coating material for metallic implants. It is considered as one of the best biocompatible coatings, ensuring a highly resorbable surface in contact with bone, while stable in contact with the metallic part of the implant. Synthetic HA is therefore successfully used to prepare thin films to cover the metallic part of the implant. Chemically, HA can be characterized by its Ca/P molar ratio, which is 1.67 for the stoichiometric compound. However, nonstochiometric HA, with a Ca/P ratio lower than 1.67, is more similar to biological apatite and, as a consequence, is more suitable for coating purposes.

Generally, HA compared to other CaPs is most insoluble above pH 4.2 (Table 10.1 in Sect. 10.1.2) and under normal physiological conditions of pH = 7.2 and is the only stable CaP compound. The solubility of HA under physiological conditions is influenced by the Ca/P ratio, the crystallinity of HA powders, particle size and porosity [20, 31].

One of the classical synthetic methods for preparation of HA proceeds by a neutralization reaction. An aqueous solution of phosphoric acid is dropped into a calcium hydroxide suspension at 85°C.

$$5\mathrm{Ca}(\mathrm{OH})_2 + 3\mathrm{H_3PO_4} \rightarrow \mathrm{Ca_5}(\mathrm{PO_4})_3\mathrm{OH} + 9\mathrm{H_2O} \tag{10.1}$$

The precipitation product has to be aged for 24 h, washed and filtered. Samples are obtained with a Ca/P ratio close to 1.67 but not very well crystallized.

In a different approach, in order to obtain well-crystallized samples, a phosphate solution [3.7 g $(\mathrm{NH_4})_2\mathrm{HPO_4}$ in 200 ml triple-distilled water (TDW)] is added slowly by dropping from a separation funnel, during 2 h, to a calcium solution [9.47 g $\mathrm{Ca}(\mathrm{NO_3})_2 \cdot 4\mathrm{H_2O}$ in 200 ml TDW].

$$\begin{aligned}(\mathrm{NH_4})_2\mathrm{HPO_4}(\ell) + {} & \mathrm{Ca}(\mathrm{NO_3})_2(\ell) + \mathrm{OH^-} \rightarrow \mathrm{Ca}_{10-x}(\mathrm{HPO_4})_x \\ & (\mathrm{PO_4})_{6-x}\mathrm{OH}_{2-x}(\mathrm{s}) + \mathrm{NH_4^+} + \mathrm{NO_3^-} + \mathrm{H^+}\end{aligned} \tag{10.2}$$

The pH range for the preparation is 6.0–11.0 and has to be maintained constant during precipitation by adding NaOH solution. The precipitation is carried out during 2 h at temperature of 85°C. Next, this temperature is raised to boiling point and the system refluxed for 2 h. The sample is then washed with TDW and dried overnight in air at 120°C.

Determination of the Ca and P content of the samples can be performed by inductive coupled plasma atomic emission spectroscopy (ICP-AES) or by other analytical methods.

HA crystallizes in a hexagonal unit cell and can be identified by its characteristic powder X-ray diffraction (XRD) pattern [32]. A typical pattern of HA is shown in Fig. 10.3. The presence of crystalline impurities can be detected by additional peaks on the pattern. The XRD pattern enables also to get information about the crystallinity of the sample.

The infrared (IR) spectrum of an HA sample is given in Fig. 10.4. It shows the IR bands of the functional groups of HA, phosphate (564, 603, 961, 1035 and 1087 cm^{-1}), OH^- (3572 cm^{-1}), carbonate (1420 and 1454 cm^{-1}) and two broader bands of water (1641 and 3432 cm^{-1}), often parts of HA, appear also on the spectrum.

Plasma spraying was the first and the only commercial method used for the synthesis of HA coatings. There are still several drawbacks of plasma-sprayed coatings related to the presence of other CaP phases, porosity and poor mechanical adhesion to the substrate. PLD has proved to be a competitive and challenging technique for the production of high-quality HA thin films.

To obtain crystalline HA films, water vapor atmosphere was mandatory because of the hydrated nature of HA. Another key parameter in PLD is the substrate temperature. It was found that depositing at temperatures less than 400°C resulted in amorphous films independent of the environment (Fig. 10.5). At temperatures above 400°C, films were either crystalline or amorphous, depending on the deposition environment.

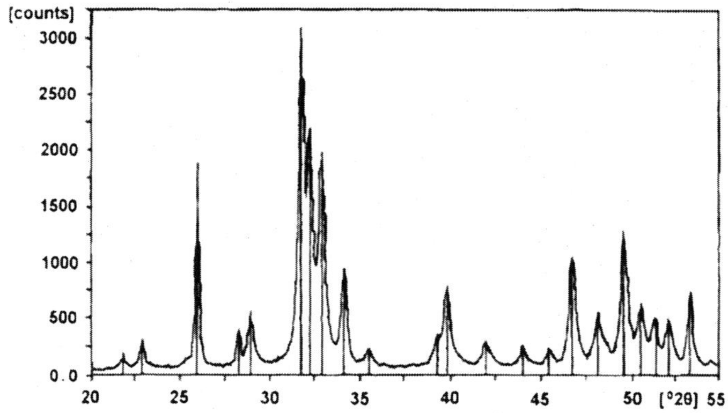

Fig. 10.3. Powder X-ray diffraction pattern of calcium hydroxyapatite

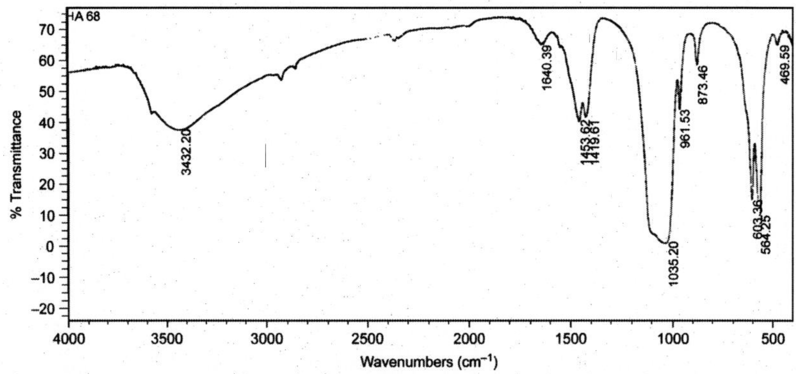

Fig. 10.4. IR spectrum of carbonate-containing calcium hydroxyapatite

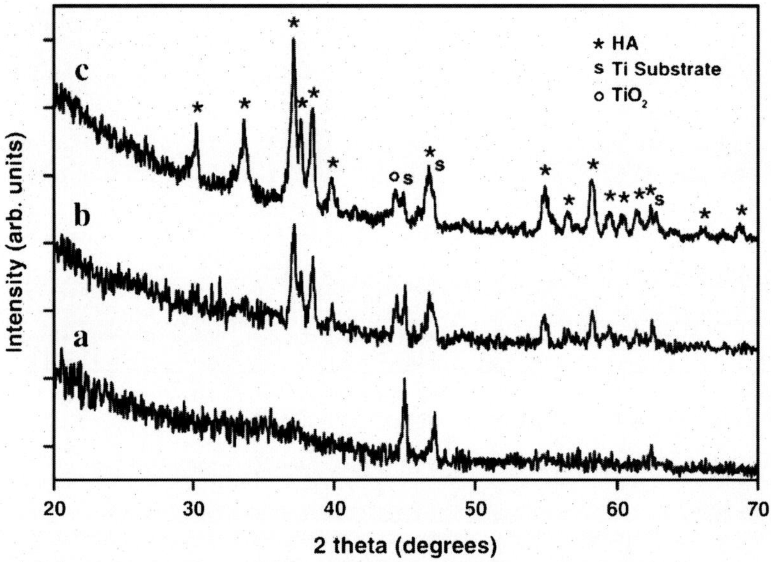

Fig. 10.5. XRD patterns of HA coatings deposited at 350°C (**a**), 400°C (**b**), and 450°C (**c**). Reproduced with permission from Materials Science and Engineering C **27**, 105 (2007)

Simple tuning of PLD parameters can deposit HA in situ in crystalline form [13], amorphous films [33] or mixture of nanocrystals and amorphous films.

Because composition and stoichiometry of the obtained nanostructures determine the resorbability rate, PLD has proved to be efficient for in situ synthesis of bioactive layers and multilayers with controlled response of biointegrability.

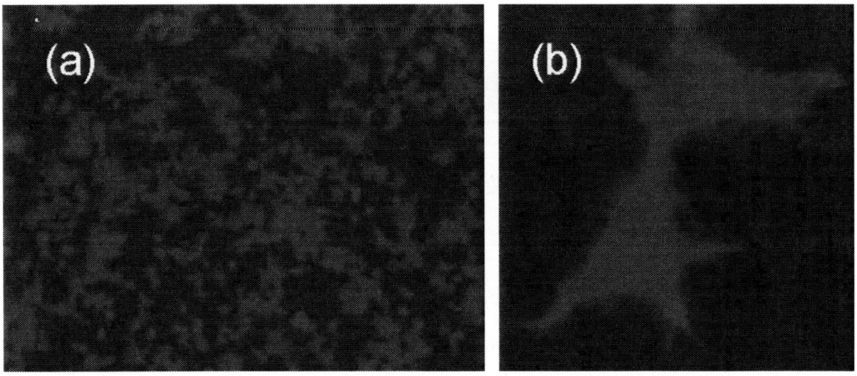

Fig. 10.6. Fluorescence microscopy images [magnifications 10× (a) and 40× (b)] of Hek293 cells cultivated on HA coatings. Cells were marked with phalloidin conjugated with Alexa Fluor 594 which stains polymerized actin filaments. Reproduced with permission from Gyorgy et al. Applied Surface Science **253**, 7981 (2007)

For understanding the biological events at the bone tissue/material interface, dedicated investigations have been performed. Hek293 (human embryonic kidney) cells were used for in vitro tests (Fig. 10.6).

Fluorescence microscopy examinations have confirmed normal morphology of the cells cultivated on HA and validated this material as an efficient surface for cell adherence. The examination of cell distribution on the surface has revealed the uniform spreading of cells grown on HA films.

10.3 Octacalcium Phosphate

The biological relevance of octacalcium phosphate, $Ca_8H_2(PO_4)_6 \cdot 5H_2O$ (OCP), is second only to that of HA, due to its possible precursor role during the deposition of carbonated apatite in the hard tissues of vertebrates. The triclinic structure of OCP displays remarkable similarities with the hexagonal structure of HA, and this phase often occurs as a transient intermediate in the precipitation of the thermodynamically more stable HA [34]. The unit cell of OCP is described as a periodic array of "apatitic layers," where Ca^{2+} and PO_4^{3-} ions occupy the same positions as in HA, and of "hydrated layers," where Ca^{2+} and PO_4^{3-} ions are more widely spaced, due to the presence of the interdispersed structural water molecules [34, 35]. Hydrated layers, about 0.8-nm thick, alternate with apatitic layers, about 1.1-nm thick, parallel to the wide (100) face.

The possible advantages of OCP with respect to crystalline HA for biomedical applications include its greater solubility and its easy hydrolysis into nanocrystalline apatite in physiological solutions [36–38]. Indeed, OCP has been demonstrated to enhance bone formation, and its conversion into apatite has been suggested to be involved in this stimulatory effect of OCP [39, 40]. OCP can be synthesized in an aqueous medium by the reaction of calcium acetate with sodium phosphate under stirring at 60°C, at a starting pH of 5.0 [38]. Due to its complex stoichiometry and structure, OCP films are quite difficult to grow by means of physical methods, and coatings were mostly deposited from supersaturated solutions and through electrochemical methods [41, 42].

We deposited OCP thin films by means of pulsed laser ablation with a pulsed UV laser source in a flux of hot water vapors. The resulting structures were submitted to heat treatment in hot water vapors for up to 6 h. The best results were obtained at a substrate temperature of 150°C during both deposition and postdeposition treatment [10]. The X-ray patterns of the coatings indicate an amorphous or poorly crystalline structure. However, OCP could be identified by the shoulder at about 4.7°, corresponding to the 100 reflection, and by the broad peak centered around 32–33° (Fig. 10.7). Accordingly, nano- and microcrystalline domains, which are dispersed throughout the layer, can be appreciated in the cross-section dark-field transmission electron microscopy (TEM) pictures of the coatings (Fig. 10.8c). The corresponding selected area electron diffraction (SAED), consistent with the structure of OCP, is given in Fig. 10.8a. The deposited material coalesces and grows perpendicular to the substrate in a tree-like structure, containing droplets of micrometric size (Fig. 10.8b), similar to a coral reef, resulting in porous films [10,43]. High-resolution TEM observation confirmed the presence of crystalline nanodomains interdispersed in the amorphous phase.

The results of in vitro tests performed with the human fetal osteoblast-like (hFOB) cell line 1.19 and murine fibroblast cell line L929 indicated that both fibroblasts and osteoblasts adhere, reach a normal morphology, proliferate and

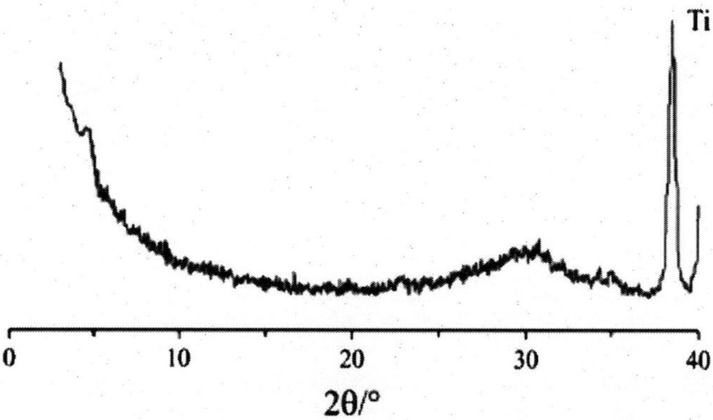

Fig. 10.7. XRD pattern of the OCP coating deposited by PLD on Ti substrate heated at 150°C and subjected to post-deposition treatment at 150°C. Both deposition and treatment were conducted in a flux of water vapors. Reproduced with permission from Socol et al., Biomaterials **25**, 2539 (2004)

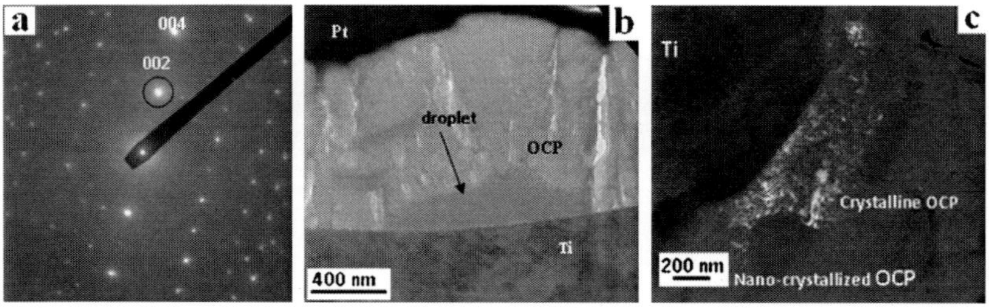

Fig. 10.8. (**a**) SAED pattern of the OCP sample, (**b**) global XTEM image and (**c**) dark field (DF) image. (**a**), (**b**) and (**c**) show the samples deposited on Ti substrate heated at 150°C and subjected to post-deposition treatment at 150°C.

remain viable when cultured on OCP coatings, supporting good biocompatibility and without any toxicity. Further tests were carried out using human primary osteoblasts (hOBs), which were cultured on OCP thin films for up to 21 days in order to evaluate cell attachment, proliferation and differentiation. The results showed that hOBs are able to adhere, remain attached and viable and proliferate on the thin films. Moreover, OCP coatings activate osteoblast metabolism and differentiation, as shown by the increased values of alkaline phosphatase (ALP) activity and transforming growth factor ß1 (TGF-ß1) level [44].

10.4 Carbonated HA and ß-TCP Doped with Mn^{2+} Coatings

In order to make synthetic HA comparable to natural bone tissues, cationic or anionic substituents have been added. Among the anionic substituents, the CO_3^{2-} content of natural apatites is ~4% in teeth and ~6% in bones. The position in which the carbonate group enters into the crystalline structure of HA is variable: with higher content B-type (CO_3^{2-} substituting in PO_4^{3-} sites) or A-type (CO_3^{2-} substituting in OH^- sites) usually in young bones.

The doping with divalent ions such as Mg, Mn, Sr and Si substituting Ca in HA can have beneficial effects on the behavior of HA in implants [45–48]. In particular, the addition of Mn^{2+} ions was found to be related to the activation of integrins, a family of receptors that mediate cellular interactions with extracellular matrix and cell surface ligands. In the presence of Mn^{2+} ions, the ligand affinity of integrin increases and cell adhesion is promoted [49].

10.4.1 Carbonated HA Doped with Mn^{2+}

The composition and morphology of carbonate apatites (CHAs) was shown to depend on preparation parameters such as pH, CO_3^{2-} concentration, excess of

PO_4^{3-} and Ca^{2+}. Thus, the crystallinity of CHA is lowered with the carbonate content [50].

CHA samples doped with Mn^{2+} ions (Mn:CHA) were prepared by precipitation, according to reaction (10.2) by adding Mn^{2+} from a $Mn(NO_3)_2$ solution at pH of 6.0, in order to avoid oxidation of Mn^{2+}. Samples without or with carbonate up to 2%, after heating up to 800°C, transform partially to β-TCP. The samples with carbonate content equal to or higher than 4% do not decompose at all during thermal treatment, but their color turns to blue above 600°C. There is spectroscopic evidence that $Mn^VO_4^{3-}$ ions – generated by oxidation of Mn^{2+} and presumably substituting CO_3^{2-} in the HA structure – are responsible for the color center. In contrast, manganese in the samples with low carbonate content remains in the (+II) oxidation state and partly migrates into Ca^{2+} positions of β-TCP. It has been shown [51] that HPO_4^{2-} ions, present in Ca-deficient HA, support the transformation of HA to β-TCP, while CO_3^{2-} ions in carbonate-containing HA prevent the incorporation of HPO_4^{2-} ions. This stabilizing effect of carbonate in the HA structure determines the thermal properties of samples with carbonate content. Electron paramagnetic resonance (EPR) spectroscopy has proved that manganese is divalent both in the as-precipitated and heated samples.

PLD Mn:CHA depositions were performed in a low flux of oxygen on chemically etched Ti substrates heated at 400°C. To improve the crystallization status and restore the OH ions, each structure was heat post-treated for 6 h in a flux of hot water vapors at the same temperature as was applied during deposition. The coatings were crystalline, and their surface displayed a granular morphology [44].

SEM analysis was performed to evaluate morphology of cells cultivated on PLD surfaces vs. reference samples (uncoated Ti and polyester). hOBs were observed to attach and spread on Ti and Mn:CHA coatings (Fig. 10.9) with some differences, even though most of the cells exhibited their phenotypic morphology: flattened, polygonal configuration, dorsal ruffles, well attached to the substrate by cellular extension. The cells grown on the pure Ti surfaces were more elongated, with a rod-like shape both after 7 and 21 days of culturing (Fig. 10.9a and b).The cells grown on Mn:CHA (Fig. 10.9c and d) coatings appeared more spread, and exhibited a polygonal configuration. At 21 days, the cells were even more flattened and well spread across the surfaces.

Dedicated bioactivity tests by ALP, CICP and TGF β1 have confirmed that the response of hOB cells cultivated on PLD Mn:CHA coatings benefited from the presence of Mn^{2+} and carbonate, counterbalancing the higher crystallinity of these structures and improving the osteoblast differentiation on coated Ti [44].

A further support of these evolutions came out after the in vivo tests on 8-month-old New Zealand female rabbits, which were carried out on Mn:CHA, OCP and HA coatings. The flat coin-shaped implants (uncoated or coated with CaPs) were placed on the cortical bone of rabbit tibia [52] and kept there for 8 weeks. The results of pullout test measurements for the CaP-coated Ti

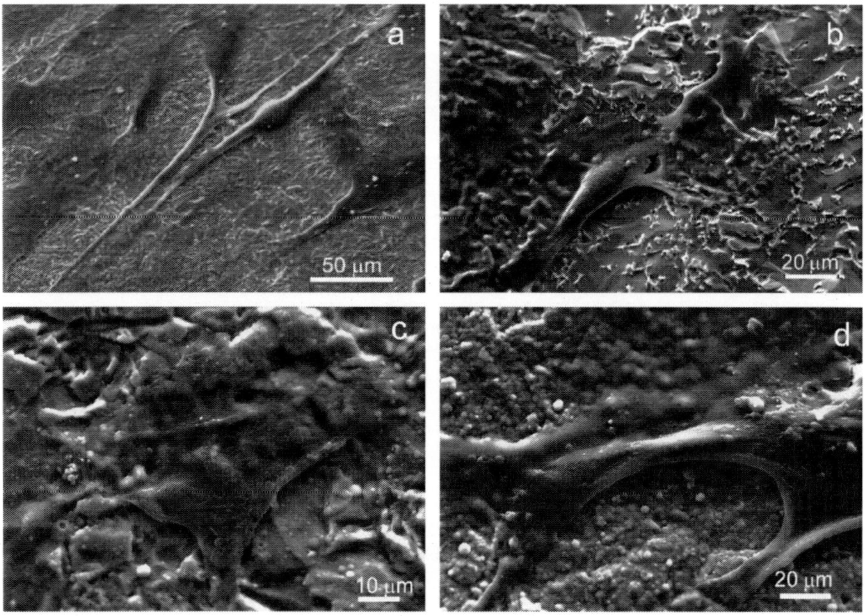

Fig. 10.9. SEM images of primary osteoblasts on (**a**, **b**) Ti, and (**c**, **d**) Mn:CHA coating. (**a**, **c**): 7 days of culturing; (**b**, **d**): 21 days of culturing. Reproduced with permission from Bigi et al. Biomaterials **26**, 2381 (2005)

compared to the uncoated Ti control implants are given in Fig. 10.10. They clearly demonstrated that all three tested groups (HA, Mn:CHA and OCP) had a significantly improved bone attachment strength (statistical significance values $p \leq 0.05$), about twice as high as that of control implants (average 5.32 N). Very important too, this strength was 25% higher for Mn:CHA (11.2 N) and 10% higher for OCP (9.11 N), as compared with HA-coated implants (8.45 N).

10.4.2 ß-Tricalcium Phosphate Doped with Mn^{2+}

ß-Tricalcium phosphate (ß-TCP) was found in many cases advantageous compared to HA [53] and was studied also when doped with Mg^{2+}, Zn^{2+} and Mn^{2+} ions or Si [45, 54, 55].

To synthesize β-TCP doped with Mn^{2+} ions (Mn: β-TCP), Mn:CHA was prepared from HA samples synthesized by method (10.2) [56]. When heated up to 800–1,100°C, most of the HA (∼75%) transforms to the rhombohedral crystal phase of TCP. As an alternative, TCP can be synthesized by a high-temperature solid-state reaction (800–1,100°C) of stoichiometric mixtures of $CaCO_3$ and $(NH_4)_2HPO_4$.

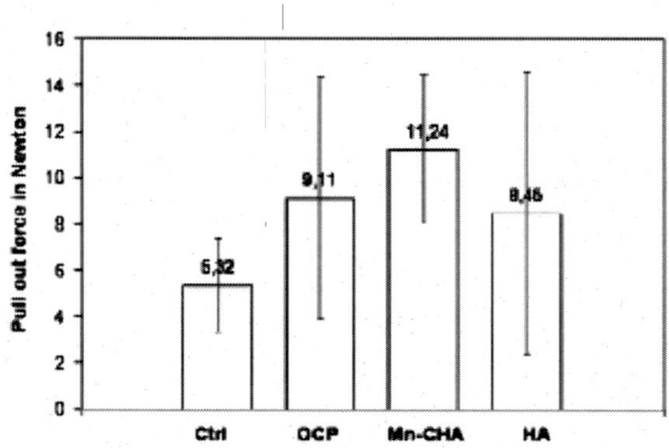

Fig. 10.10. Diagram of pullout force (tensile strength) for different types of samples. Error bars are ±SD, $n = 12$ in each group. The values are averaged over data collected for implants of the same type. Asterisks indicate a significant difference between test and control ($p < 0.05$). Ctrl denotes the control group, OCP is the octacalcium phosphate group, Mn-CHA stands for the manganese-doped carbonated hydroxyapatite group and HA is the hydroxyapatite group. Reproduced with permission from Mihailescu et al. Optoelectronics and Advanced Materials – Rapid Communications **2**, 337 (2008)

$$3\mathrm{CaCO_3\,(s)} + 2\,\mathrm{(NH_4)_2\,HPO_4\,(s)} \rightarrow \mathrm{Ca_3\,(PO_4)_2\;(s)} + 4\mathrm{NH_3\,(g)} + 3\mathrm{CO_2\,(g)} + 3\mathrm{H_2O\,(g)} \qquad (10.3)$$

Mn:β-TCP with the general composition of $\mathrm{Ca}_{3-x}\mathrm{Mn}_x\,(\mathrm{PO_4})_2$ can be prepared by the above-mentioned solid-state reaction by adding $\mathrm{Mn\,(NO_3)_2}$ to the stoichiometric mixtures. In case of $\mathrm{Ca_{2.7}Mn_{0.3}\,(PO_4)_2}$, the chemical reaction was as follows:

$$2.7\mathrm{CaCO_3\,(s)} + 2\,\mathrm{(NH_4)_2\,H\,(PO_4)\,(s)} + 0.3\mathrm{Mn\,(NO_3)_2\,(s)} \rightarrow \mathrm{Ca_{2.7}Mn_{0.3}\,(PO_4)_2} + 2.7\mathrm{CO_2\,(g)} + 4\mathrm{NH_3\,(g)} + 3\mathrm{H_2O\,(g)} + 0.3\mathrm{N_2O_5\,(g)} \qquad (10.4)$$

Samples exhibited pink color at high Mn concentration and were white at low concentrations. Samples with different x values up to the composition $\mathrm{Ca_{2.4}Mn_{0.6}\,(PO_4)_2}$ crystallize in the rhombohedral structure without reflections of any impurity phase. EPR and single-crystal XRD proved that Mn is in the divalent state and is located in the lattice sites of Mn:β-TCP.

It was shown by TEM [57] that Mn:β-TCP samples obtained by solid-state reaction via prolonged heating of the starting materials promotes the formation of bigger and better shaped crystals than in the case of Mn:β-TCP obtained by transformation of HA to TCP in which case smaller and less perfect crystals were observed. Based on the above results, it was predicted that the less perfect crystals observed for HA with the higher Mn content and for Mn:β-TCP obtained by partial transformation of HA to TCP will be good candidates for obtaining resorption sites in vivo and in vitro environments.

Mn:β-TCP thin coatings were synthesized by PLD on chemically etched Ti substrates [58]. The films were in a mostly amorphous, poorly crystalline phase and showed rather a homogeneous aspect molded to the rough relief of the Ti substrate. A Ca/P ratio of about 1.50–1.52 was measured in the deposited films.

In order to emphasize the biological role of divalent Mn^{2+} ions, interactions between two types of Mn:β-TCP ($Ca_{2.9}Mn_{0.1}(PO_4)_2$ and $Ca_{2.8}Mn_{0.2}(PO_4)_2$) coatings and mesenchymal stem cells (MSCs) during in vitro differentiation were examined. MSCs were isolated from human bone marrow extracted from a male patient. After multiplication, they were seeded at second passage on the surface of biomaterial samples and standard tissue culture material. With a view to obtain hOBs, the cells were grown in a differentiation medium. The effects of the biomaterials on cellular morphology, proliferation and spreading were examined 14 days later.

It was shown that neither of the tested materials was cytotoxic, since both of them induced cell adherence and growth over a period of 14 days in culture. The samples of 0.2 Mn-doped β-TCP showed higher potential for proliferation and increased viability (Fig. 10.11) when tested in osteoprogenitor cell culture than those with a lower Mn content. In addition, the stretched parallel pattern of actin filaments in 0.2 Mn-doped β-TCP showed an improved interaction of this biomaterial with the MSCs in comparison to 0.1 Mn-doped β-TCP (Fig. 10.12) [58].

10.5 Sr-Doped HA

Among the trace elements associated with biological apatites, strontium has attracted remarkable interest for its possible biological role. Strontium is present in the mineral phase of bone, especially at the regions of high metabolic turnover [59], and its beneficial effect in the treatment of osteoporosis is well known [60]. Strontium exerts both antiresorbing and bone forming

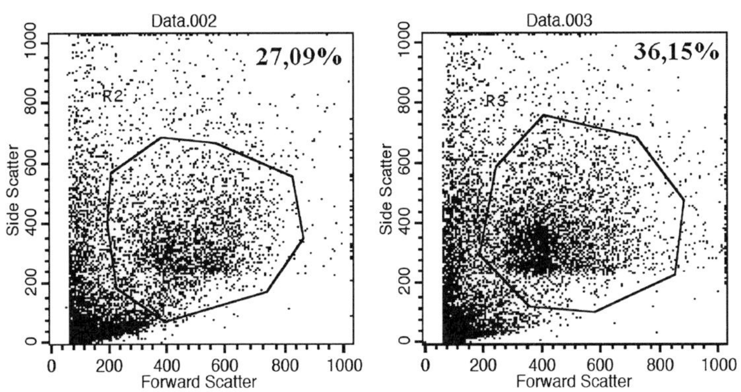

Fig. 10.11. FACS analysis of osteoprogenitor cells grown on (**a**) $Ca_{2.9}Mn_{0.1}(PO_4)_2$ and (**b**) $Ca_{2.8}Mn_{0.2}(PO_4)_2$. Reproduced with permission from Sima et al. Applied Surface Science **254**, 1155 (2007)

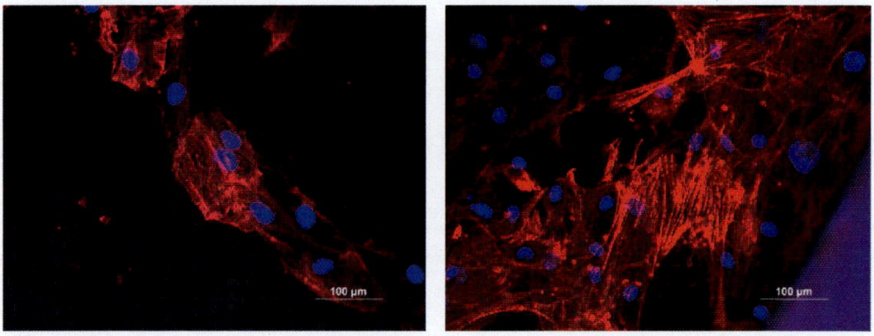

Fig. 10.12. Osteoprogenitor cell actin filament staining on (**a**) $Ca_{2.9}Mn_{0.1}(PO_4)_2$ and (**b**) $Ca_{2.8}Mn_{0.2}(PO_4)_2$ after 14 days in culture. Reproduced with permission from Sima et al. Applied Surface Science **254**, 1155 (2007)

effects in vitro, since it increases the number of osteoblasts and decreases the number and the activity of osteoclasts [60,61]. Again, strontium administration has been shown to induce a significant increase in bone mass and bone strength by a dual mechanism of action: inhibition of bone resorption and augmentation of bone formation in both normal or ovariectomized animals [62–64].

The unit cell of HA hosts 10 cations arranged in two nonequivalent positions: four at the M(1) site aligned in the column, each surrounded by 9 oxygen atoms, and 6 at the M(2) site arranged at the apexes of "staggered" equilateral triangles, each surrounded by 7 oxygen atoms [65,66]. Sr can substitute for Ca in HA structure over the whole range of composition. Structural refinements indicated that at low content, in the range of Sr concentration in the bone, the ion displays a slight unexpected preference for the smaller M(1) cationic site of the structure, whereas at greater concentrations it prefers cation site M(2) in agreement with its ionic radius [67]. Moreover, at relatively low concentration, Sr incorporation into HA induces a decrease of crystallinity, which could be responsible of the observed increased reactivity [67, 68]. Promising results on the possibility to utilize Sr-substituted HA for different applications, including bone filler and coatings of metallic implants [9, 68–71] and Sr-doped α-TCP for bone cements [72], have been reported.

We successfully grew HA thin films at different extents of strontium substitution for calcium (0, 1, 3, or 7 atom %) by means of the PLD technique. To this aim, we utilized HA and Sr-doped HA synthesized in N_2 atmosphere by dropwise addition of $(NH_4)_2HPO_4$ to $Ca(NO_3)_2 \cdot 4H_2O$ solution at pH adjusted to 10 with NH_4OH [67, 71]. The films were deposited on etched Ti substrates heated at 400°C using a UV KrF* excimer laser source in water vapor flux. The as-deposited samples were submitted to an annealing treatment in water vapor at ambient pressure for 6 h at 400°C. XRD analysis indicated that the films were made of HA as a unique crystalline phase and displayed well-defined sharp peaks in agreement with a high degree of crystallinity achieved by the technique (Fig. 10.13) [71]. However, the patterns showed a slight broadening of the diffraction peaks, which was in agreement with similar results recorded from the initial powders and was consistent with a shortening of the crystalline domains as Sr concentration increases [67].

EDS (energy dispersive spectroscopy) data indicated $[(Sr/Sr + Ca) \times 100]$ values of 0.5%, 3% and 7% for the thin films deposited from Sr1%, Sr5% and Sr10% powders, respectively, confirming that PLD is able to produce thin films with a composition close to that of the initial powders. Furthermore, the EDS maps were consistent with a homogeneous strontium distribution on the surface of the thin films. In contrast, the presence of Sr^{2+} had no significant effect on the morphology of the coatings, which displayed a

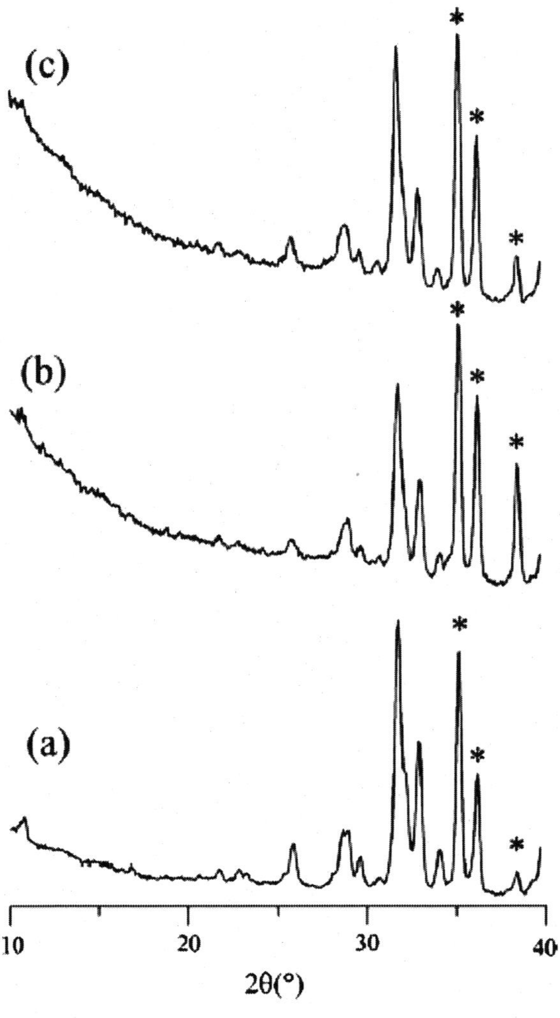

Fig. 10.13. XRD patterns of the coatings deposited on Ti substrates from HA containing (**a**) 0%, (**b**) 3% and (**c**) 7% of Sr. The reflections due to Ti are indicated. Reproduced with permission from Capuccini et al., Acta Biomaterialia **4**, 1885 (2008)

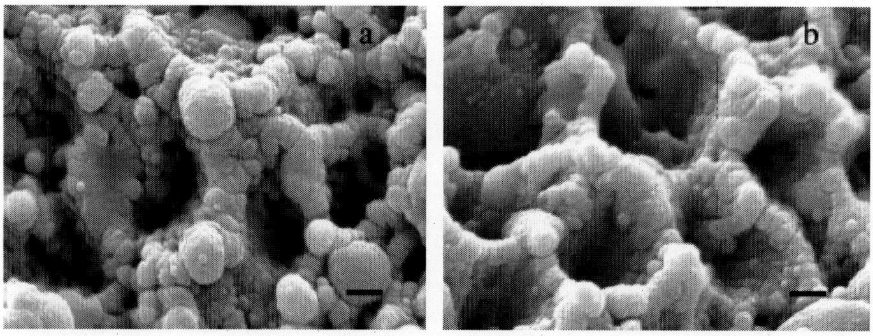

Fig. 10.14. SEM micrographs of thin films deposited from HA containing (**a**) 0% and (**b**) 7% of Sr. Scale bars = 1 μm. Reproduced with permission from Capuccini et al. Acta Biomaterialia **4**, 1885 (2008)

granular surface (Fig. 10.14). Chemical composition influenced bone cell response as shown by the results obtained on MG63 osteoblast and human osteoclast grown on the thin films at different Sr^{2+} contents [71]. In fact, osteoblast adhesion, as well as proliferation and viability (WST-1 test), increased with increasing Sr content in the coating. Osteoblast activity and differentiation were tested through the analysis of selected markers, namely alkaline phosphatase (ALP), osteocalcin (OC) and type I collagen (CICP), after 7, 14 and 21 days of culture. The results indicated that all markers of osteoblast activation and differentiation were improved in proportion to the concentration of Sr in the coating, with the statistically highest values recorded on the coatings containing 3 and 7% Sr. Moreover, considerably reduced values of osteoclast proliferation were observed on the same samples, which suggested that the presence of strontium in hydroxyapatite thin films could not only enhance the positive effect of HA coatings on osteointegration and bone regeneration but also prevent undesirable bone resorption.

10.6 Hybrid Organic–Inorganic Bionanocomposites

The introduction of biopolymers and biologically active molecules into the CaP coatings was studied with a view to synthesizing a composite similar to bone and increasing the bioactivity of the metallic implant. Successful MAPLE depositions of uniform, adherent and functional layers of biopolymers, collagen nanofibrils, fibrinogen and other proteins already were providing encouraging results [73].

10.6.1 Biopolymers–CaP

The concept of the composite HA–polymer systems, combining the biological and mechanical properties of the two components, aims to reduce the implant stiffness and to adjust its elastic modulus to that of the surrounding bone. Polymers are usually ductile, but not tough enough compared to the bony tissue. The tensile strength is around 30 MPa for the high density polyethylene (HDPE) and 50–150 MPa for the cortical bone, while their elasticity modulus is smaller than that of the bone (1 GPa for HDPE) [74, 75].

The HA–polymer composites mimic the natural bone, formed by the inorganic phase (especially nanometric CHA) and organic compounds (mainly collagen). The intimate synergy between inorganic and organic phase provides hard tissues with the beneficial features such as high fracture toughness, flexibility and strength.

The nanometric dimension of the inorganic element, of high specific surface, similar to the one in the bony apatite, is important from the point of view of the mechanical properties [74].

HA-based composites with biomedical applications include biocompatible and biodegradable polymers: polyesters such as PLA (poly(lactic acid)), PGA (poly(glycolic acid)) and their copolymers (PLGA), collagen, gelatin, chitosan, chondroitin sulfate, silk fibroin, hyaluronic acid, glycosaminoglycans and phosphorylated cellulose. Polyethylene, polysulphone [76], chitosan [77], polyethylene glycol [78] form with HA composite resorbable implants, while HA with collagen [79] or PLA [80] are developed for resorbable implants.

Most of them refer to the 3D porous scaffolds aimed to manipulate the osteoblast functionality and the guidance of the osseous growth in the desired forms. Polymeric–ceramic scaffolds of osteoinductive CaPs (HA, TCP) and biodegradable polyesters were fabricated by direct mixing or by biomimetic techniques [81–83]. Spongy osteoinductive structures of collagen–HA for osseous repair were synthesized by an alternative immersion method [79].

HA–polymer composite coatings with different contents of HA as skeletal implants were produced using a flame spray system. Films are adherent, but with a low percent of HA compared to the original raw material and nonhomogenous distribution of the HA particles inside the polymeric matrix [84]. Thick films of fibroin (tens of micrometers) and HA were deposited by alternative lamination as resorbable membranes for the guided bone regeneration (GBR) [85].

We applied MAPLE for the synthesis of hybrid nanocomposites of HA–sodium maleate (HA–NaM) copolymer on Ti substrates [86]. For preparing the frozen targets and the corresponding thin nanostructures, 0.2% (HA–NaM1) and 1% (HA–NaM2) powders containing 20% NaM were used. In parallel experiments, we synthesized HA thin coatings by PLD as control.

Interactions between the three types of coatings and MSCs were examined during in vitro differentiation. The applied procedure was described in detail in Sect. 10.4.

Cells were cultured for 24 h in direct contact with the surfaces of Ti–HA control samples and of Ti–HA–NaM1 or Ti–HA–NaM2 nanocomposites. Fixed cells were stained for actin (red), microtubules (green) and nuclei (blue). The cell morphology and actin–tubulin patterns in MSCs grown on HA and

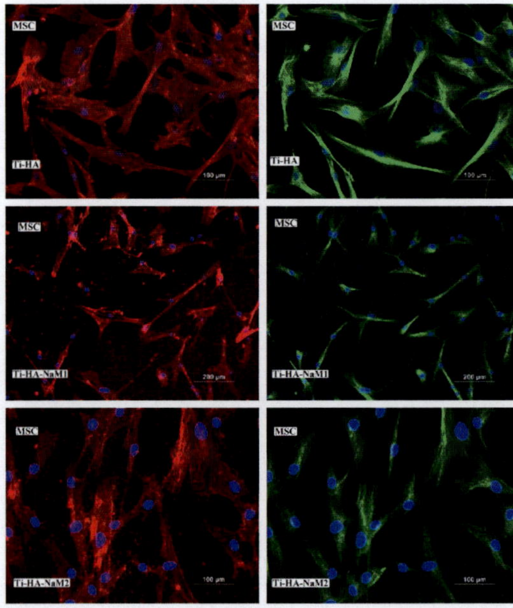

Fig. 10.15. Cytoskeletal organization and nuclei morphology of MSCs seeded on different biomaterial surfaces. Reproduced with permission from Negroiu et al. Journal of Materials Science: Materials in Medicine **19**, 1537 (2008)

HA–NaM coatings were similar to controls, suggesting that these surfaces were well tolerated by cells (Fig. 10.15). As proliferation follows cell spreading, this suggests that the polymer substantially enhanced cell adhesion to the biomaterial coating surface.

The samples of Ti–HA–NaM2 showed higher potential for proliferation and increased viability, when tested in osteoprogenitor cell culture, than did Ti–HA or Ti–HA–NaM1. In addition, the stretched parallel pattern of actin filaments in Ti–HA–NaM2 showed an improved interaction of this biomaterial with the MSCs compared to Ti–HA–NaM1 or Ti–HA.

10.6.2 Alendronate–HA

Bisphosphonates (BPs) were introduced in the 1970s for the management of disorders of bone metabolism, associated with bone loss. In particular, they are widely used for the treatment of tumor-induced hypercalcemia, Paget's disease and osteoporosis [87–90].

These chemical compounds are synthetic analogs of pyrophosphate in which the P–O–P group is replaced by the P–C–P bridge. Individual BPs are characterized by the two covalently bonded side chains attached to the central carbon atom, termed R1 and R2, which determine the efficiency of the compound. Binding to bone is enhanced when R1 is a hydroxyl group, whereas the R2 side group has some effect on binding but predominantly determines the antiresorptive potency of the BPs [88, 91–93]. The presence of nitrogen groups within the R2 side group is associated with the ability of an individual BP to inhibit farnesyl pyrophosphate (FPP) synthase, a major enzyme in the mevalonate pathway. This results in the disruption of many of the activities of the osteoclasts, including migration, attachment and resorption. Ultimately, cell death can occur via apoptosis [94].

Prolonged use of BPs, especially through intravenous preparations, has been recently associated with over suppression of bone metabolism and to possible osteonecrosis of the jaws [92]. Local administration might be helpful against the potential negative effects of their prolonged use. However, the great affinity of BPs for calcium hinders the direct synthesis of hybrid calcium phosphate crystals, as shown by the sudden precipitation of amorphous calcium alendronate that occurs when alendronate is introduced in the synthesis medium of hydroxyapatite [95].

Composite HA nanocrystals at different alendronate content, up to 7.1%wt, were synthesized in aqueous solution thanks to a strategy based on a slight modification of a classical method of synthesis of HA [95, 96]. The results of the structural refinements carried out on the composite nanocrystals were interpreted as suggesting that alendronate interacts with calcium ions through a bidentate chelation of deprotonated oxygen atoms of the BP anion without greatly affecting the crystal structure of HA [95].

In vitro tests demonstrated that alendronate is able to promote osteoblast activation and extracellular matrix mineralization processes, and to inhibit osteoclast proliferation even though incorporated in the composite nanocrystals [96]. On the basis of these promising results, we applied MAPLE to deposit thin films of alendronate–HA nanocrystals on Ti substrates. MAPLE in contrast to PLD is expected to avoid possible damage of alendronate.

Preliminary results indicate that it is possible to deposit the modified HA at low temperature. Alendronate–HA nanocrystals containing 7.1%wt of the BP display a powder XRD pattern characteristic of crystalline HA (Fig. 10.16a). The degree of crystallinity of the coatings deposited onto Ti substrates using MAPLE appears significantly reduced (Fig. 10.16b). However, the pattern clearly exhibits the peaks characteristic of HA, together with those due to Ti around 35° of 2θ (Fig. 10.16b). SEM images of the coatings (Fig. 10.17) show that the deposits consist of granules smaller and less defined with respect to those usually obtained through PLD. These data are of high biological relevance since they indicate that it is possible to use MAPLE to synthesize a coating that couples the bioactivity of HA with the local availability of alendronate.

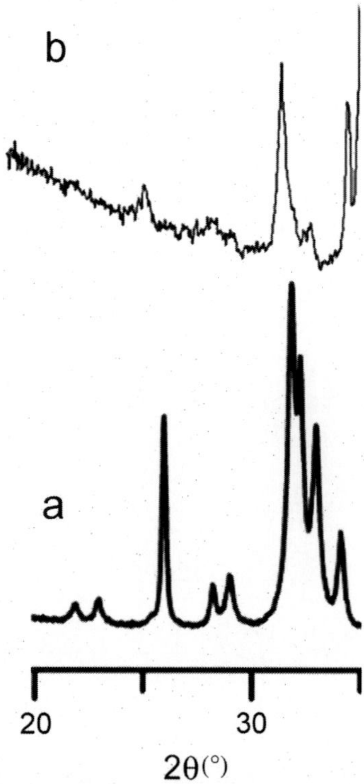

Fig. 10.16. Powder X-ray diffraction patterns of (**a**) HA nanocrystals containing 7.1wt.% of alendronate; and (**b**) MAPLE coating synthetized from the same alendronate-doped HA

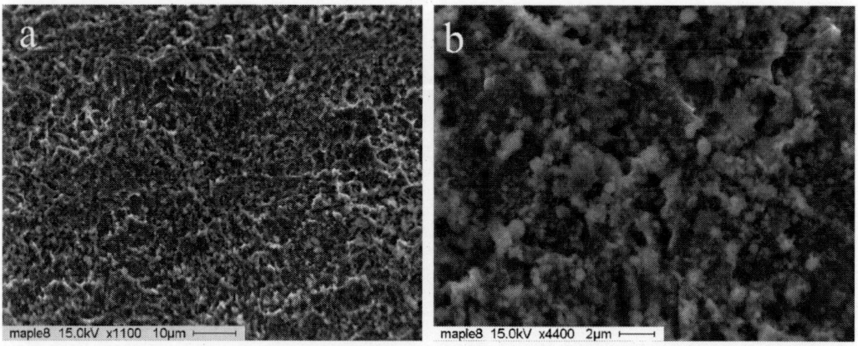

Fig. 10.17. SEM micrographs of thin films deposited from HA containing 7.1 wt.% of alendronate at two different enlargements

10.7 Conclusions

The results of the studies here reviewed clearly show that PLD and MAPLE are powerful tools to obtain CaP coatings on metallic substrates. The unique characteristics of these techniques allow depositing material with the same stoichiometry and structure as those of the target. In particular, it is possible to deposit phosphates with complex formula and crystalline structure. The availability of MAPLE further increases the variety of molecules and macromolecules, in particular heat-labile compounds, which can be co-deposited with CaPs. This is of great importance for biomedical applications, where the use of CaPs less stable than stoichiometric HA and the enrichment of the coatings with bioactive ions or molecules even in very small amounts can significantly improve the performance of the implant.

Acknowledgments

AB, INM and CR acknowledge the mobility exchange under the 15th Italian–Romanian Executive Programme of S&T Co-operation project "New Biomimetic Calcium Phosphate Coatings for Metallic Implants." IM, INM and CR acknowledge also the bilateral Agreement for Scientific Cooperation between the Israel Academy of Sciences and Humanities and Romanian Academy under the theme "Thin films and structures for medical, chemical and biological applications." All authors acknowledge with thanks the partial support of the work by EU under the contract SIMI G5RD-CT-2000–00423.

References

1. K. Duan, R. Wang, J. Mater. Chem. **16**, 2309 (2006)
2. T.J. Webster, in *Nanostructured Materials*, ed. by J.Y. Ying (Academic, New York, 2001), p. 125
3. L.L. Hench, J. Wilson, Biomater. Sci. **226**, 630 (1984)
4. W. Suchanec, M. Yoshimura, J. Mater. Res. **13**, 94 (1998)
5. R.Z. LeGeros, Clin. Orthop. Relat. Res. **395**, 81 (2002)
6. L. Sun, C.C. Berndt, K.A. Gross, A. Kucuk, J. Biomed. Mater. Res. Appl. Biomater. **58**, 570 (2001)
7. N. Hijon, M.V. Cabanas, J. Pena, M. Vallet-Regi, Acta Biomater. **2**, 567 (2006)
8. E.S. Thian, J. Huang, S.M. Best, Z.H. Barber, W. Bonfield, Mater. Sci. Eng. C **27**, 251 (2007)
9. W. Xue, H.L. Hosick, A. Bandyopadhyay, S. Bose, C. Ding, K.D.K. Luk, K.M.C. Cheung, W.W. Lu, Surf. Coating Tech. **201**, 4685 (2007)
10. G. Socol, P. Torricelli, B. Bracci, M. Iliescu, F. Miroiu, A. Bigi, J. Werckmann, I.N. Mihailescu, Biomaterials **25**, 2539 (2004)
11. R. Chiesa, E. Sandrini, M. Santin, G. Rondelli, A. Cigada, J. Appl. Biomater. Biomech. **1**, 91 (2003)

12. T. Kokubo, H. Kushitani, S. Sakka, T. Kitsugi, T. Yamamuro, J. Biomed. Mater. Res. **24**, 721 (1990)
13. V. Nelea, I.N. Mihailescu, M. Jelinek, in *Pulsed Laser Deposition of Thin Films: Applications-Led Growth of Functional Materials*, ed. by R. Eason (Wiley, New York, 2007), p. 421
14. D.B. Chrisey, A. Pique, R.A. McGill, J.S. Horwitz, B.R. Ringeisen, D.M. Bubb, P.K. Wu, Chem. Rev. **103**, 553 (2003)
15. A. Pique, in *Pulsed Laser Deposition of Thin Films: Applications-Led Growth of Functional Materials*, ed. by R. Eason (Wiley, New York, 2007), p. 63
16. I.N. Mihailescu, E. Gyorgy, in *International Trends in Optics and Photonics*, ed. by T. Asakura (Springer, Heidelberg, 1999), p. 201
17. D. Bauerle, *Laser Processing and Chemistry*, 3rd edn. (Springer, Berlin, 2000)
18. R. Eason (ed.) *Pulsed Laser Deposition of Thin Films: Applications-Led Growth of Functional Materials* (Wiley, New York, 2007)
19. R. Cristescu, I.N. Mihailescu, M. Jelinek, D.B. Chrisey, in *Functionalized Properties of Nanostructured Materials*, vol. 223, ed. by R. Kassing, P. Petkov, W. Kulisch, C. Popov (NATO Science Series II: Mathematics, Physics and Chemistry, Springer, Berlin, 2006), p. 211
20. L.L. Hench, J.M. Polak, Science **295**, 1014 (2002)
21. S.I. Stupp, G.W. Ciegler, J. Biomed. Mater. Res. **26**, 169 (1996)
22. M. Vallet Regi, J. Chem.Soc. Dalton Trans. **2**, 97 (2001)
23. S.V. Dorozhkin, J. Mater. Sci. **42**, 1061 (2007)
24. R. Cancedda, P. Giannoni, M. Mastrogiacomo, Biomaterials **28**, 4240 (2007)
25. S.V. Dorozhkin, M. Epple, Angew. Chem. Int. Ed. **41**, 3130 (2002)
26. H.A. Lowenstam, S. Weiner, *On Biomineralization* (Oxford University Press, Oxford, 1989)
27. F.C.M. Driessens, R.M.H. Verbeeck, *Biominerals* (CRC Press, Boca Raton, 1990)
28. R.Z. LeGeros, Clin. Orthop. **395**, 81 (2002)
29. R. Xin, Y. Leng, N. Wang, J. Cryst. Growth **289**, 339 (2006)
30. O. Suzuki, S. Kamakura, T. Katagiri, M. Nakamura, B. Zhao, Y. Honda, R. Kamijo, Biomaterials **27**, 2671 (2006)
31. H. Wang, J.K. Lee, A. Moursi, J. Lannutti, J. Biomed. Mater. Res. A **67**, 599 (2003)
32. International Centre of Diffraction data. Powder Diffraction File, 09–0432 (2000)
33. M. Jelínek, T. Dostálová, L. Himmlová, C. Grivas, C. Fotakis, Mol. Cryst. Liq. Cryst. **374**, 599 (2002)
34. W.E. Brown, J.P. Smith, J.R. Lehr, A.W. Frazier, Nature **196**, 1048 (1962)
35. M. Mathew, W.E. Brown, L.W. Schroeder, B. Dickens, J. Cryst. Spectrosc. Res. **18**, 235 (1988)
36. J. Zhang, G.H. Nancollas, J. Phys. Chem. **96**, 5478 (1992)
37. M. Iijima, H. Kamemizu, N. Wakamatsu, T. Goto, Y. Doi, Y. Moriwak, J. Cryst. Growth **181**, 70 (1997)
38. A. Bigi, E. Boanini, G. Cojazzi, G. Falini, S. Panzavolta, Cryst. Growth Des. **1**, 239 (2001)
39. S. Kamakura, K. Sasaki, T. Homma, Y. Honda, T. Anada, S. Echigo, O. Suzuki, J. Biomed. Mater. Res. A **83**, 725 (2007)
40. O. Suzuki, S. Kamakura, T. Katagiri, M. Nakamura, B. Zhao, Y. Honda, R. Kamijo, Biomaterials **27**, 2671 (2006)

41. P. Habibovic, J. Li, C.M. van der Valk, G. Meijer, P. Layrolle, C.A. van Blitterswijk, K. de Groot, Biomaterials **26**, 23 (2005)
42. X. Lu, Y. Leng, Q. Zhang, Surf. Coating Tech. **202**, 3142 (2008)
43. M. Iliescu, V. Nelea, J. Werckmann, I.N. Mihailescu, G. Socol, A. Bigi, B. Bracci, Thin Solid Films **453–454**, 157 (2004)
44. A. Bigi, B. Bracci, F. Cuisinier, R. Elkaim, M. Fini, I. Mayer, I.N. Mihailescu, G. Socol, L. Sturba, P. Torricelli, Biomaterials **26**, 2381 (2005)
45. E. Landi, A. Tampieri, M. Mattioli-Belmonte, G. Celotu, M. Sandri, A. Gigante, P. Fava, G. Biagini, J. Eur. Ceram. Soc. **26**, 2593 (2006)
46. E. Medvecky, R. Stulajterova, L. Parilak, J. Trpcevska, J. Durisin, S.M. Barainov, Colloids Surf. A Physicochem. Eng. Asp. **281**, 221 (2006)
47. W. Xue, H.L. Hosick, A. Bandyopadhyay, S. Bose, C. Ding, K.D.K. Luk, K.M.C. Cheung, W.W. Lu, Surf. Coating Tech. **201**, 4685 (2007)
48. E.S. Thian, J. Huang, S.M. Best, Z.H. Barber, W. Bonfield, Mater. Sci. Eng. C **27**, 251 (2007)
49. A. Armulik, G. Svineng, K. Wennerberg, R. Faessler, S. Johansson, Exp. Cell Res. **254**, 55 (2000)
50. A. Yasukava, K. Kandori, T. Ishikova, Calcif. Tissue Int. **72**, 243 (2003)
51. F. Apfelbaum, I. Mayer, J.D.B. Featherstone, J. Inorg. Biochem. **38**, 1 (1990)
52. I.N. Mihailescu, S. Lamolle, G. Socol, F. Miroiu, H.J. Roenold, A. Bigi, I. Mayer, F. Cuisinier, S.P. Lyngstadaas, Optoelectr. Adv. Mater. – Rapid Commun. **2**, 337 (2008)
53. M. Aizawa, K. Itatani, I. Okada, J. Ceram. Soc. Jpn. **113**, 245 (2005)
54. S. Kannan, A.F. Lemos, J.M.F. Ferreira, Chem. Mater. 18, 2181 (2006)
55. S. Kannan, J.H.G. Rocha, J.M.F. Ferreira, J. Mater. Chem. 16, 286 (2006)
56. J.D.B. Featherstone, I. Mayer, F.C.M. Driessens, R.M.H. Verbeeck, M. Heijligers, Calcif. Tissue Int. **35**, 169 (1983)
57. I. Mayer, F.J.G. Cuisinier, S. Gdalya, I. Popov, J. Inorg. Biochem. **102**, 311 (2008)
58. F. Sima, G. Socol, E. Axente, I.N. Mihailescu, L. Zdrentu, S.M. Petrescu, I. Mayer, Appl. Surf. Sci. **254**, 1155 (2007)
59. E. Shorr, A.C. Carter, Bull. Hosp. Jt. Dis. Orthop. Inst. **13**, 59 (1952)
60. E. Canalis, M. Hott, P. Deloffre, Y. Tsouderos, P.J. Marie, Bone **18**, 517 (1996)
61. W. Chang, C. Tu, T. Chen, L. Komuwes, Y. Oda, S. Pratt, S. Miller, D. Shoback, Endocrinology **140**, 5883 (1999)
62. P. Ammann, V. Shen, B. Robin, Y. Mauras, J.P. Bonjour, R. Rizzoli, J. Bone Miner. Res. **19**, 2012 (2004)
63. M.D. Grynpas, E. Hamilton, R. Cheung, Y. Tsouderos, P. Deloffre, M. Hott, P.J. Marie, Bone **18**, 253 (1996)
64. P.J. Marie, M. Hott, D. Modrowski, C. DePollak, J. Guillemain, P. Deloffre, Y. Tsouderos, J. Bone Miner. Res. **8**, 607 (1993)
65. K. Sudarsanan, R.A. Young, Acta Crystallogr. B **28**, 3668 (1972)
66. J.C. Elliott, *Structure and Chemistry of the Apatites and Other Calcium Orthophosphates* (Elsevier, Amsterdam, 1994)
67. A. Bigi, E. Boanini, C. Capuccini, M. Gazzano, Inorg. Chim. Acta **360**, 1009 (2007)
68. C. Capuccini, P. Torricelli, E. Boanini, M. Gazzano, R. Giardino, A. Bigi, J. Biomed. Mater. Res. A **89**, 594 (2009)
69. E. Landi, A. Tampieri, G. Celotti, S. Sprio, M. Sandri, G. Logroscino, Acta Biomater. **3**, 961 (2007)

70. A.L. Oliveira, R.L. Reis, P. Li, J. Biomed. Mater. Res. B **83**, 258 (2007)
71. C. Capuccini, P. Torricelli, F. Sima, E. Boanini, C. Ristoscu, B. Bracci, G. Socol, M. Fini, I.N. Mihailescu, A. Bigi, Acta Biomater. **4**, 1885 (2008)
72. S. Panzavolta, P. Torricelli, L. Sturba, B. Bracci, R. Giardino, A. Bigi, J. Biomed. Mater. Res. A **84**, 965 (2008)
73. M. Jelinek, T. Kocourek, J. Remsa, R. Cristescu, I.N. Mihailescu, D.B. Chrisey, Laser Phys. **17**, 66 (2007)
74. J.Y. Rho, L. Kuhn-Spearing, P. Zioupos, Med. Eng. Phys. **20**, 92 (1998)
75. H. Takadama, M. Hashimoto, Y. Takigawa, M. Mizuno, T. Kokubo, Key Eng. Mat. **254–256**, 569 (2004)
76. M. Wang, C.Y. Yue, B. Chua, J. Mater. Sci. Mater. Med. **12**, 821 (2001)
77. M. Ito, Y. Hidaka, M. Nakajima, H. Yagasaki, A.H. Kafrawy, J. Biomed. Mater. Res. **45**, 204 (1999)
78. Q. Liu, J.R. Dewijn, C.A. van Blitterswijk, Biomaterials **18**, 1263 (1997)
79. A. John, L. Hong, Y. Ikada, Y. Tabata, J. Biomater. Sci. Polym. Ed. **12**, 689 (2001)
80. N. Ignjatovic, V. Savic, S. Najman, M. Plavsic, D. Uskokovic, Biomaterials **22**, 571 (2001)
81. R. Zhang, P. X. Ma, J. Biomed. Mater. Res. **45**, 285 (1999)
82. R.C. Thomson, M.J. Yaszemski, J.M. Powers, A.G. Mikos, Biomaterials **19**, 1935 (1998)
83. R. Zhang, P.X. Ma, J. Biomed. Mater. Res. **44**, 446 (1999)
84. L. Sun, C.C. Berndt, K.A. Gross, J. Biomater. Sci. Polym. Ed. **13**, 977 (2002)
85. R. Kino, T. Ikoma, S. Yunoki, J. Biosci. Bioeng. **103**, 514 (2007)
86. G. Negroiu, R.M. Piticescu, G.C. Chitanu, I.N. Mihailescu, L. Zdrentu, M. Miroiu, J. Mater. Sci. Mater. Med. **19**, 1537 (2008)
87. O. Dyer, Br. Med. J. **326**, 243 (2003)
88. H. Fleisch, Endocr. Rev. **19**, 80 (1998)
89. M.J. Rogers, J.C. Frith, S.P. Luckman, F.P. Coxon, H.L. Benford, J. Monkkonen, S. Auriola, K.M. Chilton, R.G.G. Russell, Bone **24**, 73S (1999)
90. R.G.G. Russell, M.J. Rogers, Bone **25**, 97 (1999)
91. S.B. Woo, J.W. Hellstein, J.R. Kalmar, Ann. Intern. Med. **144**, 753 (2006)
92. R. Rizzoli, N. Burlet, D. Cahall, P.D. Delmas, E.F. Eriksen, D. Felsenberg, J. Grbic, M. Jontell, M. Landesberg, A. Laslop, M. Wollenhaupt, S. Papapoulos, O. Sezer, M. Sprafka, J.Y. Reginster, Bone **42**, 841 (2008)
93. S.M. Ott, J. Clin. Endocrinol. Metab. **86**, 1835 (2001)
94. R.G.G. Russell, Bone **40**, S21 (2007)
95. E. Boanini, M. Gazzano, K. Rubini, A. Bigi, Adv. Mater. **19**, 2499 (2007)
96. E. Boanini, P. Torricelli, M. Gazzano, R. Giardino, A. Bigi, Biomaterials **29**, 790 (2008)

11

Laser Direct Writing of Idealized Cellular and Biologic Constructs for Tissue Engineering and Regenerative Medicine

Nathan R. Schiele, David T. Corr, and Douglas B. Chrisey

11.1 Conventional Tissue Engineering

Conventional tissue engineering typically involves homogenously seeding cells into a scaffold, then manipulating the scaffold either mechanically, using bioreactors, or chemically, using growth factors, in an attempt to tailor the mechanical and biological properties of the engineered tissue. The material composition of the scaffold gives the construct its initial strength; then the scaffold either remodels or dissolves when implanted in the body. An ideal tissue replacement scaffold would be biocompatible, biodegradable, implantable, and would match the strength of the tissue it is replacing, and would remodel by natural mechanisms [1]. Finding or creating scaffold materials that meet all these specifications while providing an environment for cell attachment and proliferation is one of the main goals of conventional tissue engineering. Popular current scaffold materials include poly-l-lactic acid (PLLA) [2] and collagen [3]. Typically, the utilization of scaffolds in tissue engineering employs a top-down approach in which cells are seeded homogenously into the scaffold, then incubated in vitro prior to implantation. Scaffold properties, such as geometric dimensions (e.g., thickness) and cellular in-growth, are limited by the diffusion of nutrients, since these scaffolds do not incorporate vascular structures to transport nutrients and remove wastes deep into the scaffold as in native tissue [4]. Although seeded scaffolds have proven successful in some cases, there remains the need to have greater control of cell placement as well as the placement of additional features such as vascular structures, multiple cell types, growth factors, and extracellular matrix proteins that will aid in the fabrication of the next generation of engineered tissues.

11.2 History of Cell Patterning and Direct Writing Biomaterials

The direct placement of cells and control of their locations is essential to truly understanding cell–cell interactions, and to create tissue replacements that will ultimately best mimic native tissues. Recent developments in biological material patterning techniques, such as micropatterning [5, 6], photolithography [7], ink jet printing [8], microfluidic patterning [9], laser forward transfer [10–12], dip pen nanolithography [13], and matrix-assisted pulsed laser evaporation direct write (MAPLE DW) [14–18], offer new ways to precisely construct cellular cultures, and thus avoid the uncertainties introduced by manually pipetting cell suspensions onto substrates. Table 11.1 displays various patterning techniques and their capabilities and limitations.

Cellular micropatterning employs microlithography and nanolithography techniques to create stencils, which are then used to stamp a thin adherent

Table 11.1. Capabilities of various cell and biological material patterning techniques

	Photolithography	Micro-stencil	Dip pen nanolithography	Microfluidic patterning	Ink-jet printing	Laser-induced forward transfer	MAPLE DW
Capable of patterning living cells on substrates	×	×		×	×	×	×
Capable of patterning active proteins on substrates	×	×	×	×	×	×	×
Capable of transferring different biomaterials side-by-side			×	×	× (only with multiple dispensers)	×	×
Capable of transferring multiple cell types side-by-side				×	× (only with multiple dispensers)	×	×
Optical imaging							×
Selectively captures and immobilizes cells							×
CAD/CAM compatible					×	×	×
Single-cell resolution							×

micropattern surface onto a culture dish. Cells are applied in bulk, allowed to attach, and then are washed off, resulting in cell attachment only in the micropatterned area. Additional cells can then be added to this cell pattern to obtain a coculture; however, there is no pattern to guide the adherence of this second cell type since a second adherent surface cannot be applied once cells have been seeded. Thus, micropatterning cannot direct the adherence of multiple cell types simultaneously, nor can it establish precise integrated patterns of heterogeneous cell types. Therefore, while it offers great control of cells in culture, micropatterning does not lend itself to precise cocultures or multi-cultures.

Photolithography can be used to create patterns of biologically active proteins that allow for cell attachment. The biologic photolithography process begins with the creation of a substrate by first spin-coating a photoresist onto the surface of a substrate, then exposing the surface to a UV light through a stencil. The stencil ensures that only areas of interest are exposed to the UV light. A mask of cured photoresist is left on the surface of the receiving substrate. Biologically active cell attachment factors can then be added which will reside solely in the pattern specified by the photoresist mask. Once the cured photoresist is removed, cells are added on top of the substrate and will only adhere to the areas that contain the attachment factors. This process can produce very intricate two-dimensional patterns, but cannot be used for more than one layer of cells. Furthermore, each bio-photolithography application is limited to attaching a single cell type to a single cell attachment factor.

An additional approach, modified ink jet bioprinting, controls cell patterning through the regulated dispensing of cells, suspended in solution, to precise locations on the substrate. This technique excels at printing intricate patterns, and thus allows rapid pattern generation with multiple cell types. However, since the cells are suspended in solution, this process precisely controls the amount of solution printed, not the amount of cells. The density of cells in ink-jet printing is governed by the statistical probability that cells are contained in the delivered fluid volume, and thus the number of cells within the delivered fluid volume is uncertain, and their precise positioning is left to chance. As a result, ink jet bioprinting is not able to directly control the number of cells printed, nor their precise location, and thus cannot create the desired precise cocultures. In addition, the spot size is also limited, which has a direct effect on the number of cells that can be printed. The smaller the spot, the greater the pressure build-up within the ink-jet cartridge, and thus greater stresses experienced by the cells. This increased stress can be damaging to the cells.

Laser-induced forward transfer makes use of a pulsed laser to transfer spots from a thin liquid film on a ribbon to a receiving substrate. Biologic materials, including cells, can be suspended in a liquid and placed on a ribbon. Although very similar to matrix assisted pulsed laser direct write, the liquid suspension used in laser forward transfer makes the targeting of single cells quite difficult, since cells are not attached in a rigid location.

Dip pen nanolithography employs an atomic force microscope (AFM) to deposit biological materials to a substrate at the nano-scale. The AFM tip is used to deposit the various materials which have included, small organic molecules, DNA, proteins, peptides, along with many other nonbiologic materials [13]. Although this technique provides very high resolution and the ability to operate at the nano-scale, dip pen nanolithography cannot be used to transfer live cells. Despite this limitation, dip pen nanolithography may be used to lay the foundation for live cells by depositing proteins or peptides.

Microfluidic patterning is very similar to the modified ink jet printing technique in that cells must be suspended in fluid or an extracelluar matrix. The number of cells patterned depends on the volume of deposited material and the cell density. One recent example of this technique utilized a syringe to produce three-dimensional channels of cells suspended in ECM [9]. This process is limited in that the actual number of transferred cells is unknown, and the physical dimensions of the channels are bounded by the pressures generated within the syringe. These limitations are very similar to those of modified ink jet printing.

11.3 Matrix-Assisted Pulsed Laser Evaporation Direct Write

Many of the aforementioned shortcomings or limitations can be overcome using laser direct writing (Table 11.1). Matrix-assisted pulsed laser evaporation direct write (MAPLE DW), a laser direct writing process with CAD/CAM capabilities, can be used to build tissues on a cell-by-cell basis using a bottom-up approach. With this technique a scaffold can be built around the cell, giving complete control over geometry and location. Multiple cell types can be placed in desired locations along with growth factors and even vascular structures. This technology allows for the creation of engineered tissues that truly mimic tissue in vivo. To date, many different cell types [18–21], extracellular matrix (ECM) [19], proteins [14,17,22], and other biomaterials [14,16,23] have all been written using the MAPLE DW process.

MAPLE DW was originally developed for processing mesoscopic patterns of electronic components. Thick films of Ag, $BaTiO_3$, $SrTiO_3$, and Y_3Fe_5 O_{12} have been transferred to various substrates to build different types of electrical components that include parallel-plate and interdigitated capacitors, flat inductors, conducting lines, resistors, and chemoresistive gas sensors [24]. These components were able to be processed with micron resolution. This technique was translated to the processing of patterns of biological materials and cells, and in this chapter, we demonstrate the power of MAPLE DW in applications in tissue engineering, regenerative medicine, and cancer research.

The MAPLE DW system used in these experiments incorporates a UV excimer laser, and motorized ribbon and receiving stages all of which are computer controlled. The laser energy is transmitted to the ribbon through

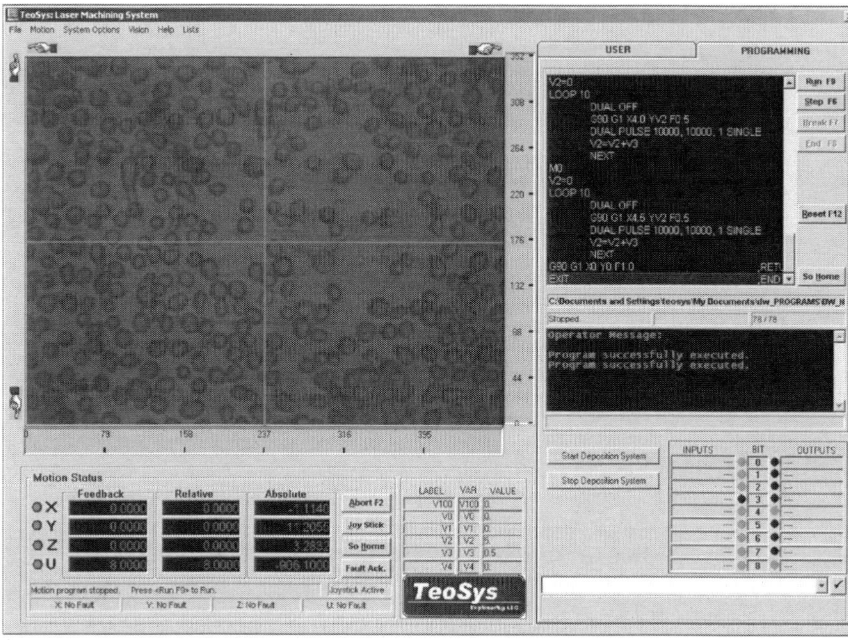

Fig. 11.1. Graphical interface for TeoSys laser, indicating trypsinized cells on the ribbon during transfer

a series of mirrors, two irises to set the spot size, and lastly though a UV objective to focus the beam to spots as small as ~2 μm. The x–y motorized stages and laser firing are controlled by visual basic source code. The user interface, optical vision system showing the quartz ribbon with trypsinized cells, and laser firing and motion code can be seen in Fig. 11.1.

Patterns used for cell deposition can be drawn in a commercial two-dimensional CAD package, and converted into motion and laser firing code. User specified code can be written in a.PRG format without needing a CAD file. Our MAPLE DW system utilizes an ArF excimer laser (TeoSys, Crofton, MD) operating at 193 nm. This system provides a spot size that can be varied from approximately 2 μm to 400 μm. The laser has an intra-cavity variable aperture with fourteen different laser spot diameter settings. This feature yields the extremely small spot size and the near Gaussian beam distribution. An in situ energy meter is used to record the energy of every laser shot in order to track shot-to-shot repeatability and ensure that the energy delivered to the ribbon is appropriate. If the energy is too high, the cells may be destroyed. Yet, if the energy is too low, the cells will not be successfully transferred from the ribbon to the receiving substrate. A computer-controlled variable attenuator can also be added into the laser path to further control the fluence that is imparted to the ribbon without affecting the spot size. If the fluence is appropriate, it has been shown that cells can be transferred using a UV laser with little to no damage to their DNA, or membrane [20,25]. In our MAPLE DW process, cells are attached to the ribbon through a thin biological polymer that allows for cell attachment and a volatilization zone of laser interaction.

In order to perform cell cocultures and three-dimensional tissue construction, the ribbon stage must be controlled separatly from the control of the receiving stage. This provides independent movement of the ribbon and receiving substrate, which is essential for the building of three-dimensional structures. The ribbon stage is computer-controlled and motorized in the x and y directions. Moving to a new area on the ribbon after each transfer allows multiple cells to be stacked at a single location on the receiving substrate. Interchanging ribbons during this process will allow for ECM to be stacked between cells to create three-dimensional tissues built on a cell-by-cell basis.

The receiving platform in our TeoSys MAPLE DW system contains an x-y-z motorized, computer-controlled vacuum stage. This stage securely holds the receiving substrate with a vacuum chuck while giving it independent control from the ribbon. Translation in the z-direction is necessary to establish proper registry between the ribbon and receiving substrate. Using the z-translation, the receiving substrate can be positioned within 50 μm of the ribbon. This allows x-y-z control of cell placement, to aid in tissue construction and improve registry between the targeted spots on the ribbon and their corresponding placement on the receiving substrate.

In our MAPLE DW system, a CCD camera shares the optical path with the laser as it passes through the final objective. This allows the user to visualize the cells on the ribbon prior to transfer, as well as following transfer. This optical set up is truly unique, and offers a significant advantage over many other cell deposition systems. It allows specific cell targeting by the user, as well as visual verification of all transfers, in real time. A full schematic of the MAPLE DW system is shown in Fig. 11.2. This optical configuration allows machine vision to be incorporated into the system. The MAPLE DW system is able to visualize exactly what material will be deposited, pretransfer. Visualization is achieved in real-time, providing the necessary requirements for a high-throughput automated machine vision system. To accomplish such high throughput automation, the machine vision system will need to image the ribbon and rapidly complete a blob segmentation to separate the background of the image from the cells, and then determine x–y coordinates for the centroids of each individual cell. The system will then select optimal cells for deposition. An optimal cell will be slightly separated from other cells especially for single-cell depositions. The cells that are touching, overlapping or too close to another cell will be eliminated from consideration for deposition. Once optimal cells have been targeted, and their locations recorded, the machine vision system will automatically move the ribbon into position in relation to the receiving substrate and the laser will be pulsed. With the automated

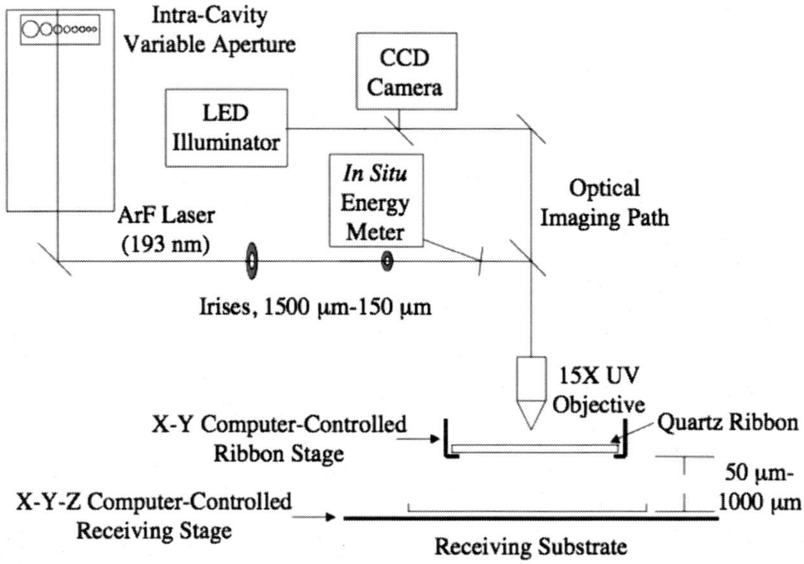

Fig. 11.2. Schematic of the TeoSys laser system showing the laser and optical beam paths. The ribbon and receiving stages as well as the set-up for cell and biological material transfer are also shown

deposition of cells, but more importantly single-cells, a high-throughput process for building combinatorial libraries of idealized cellular constructs and engineered tissue replacements can be developed.

11.4 Preparation of a Ribbon for Direct Write of Cells

Our MAPLE DW process for cell direct writing begins with attaching the desired cells or other biologic material to be transferred to the quart ribbon. To prepare the ribbon to accept cells, a UV transparent quartz ribbon is mounted on a bench-top spin coater, and 0.3 ml of Matrigel® GFR basement membrane matrix (Growth Factor Reduced, BD Biosciences, Bedford, MA) is thawed at 0°C and applied to the ribbon using chilled pipettes, while spinning at 1,000 rpm, for 20 s. The Matrigel® coated ribbon is then placed in a standard cell culture incubator (37°C, 5% CO_2, 95% RH) for 5 min and allowed to polymerize. The receiving substrate, a 100-mm diameter Petri dish, is uniformly coated with Matrigel®, by hand, using chilled pipettes. Prior work has shown that a receiving substrate of at least 40-μm thickness results in nearly 100% cell viability post transfer [20].

The cells to be transferred are grown in tissue culture flasks, trypsinized, centrifuged, and re-suspended in fresh culture media warmed to 37°C. Media constituents and concentrations are adjusted for each individual cell type. An appropriate cell-seeding density is chosen based on the goal of the

MAPLE DW. To transfer with single-cell resolution, the ribbon is seeded with a lower density, whereas to transfer groups of cells (around 20–40 cells, determined by the spot size) the ribbon is seeded to achieve near 100% confluence. Approximately 2 ml of suspended cell solution is pipetted over the Matrigel®-coated ribbon. The ribbon with cells is then incubated for a prescribed duration (from 5 min to 1 h, depending on cell type) to allow cells to begin adhering to the Matigel® coating. In this process the cells need to be loosely attached to the ribbon, and initial cell adhesion is visually confirmed on an inverted optical microscope (Carl Zeiss, Inc., Göttingen, Germany) using phase contrast and a 10× objective. Cells are deemed ready for transfer if they remain attached to the ribbon when the microscope stage is quickly jogged yet still retain their balled-up trypsinized morphology. If, however, the cells exhibit their normal attached morphology, then the cells are too tightly adhered to the ribbon, and the ribbon must be discarded. Next, the cell-seeded ribbon is tilted to remove excess culture media and any unattached cells, blotted dry, inverted, and mounted in the ribbon-holding stage. The receiving substrate is then secured via a vacuum chuck on an x-y-z computer controlled motorized stage, and raised into position using the z-translator to obtain a minimal substrate-ribbon distance while ensuring that the cells on the ribbon do not touch the substrate. The laser fires causing volatization of the Matrigel® at the ribbon interface, transforms the material to the receiving substrate, while the notion stages move to produce the user specified pattern. After cell transfer, the receiving substrate is incubated for 10 min, then warmed fresh culture media is added, and the substrate is returned to the incubator. This procedure has proven successful for multiple cell types and is translatable to many applications. By incorporating multiple cell types on multiple ribbons, various combinations of heterogeneous cellular arrays can be formed to study a wide range of cell behaviors: from cancer metastasis to neural stem cell differentiation.

11.5 Combinatorial Libraries of Idealized Constructs

Combinatorial experimentation has been used extensively in the pharmaceutical industry to rapidly conduct large scale testing on thousands of compounds to identify combinations of high biological activity [26]. Although, this high-throughput testing is not currently widely used in tissue engineering and regenerative medicine, combinatorial libraries of idealized constructs could lead to rapid advancements in the field due to high-throughput screening and testing. This type of analysis requires the rapid construction of precise cellular constructs to test multiple parameters, as well as their interaction. Due to its cellular precision and full CAD/CAM control, MAPLE DW is an ideal platform for conducting combinatorial testing on both two- and three-dimensional cellular constructs.

The ability to control cell placement and produce idealized cellular constructs is essential for understanding and controlling intercellular processes and ultimately for producing engineered tissue replacements. Much of the current focus of tissue engineering has shifted from a top-down approach with homogenously introducing cells in a preformed scaffold material, to a bottom-up approach involving placing cells in very specific locations and directly integrating the scaffold with the cells. Using MAPLE DW osteoblast-like cells have been co-transferred along with hydroxyapatite to create cell-ceramic composite scaffolds [21]. Placement of cells into and around scaffold materials can be achieved with precision, as well as in multiple layers to create three-dimensional structures. Controlling cell proximity and placement with regard to homogenous or heterogeneous cell types can be used to adjust or control lineage in neural progenitor cells, stem cells, or collagen production in fibroblast cells. To discover ideal cell proximities for engineered tissue replacements, cells can be directly transferred with great precision using MAPLE DW into an array of gradient proximities, and then assayed to find which combination leads to greatest bioactivity. Finding the correct combination of cell locations as well as biochemicals can be done rapidly by essentially conducting hundreds of experiments all at a time, using large arrays. Combinatorial libraries of idealized cellular constructs can be used to increase tissue engineering throughput. Arrays of cells can be produced and exposed to varying cell types at indexed proximities and geometric configurations, with the addition of biochemicals, growth factors, and matrix proteins to explore ideal combinations. Such libraries can be created to investigate cell interactions when different drugs are introduced, in both homogenous arrays as well as in coculture. These idealized constructs can allow for high-throughput experimentation to enable faster discoveries and ensure that no negative side-effects are detected. When coupled with multiple cell types in co-cultures the similarly to in vivo conditions increases, resulting in constructs with more realistic morphological histologic representation, and thus greater physiologic relevance.

11.6 Current MAPLE DW for Tissue Engineering, Regenerative Medicine, and Cancer Research

Many different mammalian cell types have been transferred using MAPLE DW. Some of our most recent work involves human dermal fibroblast cells, mouse C2C12 myoblast cells, rat neural stem cells, bovine pulmonary artery endothelial cells (BPAEC), and human breast caner cells [19].

11.7 Musculoskeletal Tissue Engineering

Fibroblast cells are crucial in the human body for maintenance of the extracellular matrix and they play a critical role in wound healing of soft tissues such as skin [27]. Directing fibroblast growth is an important step in tissue engineering where the focus has started to move from a top-down approach of homogeneously introducing cells into a preformed scaffold, to a bottom-up approach in which the tissue construct is built on a cell-by-cell basis. Such a bottom-up approach has the ability to manipulate specific cell environments through location, proximity, and geometry. The directing of cell proliferation to encourage organized tissue formation can provide tissue engineers a means of controlling the architectural and mechanical properties of soft tissue scaffolds. The fact that fibroblast cells produce collagen can be exploited for the construction of engineered tissue scaffolds. Collagen is the building block for many soft tissues such as skin, tendon, and ligament. Encouraging collagen production by fibroblasts, coupled with their controlled placement, can lead to tissue replacements that have produced their own collagen scaffolds. This approach to functional tissue engineering represents a novel direction for the development of replacement tissues. Recent work in human dermal fibroblasts has demonstrated their successful transfer from Matrigel®-coated ribbons to Matrigel®-coated receiving substrates (Fig. 11.3), thereby indicating their utility in MAPLE DW.

Myoblast cells, under correct conditions, can form myobtubes and serve as the basis for muscle. C2C12 Mouse myoblast cells have also been successfully

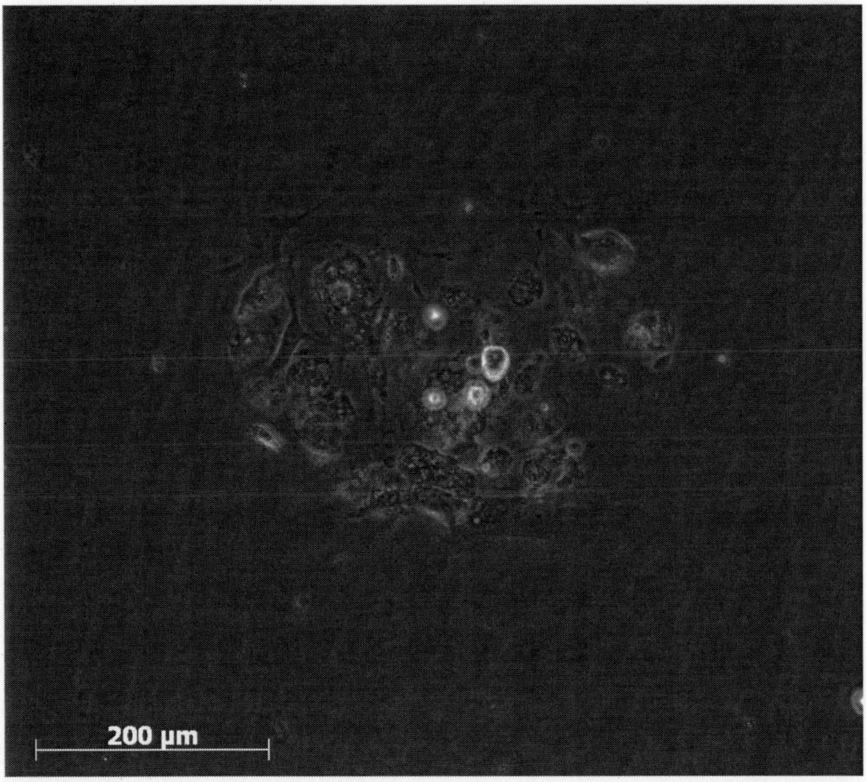

Fig. 11.3. Fibroblast cells 4 h after MAPLE DW. The cells can be seen beginning to spread out and exhibit normal morphology

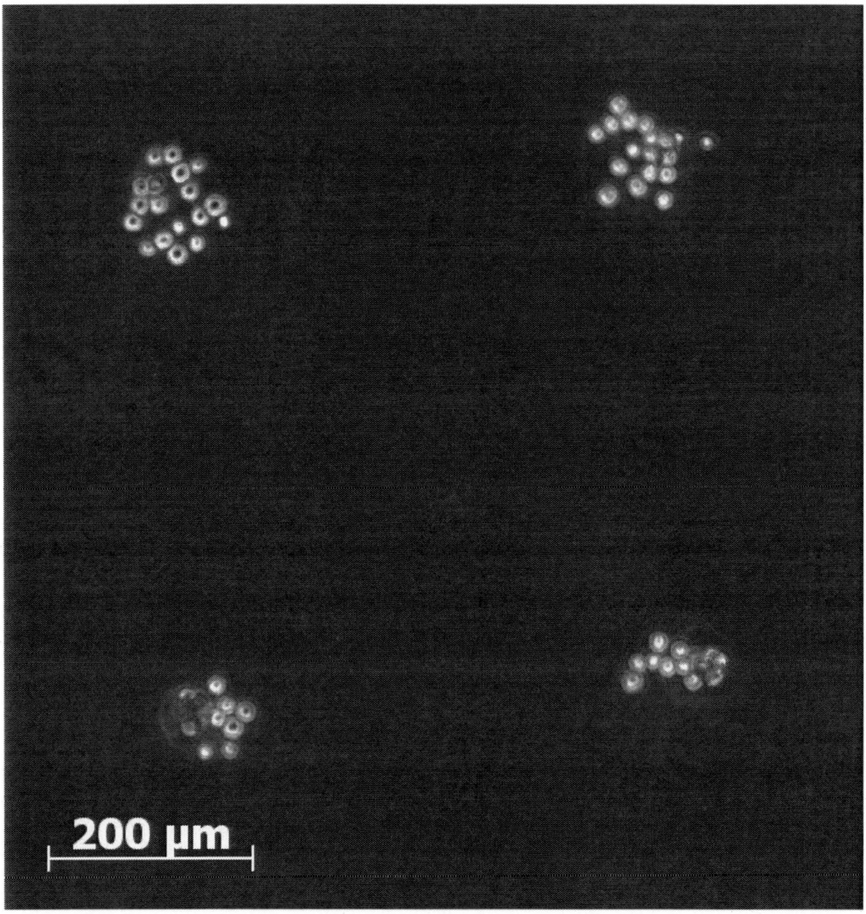

Fig. 11.4. A 2 × 2 spot array of myoblast cells directly after transfer

transferred from Matrigel®-coated ribbons to Matrigel®-coated receiving substrates using MAPLE DW (Fig. 11.4). Using this technology, cells can be directly written into patterns that encourage the myoblasts to align with a specified geometry and form myotubes, which can eventually be used to produce functioning muscle tissue.

11.8 Breast Cancer Metastasis

Controlling the location of cancer cells in relation to normal, healthy cells is an important challenge that lies at the core of many studies in cancer. Such cellular control would enable studies into the chemical signals secreted by noncancerous cells and how they react within a certain proximity to cancer cells. Fibroblast cells are known to affect the progression of breast cancer

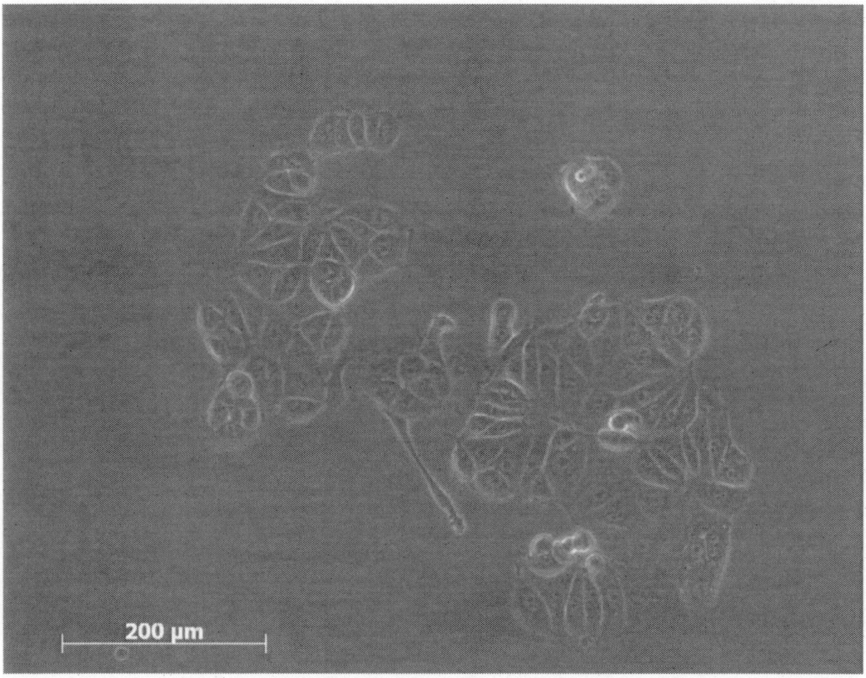

Fig. 11.5. Breast cancer cells, 6 days post transfer

in co-inoculation [28]. By precisely controlling the number and location of fibroblast cells in co-culture with breast cancer cells, a targeted investigation of the changes in normal cells with respect to distance and number could be conducted. This would provide unprecedented insight into the fundamental aspects of cancer progression, such as what factors enable cancer metastasis and which cellular signals are being secreted by normal cells that influence, either attenuate or encourage metastasis. Human breast cancer cells transferred by the MAPLE DW process are shown in Fig. 11.5.

11.9 The Neural Stem Cell Niche

The niche of neural stem cells in vivo is a local microenvironment that maintains the self-renewal and differentiation capabilities of neural stem cells [29]. The ability to create this neural stem cell niche in vitro is essential to the development of tissue replacements for the nervous system, as well as giving a better understanding to the role that surrounding vascular structures play in neural stem cell progression. Vascular endothelial cells secrete factors that affect neural stem cells and this interaction can be controlled by selectively placing endothelial cells in various proximities to neural stem cells. Figs. 11.6 and 11.7 show neural stem cells and BPAEC that have been transferred using MAPLE DW.

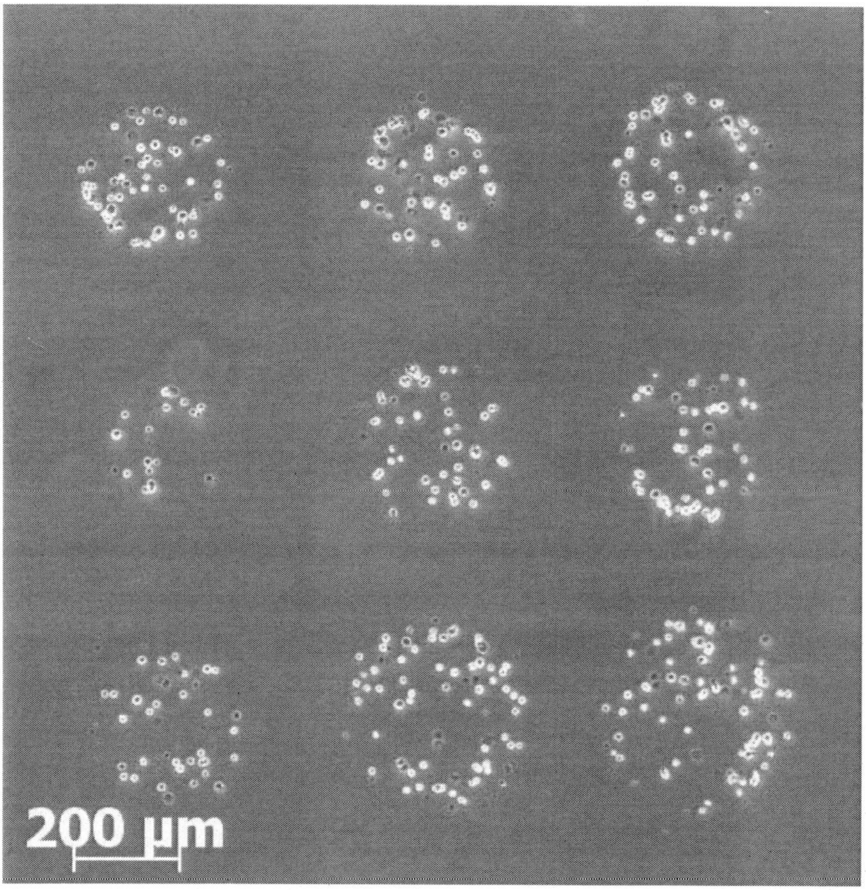

Fig. 11.6. A 3 × 3 array of neural stem cells directly after transfer

11.10 Extracellular Matrix

In addition to cells, MAPLE DW has demonstrated the ability to transfer other biological materials, such as Matrigel®, with high reproducibility and uniformity, as shown in Fig. 11.8. Precise, uniform transfer of extracellular matrix is critical to tissue engineering applications of MAPLE DW, since it can be used in combination with cells to construct three-dimensional structures. Engineered tissues can be constructed using controlled placement of multiple cell combinations and other biological materials such as extracellular matrix. This heterogeneous construction technique allows precisely engineered tissue replacements to be built that truly mimic native tissues in both structure and composition.

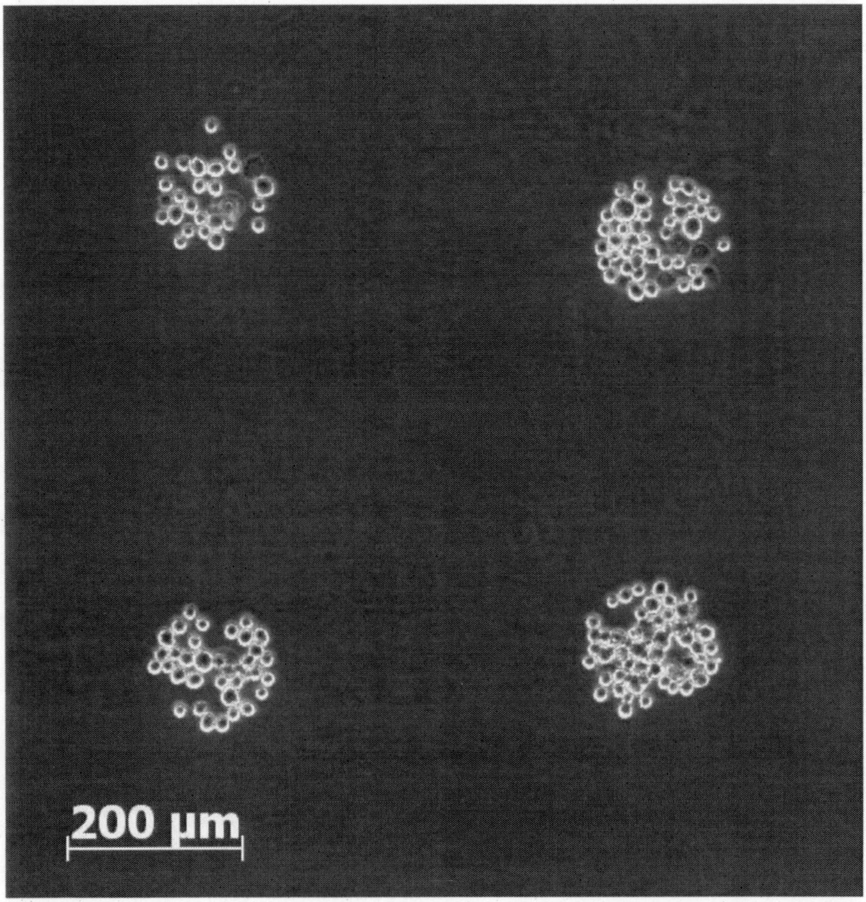

Fig. 11.7. A 2 × 2 array of bovine pulmonary artery endothelial cells (BPAEC) directly after transfer

11.11 Reproducibility and Repeatability

Reproducibility in biological environments is something for which many researchers strive. MAPLE DW possesses high reproducibility due, in part, to the strict requirements imposed by its original applications in the field of microelectronics. In our system, the variability in laser energy was measured in situ using an energy meter, indicating shot-to-shot energy variations (µJ) less than 7%. Positional reproducibility as identified by the resolution of the computer controlled motorized ribbon and receiving stages, is on the order of roughly 1 µm. The extension of MAPLE DW to cellular applications introduces other factors that affect system variability, such as surface quality and thickness of the compound layer used to attach the cells to the ribbon. Variations in surface quality or thickness in the ribbon's cell attachment

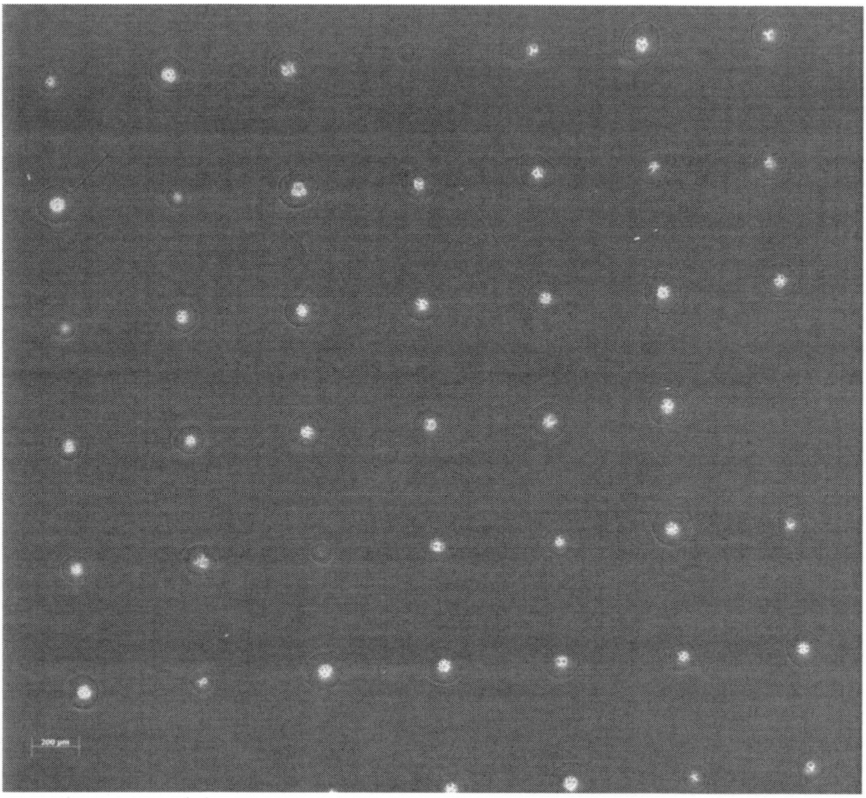

Fig. 11.8. Matrigel® transferred by MAPLE DW showing the reproducibility and repeatability of the transferred spots on a micro and macro scale

compound (e.g., biopolymer, ECM) can cause changes in the distance of cells from the quartz ribbon surface. These changes can cause shifts in the focus of the cells from the laser, and introduce artifact into the transfer process, such as increasing spot size. By utilizing spin-coating techniques to place the Matrigel® or other extracellular matrix compounds onto the ribbon and receiving substrate, the exact thicknesses of material on these surfaces can be controlled, in a reproducible manner. Furthermore, spin-coating produces a biopolymeric layer with a smooth, consistent, uniform surface, suitable for both cell attachment and laser transfer. This combined precision provides a level of biological reproducibility not previously seen.

Uniformity of the transferred biological materials can be further increased through the use of a biologically inert, but laser activated, dynamic release layer. Materials that have strong interactions with UV light can be incorporated into the ribbon assembly as a thin layer between the quartz ribbon and the biopolymer necessary for cellular attachment. The result is a layer of material that absorbs the UV light transmitted through the quartz ribbon, and reacts strongly. One such material, triazene, is biologically inert, and has been utilized to create a dynamic release layer for cellular application of MAPLE DW [30]. Triazene has been spin coated (approximately 50-nm thick) on to the quartz ribbon to create an efficient absorber of UV laser energy with no negative cellular effects. Biopolymers that allow for cell attachment can be easily spin coated onto the triazene layer. In the MAPLE DW process, the triazene layer connecting the biopolymer to the quartz ribbon is efficiently volatized by the laser pulse, and the biopolymer and attached cells are freely detached from the quartz ribbon. Incorporation of a dynamic release layer material, such as triazene, can further increase the reproducibility of the depositions as well as reduce any stray UV light that makes it to the cell.

11.12 Conclusions

The ability to accurately transfer single cells or groups of cells, and control their placement and proximity to additional cell types, is necessary to explore the many fundamental aspects of tissue engineering and regenerative medicine research. By controlling cell type and proximity one can prescribe the types of cells in communication, as well as the manner in which they communicate. When touching, cells communicate through direct cell–cell interaction, when they are close but not in direct contact they use paracrine signaling, and when the distance between cells gets sufficiently large, communication is established through endocrine signaling. Since the response of cells in culture will be greatly influenced by these factors, it is crucial to establish a precise multicell culture representative of the desired tissue, structure or pathology to be studied, to gain truly relevant insight from culture. MAPLE DW provides the technology and reproducibility required for this level of research.

In addition to cells, the other building blocks of tissues (e.g., extracellular matrix, and proteins) can also be transferred with MAPLE DW. Thus, utilizing these materials with the identification and specification capabilities offered by machine vision, true three-dimensional tissues can be engineered and generated. Cells and extracelluar matrix can be stacked in alternating layers, like bricks and mortar, to build tissues. Three-dimensional constructs have been shown to have significant advantages over their two-dimensional counterparts, with respect to structure, function, application, and cell signaling [31]. Some of the challenges of three-dimensional tissue fabrication can be overcome using MAPLE DW since it provides the ability to precisely control both the location and number cells as well as scaffold materials. Machine vision will allow for accurate placement of the proper constituents. By utilizing multiple cell types on multiple ribbons, features such as vascular structures can be built *into* the thickness of a deposited tissue. This would provide a significant advancement in tissue engineering, since the limiting factor of in vitro tissue growth is diffusion of nutrients into the tissue. If vascular structures can be built into the in vitro tissue, the dimensions of engineered constructs would no longer be diffusion limited.

There are some limitations to the system, especially with regard to the biologic cell attachment polymer. Matrigel® does not provide the ideal ribbon or receiving surface treatments. Matrigel® can have variations from lot to lot with regard to viscosity and composition. It also has unknown amounts of various growth factors that may affect different cell types differently, independent of the variables being studied. Also Matrigel® is relatively expensive and difficult to work with since it changes from a solid below 0°C, to a liquid at 0°C, to a gel at anything above 4°C. This requires precise temperature controls during spin coating so that the Matrigel® remains a liquid to achieve a thin, uniformly spin layer.

11.13 Future Directions

Improved materials, for both the receiving substrate and the cellular attachment to the ribbon, need to be created or found. Such a material must be inert, which means that it cannot affect the cells in any way through chemical signals such as growth factors. It must also be easily applied via spin-coating, in a reproducible manner, and must provide adequate energy dissipation, or "cushion," to the cells during transfer. Furthermore, it must contain the proper moisture content so that the cells do not dry out during deposition. As a receiving substrate, the material must not interfere with normal cell–cell interaction as well as cellular proliferation. MAPLE DW has proven to be a platform that can be translated to biomaterials and multiple cell types with single-cell resolution for high throughput experimentation. This technology allows for precise control over cell placement for the development of engineered tissues and combinatorial arrays of idealized constructs. Building three-dimensional tissues on a cell-by-cell basis from the ground-up has become feasible. With the development of direct writing of vascular features into a three-dimensional tissue, there will no longer be dimensional limitations due to diffusion. Furthermore, with the incorporation of machine vision and multiple ribbons with multiple cell types and biomaterials, it may be possible to engineer in vitro microenvironments that replicate those observed in vivo. Such achievements would greatly shape the future directions in tissue engineering and regenerative medicine research.

References

1. M.A.K Liebschner, M.A. Wettergreen, in *Topics in Tissue Engineering*, vol. 1, ed. by N. Ashammakhi, P. Ferretti (2003)
2. G. Wei, Q. Jin, W.V. Giannobile, P.X. Ma, Biomaterials **28**, 2087 (2007)
3. M. Jager, T. Feser, H. Denck, R. Krauspe, Ann. Biomed. Eng. **33**, 1319 (2005)

4. H.C. Ko, B.K. Milthorpe, C.D. McFarland, Eur. Cell. Mater. **14**, 1 (2007)
5. Y.S. Zinchenko, R.N. Coger, J. Biomed. Mater. Res. A **75**, 242 (2005)
6. A. Folch, B.H. Jo, O. Hurtado, D.J. Beebe, M. Toner, J. Biomed. Mater. Res. **52**, 346 (2000)
7. S. Rohr, R. Fluckiger-Labrada, J.P. Kucera, Pflugers Arch. **446**, 125 (2003)
8. E.A. Roth, T. Xu, M. Das, C. Gregory, J.J. Hickman, T. Boland, Biomaterials **25**, 3707 (2004)
9. W. Tan, T.A. Desai, Tissue Eng. **9**, 255 (2003)
10. M. Duocastella, M. Colina, J.M. Fernandez-Pradas, P. Serra, J.L. Morenza, Appl. Surf. Sci. **253**, 7855 (2007)
11. M. Duocastella, J.M. Fernandez-Pradas, P. Serra, J.L Morenza, Appl. Phys. A **93**, 453 (2008)
12. M. Colina, P. Serra, J.M. Fernandez-Pradas, L. Sevilla, J.L. Morenza, Biosens. Bioelectron. **20**, 1638 (2005)
13. K. Salaita, Y. Wang, C.A. Mirkin, Nat. Nanotechnol. **2**, 145 (2007)
14. V. Dinca, A. Ranella, M. Farsari, D. Kafetzopoulos, M. Dinescu, A. Popescu, C. Fotakis, Biomed. Microdevices **10**, 719 (2008)
15. D.B. Chrisey, Science **289**, 879 (2000)
16. B.R. Ringeisen, D.B. Chrisey, A. Pique, H.D. Young, J. Jones-Meehan, R. Modi, M. Bucaro, B.J. Spargo, Biomaterials **23**, 161 (2002)
17. C.Z. Dinu, V. Dinca, J. Howard, D.B. Chrisey, Appl. Surf. Sci. **253**, 5 (2007)
18. T.M. Patz, A. Doraiswamy, R.J. Narayan, W. He, Y. Zhong, R. Bellamkonda, R. Modi, D.B. Chrisey, J. Biomed. Mater. Res. B Appl. Biomater. **78**, 124 (2006)
19. N.R. Schiele, R.A. Koppes, D.T. Corr, K.S. Ellison, D.M. Thompson, L.A. Ligon, T.K.M. Lippert, D.B. Chrisey, Appl. Surf. Sci. in press (2008)
20. B.R. Ringeisen, H. Kim, J.A. Barron, D.B. Krizman, D.B. Chrisey, S. Jackman, R.Y. Auyeung, B.J. Spargo, Tissue Eng. **10**, 483 (2004)
21. A. Doraiswamy, R.J. Narayan, M.L. Harris, S.B. Qadri, R. Modi, D.B. Chrisey, J. Biomed. Mater. Res. A **80**, 635 (2007)
22. B.R. Ringeisen, P.K. Wu, H. Kim, A. Pique, R.Y. Auyeung, H.D. Young, D.B. Chrisey, D.B. Krizman, Biotechnol. Prog. **18**, 1126 (2002)
23. D.B. Chrisey, A. Pique, R.A. McGill, J.S. Horwitz, B.R. Ringeisen, D.M. Bubb, P.K. Wu, Chem. Rev. **103**, 553 (2003)
24. A. Pique, D.B. Chrisey, R.C.Y. Auyeung, J. Fitz-Gerald, H.D. Wu, R.A McGill, S. Lakeou, P.K. Wu, V. Nguyen, M. Duignan, Appl. Phys. A **69**, S279 (1999)
25. B. Hopp, T. Smausz, N. Kresz, N. Barna, Z. Bor, L. Kolozsvari, D.B. Chrisey, A. Szabo, A. Nogradi, Tissue Eng. **11**, 1817 (2005)
26. J.N. Cawse, Acc. Chem. Res. **34**, 213 (2001)
27. P. Martin, Science **276**, 75 (1997)
28. M. Yashiro, K. Ikeda, M. Tendo, T. Ishikawa, K. Hirakawa, Breast Cancer Res. Treat. **90**, 307 (2005)
29. K. Barami, J. Clin. Neurosci. **15**, 5 (2008)
30. A. Doraiswamy, R.J. Narayan, T. Lippert, L. Urech, A Wokaun, M. Nagel, B. Hopp, M. Dinescu, R. Modi, R.C.Y. Auyeung, D.B. Chrisey, Appl. Surf. Sci. **252**, 4743 (2006)
31. J. Lee, M.J. Cuddihy, N.A. Kotov, Tissue Eng. Part B Rev. 14, 61 (2008)

12

Ultrafast Laser Processing of Glass Down to the Nano-Scale

Koji Sugioka

Summary. Ultrafast lasers can induce strong absorption in materials and even in transparent materials, due to nonlinear multiphoton absorption. By using this phenomenon, surface microstructuring and dicing of glass are successfully demonstrated. When the ultrafast laser is focused inside a transparent material with adequate pulse energies, absorption can be confined to a region near the focus point allowing for internal processing of the transparent material such as three-dimensional (3D) optical waveguide writing and fabrication of micro-optical components and microchannels buried inside the glass. Another important feature of ultrafast lasers is the suppression of heat diffusion to the surroundings of the processed area, which makes nanoscale fabrication possible. In addition, nonlinear multiphoton absorption can further improve the spatial resolution beyond that of the laser. In this chapter, the features of ultrafast laser processing are first described and clarified. Then, some relevant topics of glass processing including nanoscale fabrication are reviewed.

12.1 Introduction

The rapid development of ultrafast lasers, such as femtosecond (fs) lasers, is opening up new avenues for material processing, and ultrafast lasers are becoming a common tool for various applications. Ultrafast lasers are now used for several practical applications including photomask repair [1] and ink jet nozzle drilling [2]. The extremely short pulse width of an ultrafast laser of several tens femtoseconds to a few picoseconds minimizes the heat-affected zone in the processed region, allowing for high-quality microfabrication of soft materials such as biological tissues [3] as well as hard or brittle materials such as semiconductors and insulators [4]. Such suppression of heat diffusion to the surroundings of the processed areas can improve the spatial resolution of processing to the nanoscale [5]. Meanwhile, the extremely high peak powers generated can induce strong absorption even in transparent materials due to the nonlinear multiphoton absorption as discussed later. This phenomenon allows for surface microstructuring and dicing of transparent materials such as glass [6]. By focusing the laser beam inside the glass with moderate pulse energy, the multiphoton absorption can be confined to a region near the focus point, and, thereby, internal modification of transparent materials can be realized. One of the most active research areas in this field is the writing of optical waveguides buried in glass by refractive index modification [7, 8]. Refractive index modification can also be applied to the fabrication of other buried optical microcomponents including optical couplers and splitters [9], volume Bragg gratings [10], diffractive lenses [11], and waveguide lasers [12]. Another interesting research area is the formation of three-dimensional (3D) hollow microstructures by internal modification followed by selective etching in wet chemicals, with the final products being microchips used for chemical analysis and biomedical inspections, such as in optofluidics and micro-total analysis systems (μ-TAS) [13, 14]. In addition to the suppression of heat diffusion, nonlinear multiphoton absorption can also improve the spatial resolution [15]. The formation of nanovoids [16] and nanochannels [17] inside glass has been successfully demonstrated.

In this chapter, the features of ultrafast laser processing are first explained and clarified, and important factors to achieve nanoscale resolution are discussed. Some relevant topics of glass surface and volume processing down to the nanoscale are then introduced.

12.2 Features of Ultrafast Laser Processing

12.2.1 Minimal Thermal Influence

One of the most important features of ultrafast laser processing is the suppression of heat diffusion to the surroundings of the processed region, due to the extremely short pulse width of several tens of femtoseconds to a few picoseconds, allowing for high-quality microfabrication of soft materials as well as hard or brittle materials. When the pulse width of the laser is shorter than the electron–phonon coupling time in laser–matter interactions, thermal diffusion to the inside of a material can be almost ignored. In the case of most metals, the electron–phonon coupling time is in the order of picoseconds [18], which is sufficiently long compared to the pulse width of ultrafast lasers. In this regime, the thermal diffusion length l_{d}, when the material is heated to around the melting point T_{im} by ultrafast laser irradiation, is given by:

$$l_{\mathrm{d}} = \left[\frac{128}{\pi}\right]^{1/8} \left[\frac{DC_{\mathrm{i}}}{T_{\mathrm{im}}\gamma^2 C'_{\mathrm{e}}}\right]^{1/4}, \tag{12.1}$$

where D is the heat conductivity, C_{i} is the lattice heat capacity, C'_{e} is given by $C'_{\mathrm{e}} = C_{\mathrm{e}}/T_{\mathrm{e}}$ (C_{e} is the electron heat capacity and T_{e} is the electron temperature), and γ is the electron–phonon coupling constant [19]. For example, when copper is heated up to its melting point of $T_{\mathrm{im}} = 1,356\,\mathrm{K}$ by an ultrafast laser, l_{d} is calculated to be 329 nm [20].

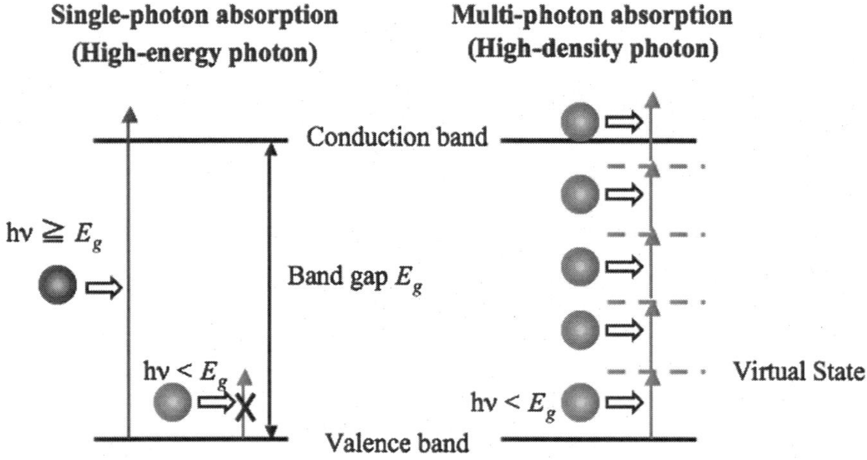

Fig. 12.1. Electron excitation process in materials by single and multiphoton absorption

On the other hand, when the pulse width of the laser τ is much longer than the electron–phonon coupling time, l_d can be roughly calculated as:

$$l_d = \sqrt{\kappa\tau}. \tag{12.2}$$

Here, κ is the thermal diffusivity. For copper, l_d is estimated to be 1.5 μm for τ = 10 ns. Thus, an ultrafast laser can clearly reduce the thermal diffusion length, which means that it can minimize the heat-affected zone at the processed region.

12.2.2 Multiphoton Absorption

Another important feature of ultrafast laser processing is that strong absorption can be induced even in materials that are transparent to the femtosecond laser beam, by the phenomenon of nonlinear multiphoton absorption. Figure 12.1 explains single and multiphoton absorption based on the electron excitation process. The normal absorption process is linear single-photon absorption. When light, whose photon energy is larger than the band gap of a specific material, is incident on the material, it is absorbed by the material and an electron is excited from the valence band to the conduction band by a single photon. On the other hand, light whose photon energy is smaller than the band gap cannot excite electrons, so no absorption occurs at the stationary state. However, when extremely high-density photons are incident on the material, an electron can be excited by multiple photons even if the photon energy is smaller than the band gap. This phenomenon is referred to as multiphoton absorption. Such extremely high-density photons inducing multiphoton absorption can be easily obtained using ultrafast lasers due to the ultra-short

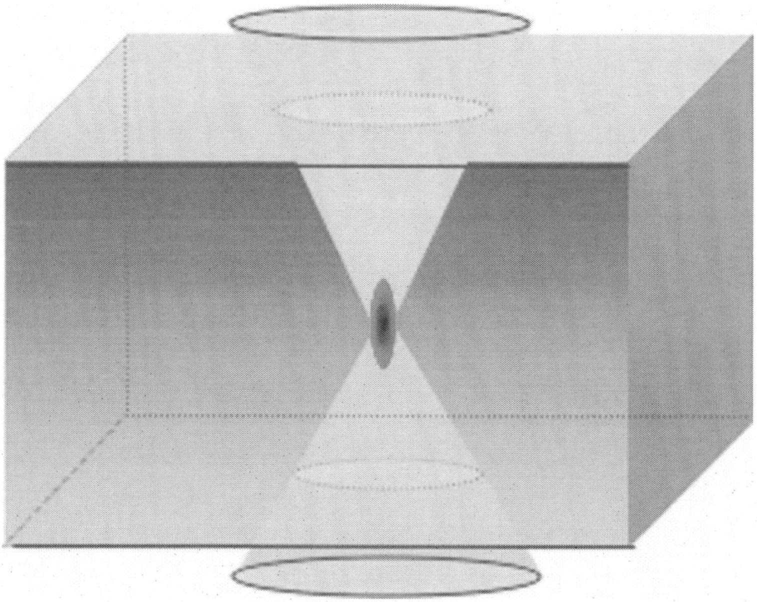

Fig. 12.2. Schematic image of internal modification in a material by multiphoton absorption using an ultrafast laser

pulse width. Thus, ultrafast lasers can induce strong absorption even in transparent materials, thereby allowing for high-quality microprocessing of glass materials.

12.2.3 Internal Modification

Multiphoton absorption is a nonlinear process and can be induced only at intensities above a threshold which is dependent of both the material and the pulse width. When the femtosecond laser beam is focused inside a transparent material with adequate pulse energy, as shown in Fig. 12.2, absorption can be confined to a region near the focus point inside the material. Thus, internal modification and fabrication of transparent materials can be performed, and these are possible only using ultrafast lasers. The internal modification can be applied to 3D optical waveguide writing, and fabrication of micro-optical components and microchannels buried inside glass, as discussed later.

12.3 Spatial Resolution of Ultrafast Laser Processing

Ultrafast lasers can suppress heat diffusion to the surrounding of the processed area, as discussed in Sect. 12.2.1, which presents an advantage for yielding higher spatial resolutions. When a 10-ns pulse laser is irradiated on to Cu with a spot size the same as the laser wavelength (typically several hundred nm to 1 μm), the processed region becomes larger than the spot size due to the thermal diffusion length of 1.5 μm. On the other hand, since thermal diffusion can be almost ignored for the ultrafast laser irradiation, it is unlikely that the processed region extends beyond the spot size.

The use of nonlinear multiphoton absorption can further improve the spatial resolution. Ideally, the ultrafast laser beam has a Gaussian profile of the spatial intensity, as shown by the thick dashed line in Fig. 12.3. For single photon absorption, the spatial distribution of the laser energy absorbed by the material corresponds to this beam profile. However, for multiphoton absorption, the distribution of the absorbed energy becomes narrower as the order (n) of the multiphoton absorption increases, since the effective absorption coefficient for n-photon absorption is proportional to the nth power of the laser intensity. Therefore, the effective beam size ω for the n-photon absorption is expressed by:

$$\omega = \omega_0/\sqrt{n}, \tag{12.3}$$

for the actual spot size ω_0 of the focused laser beam. Figure 12.3 shows the spatial distributions of the laser energy absorbed by transparent materials in two- (solid line) and three-photon (thin dashed line) absorption. From (12.3), a spatial resolution much smaller than the wavelength can be expected for multiphoton absorption. In addition, when a threshold in the laser intensity exists above which a reaction only takes place after absorption, the fabrication resolution can be further improved by adjusting the laser intensity. For example, if the threshold intensity for the reaction corresponds to the solid and straight line in Fig. 12.3, the fabrication width can be reduced to 2/5th that

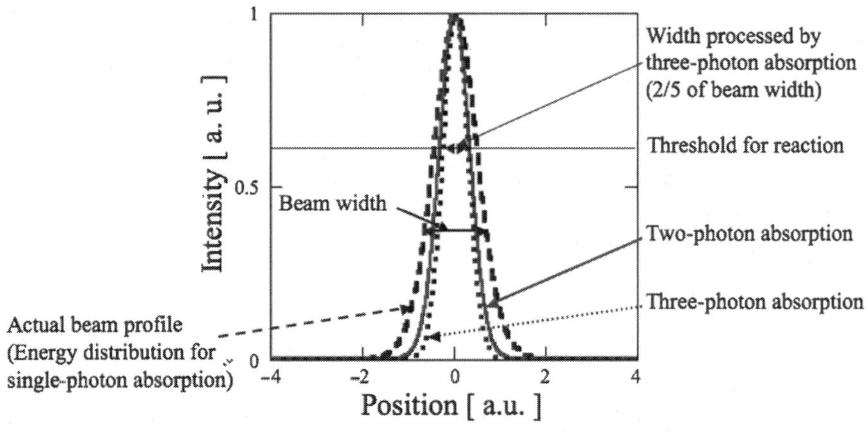

Fig. 12.3. Actual beam profile (*thick dashed line*) and spatial distributions of laser energy absorbed by transparent materials by two- (*solid line*) and three-photon (*thin dashed line*) absorption. The solid and straight line corresponds to the threshold intensity of the reaction

of ω_0. Thus nonlinear multiphoton absorption can overcome the diffraction limit of a laser beam and thereby achieve subdiffraction-limit resolution. In two-photon photopolymerization and photolithography a typical resolution of around 100 nm is achieved [15, 21], using this feature of ultrafast lasers, which can be further improved to 30 nm by very carefully adjusting the laser intensity [22].

12.4 Surface Micromachining

As discussed in Sect. 12.2.1, the suppression of heat diffusion for ultrafast lasers allows for high-quality and high-precision microfabrication even of brittle materials such as glass. In the ablation mechanism of glass by the ultrafast laser, multiphoton absorption, as described in Sect. 12.2.2, generates free electrons in the conduction band at first. These generated free electrons are accelerated by the strong electric field of the ultrafast laser and collide with the surrounding atoms and ions, so that additional electrons are generated by avalanche ionization. Finally, the generation of excessive free electrons induces surface break-up (Coulomb explosion), resulting in ablation. Since the ablation plasma is produced in several tens to several hundred picoseconds after the laser pulse irradiation which is longer than the pulse width of ultrafast laser, the pulse energy of ultrafast laser ablation is effectively deposited into the material without shielding the latter part of the laser pulse by the plasma (plasma shielding). This is another advantage of ultrafast laser processing.

Figure 12.4 shows a scanning electron microscope (SEM) image of glass microstructured by ultrafast laser ablation. The edges and side walls of the fabricated structure are sharp and smooth, respectively. Furthermore, no damage or cracks are observed around the processed region.

12.5 Internal Modification of Refractive Index

Ultrafast lasers can be used to perform internal modification of glass, as described in Sect. 12.2.3. In 1996, Davis et al. reported permanent change of the refractive index and optical waveguide writing inside glass by an ultrafast laser [7]. Currently, many researchers are investigating writing of optical waveguides embedded in various glasses such as fused silica, borosilicate glass, and chalcogenide glass, among others. The detailed mechanism of refractive index modification is still under investigation; however, a thermal model [23] and a color center model [7] have been proposed. The diameter of the core of the written optical waveguides, whose refractive index is increased by 10^{-4}–10^{-2}, ranges from 2 to 25 μm depending on the writing parameters used. The propagation loss of an optical waveguide written in fused silica was evaluated to be ca. 0.8 dB cm^{-1} [24].

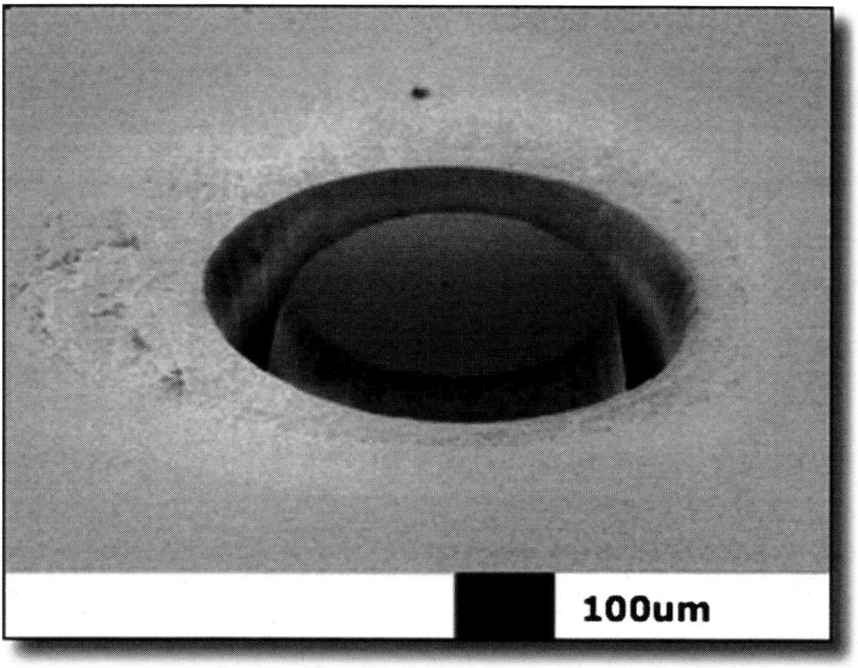

Fig. 12.4. SEM image of glass microstructured by ultrafast laser ablation (courtesy of M. Gower (Exitech Ltd.))

Refractive index modification was applied to the fabrication of 3D optical micro-devices such as optical couplers and splitters [9], volume Bragg gratings [10], diffractive lenses [11], and distributed feedback (DFB) lasers [12]. A 1×3 optical splitter was fabricated in fused silica by 3D integration of three optical waveguides, as shown in Fig. 12.5 [25]. In this case, the sample was moved perpendicular to the laser beam axis for the optical waveguide writing (side writing). Figure 12.6 shows the near-field distribution of light with 1.05 μm wavelength (fiber-coupled laser diode) guided by the fabricated structure. The splitting ratio (32:33:35) for the three beams is almost equal.

In addition to multiphoton absorption, ultrafast lasers can also induce other non-linear optical effects of self-focusing of the laser beam in transparent materials due to the Kerr effect, when the laser beam is focused by a lens with a low numerical aperture (NA) of around 0.05–0.1. This phenomenon induces a refractive index change with 10–500 mm length along the propagation direction of laser beam in silicate glass. The translation of the laser beam perpendicular to the optical axis forms a layer of refractive-index change. Stacking the layers periodically allows for the fabrication of volume gratings. A diffraction efficiency of 74.8% was achieved for the fabricated structure with a grating period of 3 μm and a thickness of 150 μm [26]. Meanwhile, two-beam interference of ultrafast lasers, in which two laser beams split from a single

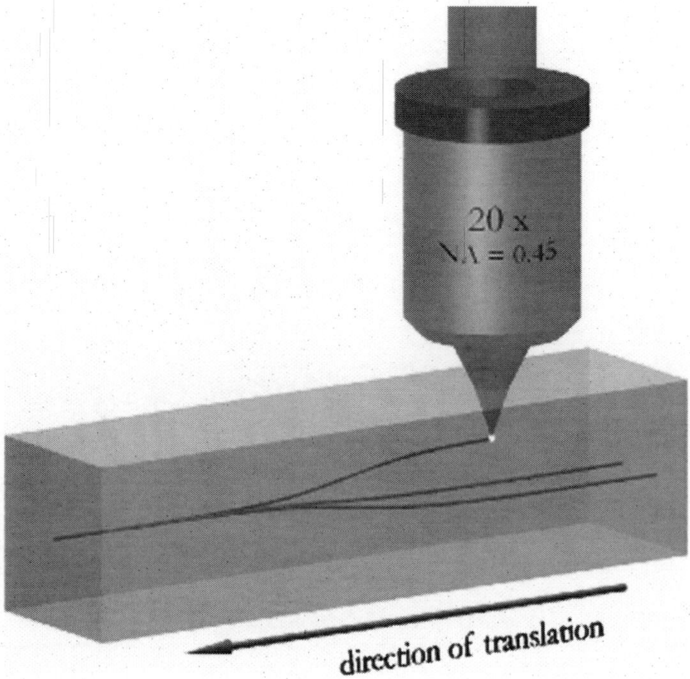

Fig. 12.5. Schematic illustration of the writing scheme of optical waveguides and a fabricated 3D optical splitter (courtesy of S. Nolte)

ultrafast laser, must spatiotemporally overlap in a single pulse irradiation, encoding the volume gratings even inside glass [27]. A period of the grating as narrow as 1 μm or less was realized by this technique.

Self-focusing of ultrafast lasers can be utilized for the fabrication of multilevel phase-type diffractive lenses by the 3D translation of the laser beam inside a glass. Figure 12.7 shows the schematic design (a) and optical microscope images of a four-level diffractive lens (side (b) and top (c)), and a spot image of a He-Ne laser focused by the fabricated lens (d). The efficiency and focused spot size were evaluated to be 56.9% and 9.9 μm, respectively [28].

Optical waveguides written in active materials, such as in the case of erbium-ytterbium-doped phosphate glass by an ultrafast laser, acts as a compact and efficient single longitudinal mode laser at 1.5 μm wavelength providing up to 55 mW maximum output power [12]. By encoding refractive index-modulated volume-type gratings in the written optical waveguide using a single interfered femtosecond laser pulse, a distributed feedback (DFB) laser was realized in lithium fluoride (LIF) [29].

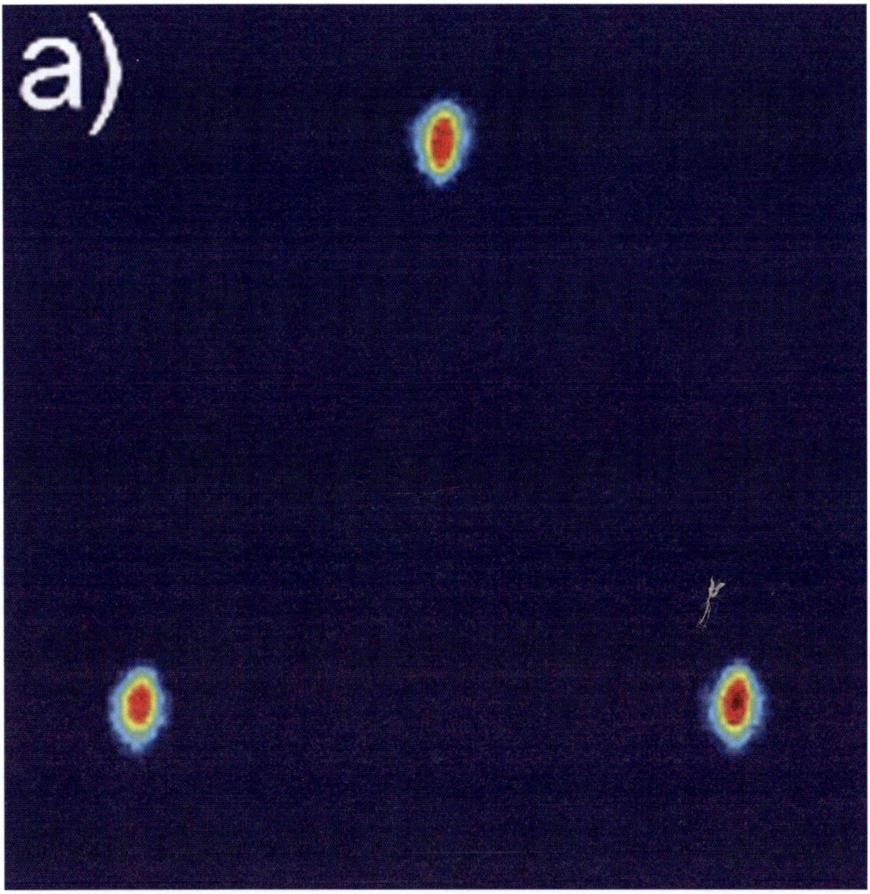

Fig. 12.6. Near-field intensity distribution at 1.05 μm measured at the exit of the splitter. The split beams are separated by 100 μm (courtesy of S. Nolte)

12.6 Fabrication of 3D Hollow Structures

The internal modification in materials by ultrafast lasers induced a change not only in the refractive index but also the chemical properties of the materials. Ultrafast laser direct writing followed by chemical wet etching in a dilute HF acid solution formed 3D hollow microstructures in fused silica, including 3D channels as narrow as 10 μm in diameter inside the volume with any angle of interconnection between the channels and a high aspect ratio [30]. Photosensitive glass, which is a lithium aluminosilicate glass doped with trace amounts of silver and cerium, is a more attractive material for this application due to high efficiency and high throughput of processing as well as much smoother etched surfaces, although a thermal treatment is necessary before the wet etching.

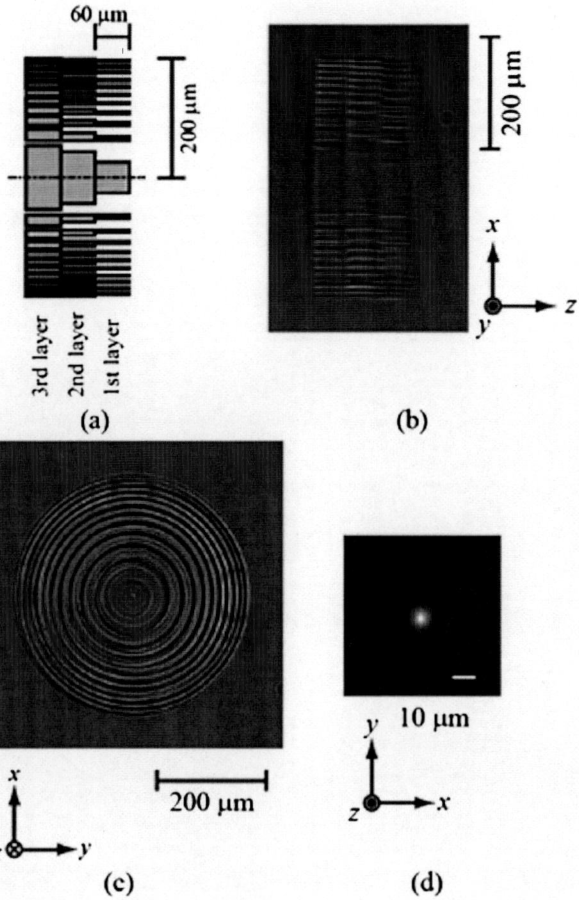

Fig. 12.7. Schematic design (**a**) and optical microscope images of a four-level diffractive lens (side (**b**) and top (**c**)), and a spot image of a He–Ne laser focused by the fabricated lens (**d**) (courtesy of W. Watanabe)

Figure 12.8 shows a prototype of a microfluidic device fabricated in photosensitive glass by the present technique. This device contains a microplate detached from the glass matrix that can move in the microfluidic chamber and can switch the flow direction of liquid samples.

The present technique of fabricating 3D hollow microstructures embedded in photosensitive glass can also be used to fabricate microoptics components such as mirrors, beam splitters and lenses. The integration of microfluidics and microoptics can be used to realize a microfludic dye laser, which is a tunable light source in the visible range. Another interesting application of the fabricated 3D hollow microstructures is the manufacture of microchips designed for dynamic observations of microorganisms, which has been referred to as a *nano-aquarium* [31]. The *nano-aquarium* has several advantages over conventional

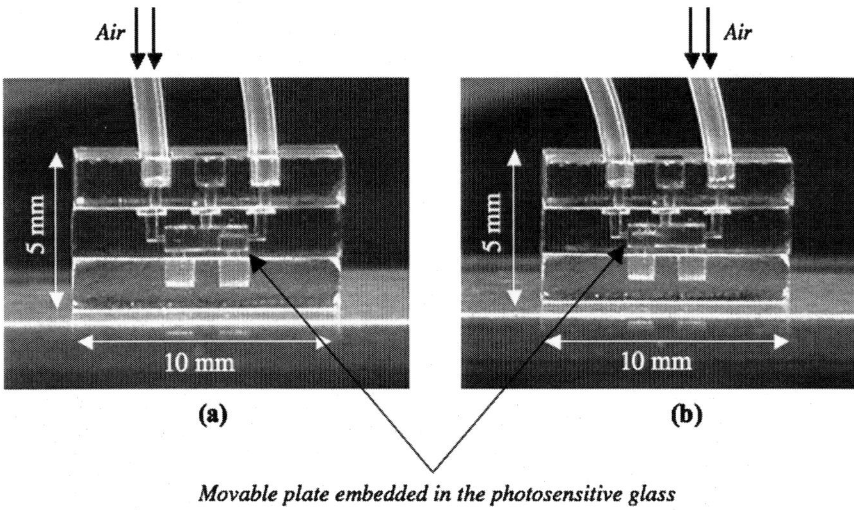

Fig. 12.8. Prototype of a microfluidic device with a freely movable microplate which can switch the flow direction of liquid samples. (**a**) By infusing compressed air from the left opening of the top, the microplate moves to the right side. (**b**) As the compressed air is infused from the right opening of the top part, the microplate moves to the left side

observation methods, namely, it can significantly shorten observation times, and can be used for 3D observations. In addition, the microorganisms can be easily stimulated during the observation.

12.7 Integration of Optical Waveguide and Microfluidics for Optofluidics Applications

Optical waveguides and microfluidics can be easily integrated in a single glass chip using a single ultrafast laser system for the manufacture of microchips for biochemical analyses and medical inspections. Such integrated microchips are often referred to as optofluidics. Microchannels integrated with optical waveguides in fused silica were used to successfully demonstrate site-selective excitation of fluorescence in several different points in a microchannel [32].

Figure 12.9 shows a schematic illustration of 3D integrated optofluidics for photonic biosensing, in which one optical waveguide 6 mm in length is connected to a microfluidic chamber of $1.0 \times 1.0 \times 1.0\,\mathrm{mm}^3$ volume and also two microlenses with 0.75 mm curvature radius are arranged on the left side of the microchamber and at the opposite side from the optical waveguide across the microchamber at a distance of 300 µm. This device can be used for fluorescence and absorption measurements. The inset shows an optical microscope image of the fabricated microchip. Experimental demonstration of photonic biosensing

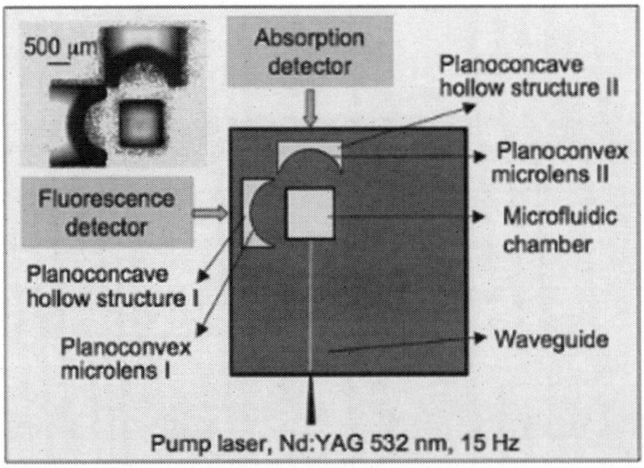

Fig. 12.9. Schematic configuration of optofluidics in which microoptics, such as microoptical planoconvex lenses and an optical waveguide, are integrated with a microfluidic chamber in a single glass chip. The optical microscope image of the top view of the fabricated microchip is shown in the upper left corner

using the integrated microchip has revealed that fluorescence analysis and absorption measurement of liquid samples can be performed with efficiencies enhanced by factors of 8 and 3, respectively [33].

12.8 Nanofabrication

When an ultrafast laser is focused by an objective lens with NA over 1.0, the spot size at the focus point can be almost the same as its wavelength. According to (12.3), the effective beam size becomes much smaller than the wavelength in the case of multiphoton absorption. In fact, nanodots 400 nm in diameter were formed in fused silica by refractive index modification using the ultrafast laser with 800 nm wavelength [34]. Increasing the laser intensity forms nanovoids with sizes as small as 200–300 nm in diameter in the glass. The 3D arrangement of nanovoids inside fused silica allows for 3D optical data storage with an ultra high density of Gbits cm^{-3} [35]. Meanwhile, in one study, by arranging the nanovoids periodically, a photonic crystal was fabricated [36]. The advantage of this technique is that introduction of defect structures into the photonic crystal is very easy, which is very attractive for photonic crystal research and applications.

Super-resolution beyond the diffraction limit of a laser beam can also be obtained in ablation. Three-dimensional nanofluidic channel networks were formed in fused silica by ultrafast laser ablation. In this case, the ultrafast laser beam at 527 nm wavelength was focused through a high NA objective lens from the rear surface of a glass substrate immersed in water, as shown in Fig. 12.10 [17]. The water plays an important role to remove ablated materials

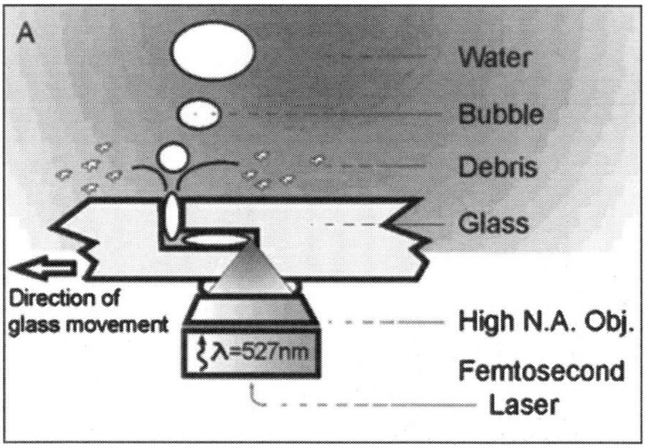

Fig. 12.10. Schematic illustration of the scheme for nanochannel fabrication in fused silica by ultrafast laser ablation in water (courtesy of A. Hunt)

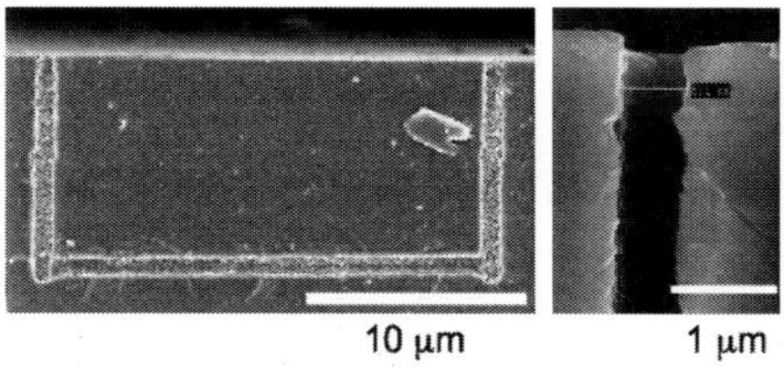

Fig. 12.11. SEM photographs of nanochannels embedded in fused silica (**a**) and a closeup view (**b**) (courtesy of A. Hunt)

from the formed nanochannels. By scanning the glass substrate through the laser focus, embedded 3D nanochannels with a structure with an extremely small diameter (<700 nm) and a relatively long length (>200 μm) can easily be produced free of debris as shown in Fig. 12.11.

Another interesting topic in the field of ultrafast laser nanostructuring is the formation of periodic nanostructures with much smaller periods than expected from the interference between incident and scattered waves. In fact, multiple pulse irradiation of a linearly polarized femtosecond laser beam at fluences around the ablation threshold can produce nano-ripple structures on various materials with a period ranging from one-half to one-tenth of the laser wavelength [37]. Periodic nanostructures can even be fabricated inside silica glass [38]. The period of the formed grating structures can be controlled from 140 to 320 nm by changing the pulse energy and the number of pulses. The formation mechanism was explained in terms of interference between the incident light field and the electric field of the bulk electron plasma wave, inducing periodic modulation of the electron plasma concentration.

12.9 Conclusions

Ultrafast lasers are now becoming common tools for micro- and nano-scale structuring, and they are being used for high quality and high precision microfabrication of various materials. They have already been used for commercial applications such as photomask repair and ink jet nozzle drilling. In addition, the technique is powerful for the direct formation of 3D microstructures inside glass since such a process is only possible by using an ultrafast laser. Spatial resolution of around 200 nm in fabrication of structures in glass, and around 100 nm in photopolymerization and 3D lithography has been achieved by using the nonlinear absorption effect. Further challenges can improve the resolution below 100 nm.

References

1. R. Haight, D. Hayden, P. Longo, T.E. Neary, A. Wagner, J. Vac. Sci. Technol. **B17**, 3137 (1999)
2. C.H. Chen, X.B. Liu, in *Proceedings of ICALEO 2005* (Laser Institute of America, Jacksonville, FL, 2005), M401
3. M.F. Yanik, H. Cinar, H.N. Cinar, A.D. Chisholm, Y.I. Jin, A. Ben-Yakar, Nature, **432**, 822 (2004)
4. N. Barsch, K. Korber, A. Ostendorf, K.H. Tonshoff, Appl. Phys. A **77**, 237 (2003)
5. Y. Nakata, T. Okada, M. Maeda, Appl. Phys. Lett. **81**, 4239 (2002)
6. P. Rudolph, J. Bonse, J. Kruger, W. Kautek, Appl. Phys. A **69**, 763 (1999)
7. K.M. Davis, K. Miura, N. Sugimoto, K. Hirao, Opt. Lett. **21**, 1729 (1996)
8. T. Gorelik, M. Will, S. Nolte, A. Tünnermann, U. Glatzel, Appl. Phys. A **76**, 309 (2003)
9. W. Watanabe, T. Asano, K. Yamada, K. Itoh, J. Nishii, Opt. Lett. **28**, 2491 (2003)
10. L. Sudrie, K.A. Winick, J. Lightwave Technol. **21**, 246 (2003)
11. E. Bricchi, J.D. Mills, P.G. Kazamsky, B.G. Klappauf, J.J. Baumberg, Opt. Lett. **27**, 2200 (2002)
12. G.D. Valle, S. Taccheo, R. Osellame, A. Festa, G. Cerullo, P. Laporta, Opt. Exp. **84**, 3190 (2007)
13. A. Marcinkevicius, S. Juodkazis, M. Watanabe, M. Miwa, S. Matsuo, H. Misawa, Opt. Lett. **26**, 277 (2001)
14. K. Sugioka, Y. Cheng, K. Midorikawa, Appl. Phys. A **81**, 1 (2005)
15. S. Kawata, H.B. Sun, T. Tanaka, K. Takada, Nature **412**, 697 (2001)
16. S. Juodkazis, H. Misawa, T. Hashimoto, E.G. Gamaly, B. Luther-Davies, Appl. Phys. Lett. **88**, 201909 (2006)
17. K. Ke, E.F. Hasselbrink, A.J. Hunt, Anal. Chem. **77**, 5083 (2005)
18. S.I. Anisimov, B. Rethfeld, Proc. SPIE **3093**, 192 (1997)
19. P.B. Corkum, F. Brunel, N.K. Sherman, T. Srinivasan-Rao, Phys. Rev. Lett. **61**, 2886 (1988)
20. M. Fujita, M. Hashida, Oyo Butsuri **73**, 178 (2004) (in Japanese)
21. S. Maruo, H. Inoue, Appl. Phys. Lett. **89**, 144101 (2006)

22. S. Juodkazis, V. Mizeikis, K.K. Seet, M. Miwa, H. Misawa, Nanotechnology **16**, 846 (2005)
23. J.W. Chan, T.R. Huser, S. Risbun, D.M. Krol, Opt. Lett. **26**, 1726 (2001)
24. M. Will, S. Nolte, B.N. Chichkov, A. Tuennermann, Appl. Opt. **41**, 4360 (2002)
25. S. Nolte, M. Will, J. Burghoff, A. Tuennermann, Appl. Phys. A **77**, 109 (2003)
26. K. Yamada, W. Watanabe, K. Kintaka, J. Nishii, K. Itoh, Jpn. J. Appl. Phys. **42**, 6916 (2003)
27. K. Kawamura, M. Hirano, T. Kamiya, H. Hosono, Appl. Phys. Lett. **81**, 1137 (2002)
28. K. Yamada, W. Watanabe, Y. Li, K. Itoh, Opt. Lett. **29**, 1846 (2004)
29. K. Kawamura, M. Hirano, T. Kurobori, D. Takamizu, T. Kamiya, H. Hosono, Appl. Phys. Lett. **84**, 311 (2004)
30. A. Marcinkevicius, S. Juodkazis, M. Watanabe, M. Miwa, S. Matsuo, H. Misawa, J. Nishii, Opt. Lett. **26**, 277 (2001)
31. Y. Hanada, K. Sugioka, H. Kawano, I.S. Ishikawa, A. Miyawaki, K. Midorikawa, Biomed. Microdevices **10**, 403 (2008)
32. R. Osellame, V. Maselli, R.M. Vazquez, R. Ramponi, G. Cerullo, Appl. Phys. Lett. **90**, 231118 (2007)
33. Z. Wang, K. Sugioka, K. Midorikawa, Appl. Phys. A **93**, 225 (2008)
34. K. Hirao, Oyo Butsuri **67**, 950 (21998) (in Japanese)
35. E.N. Glezer, M. Milosavljevic, L. Huang, R.J. Finlay, T.H. Her, J.P. Callan, E. Mazur, Opt. Lett. **21**, 2023 (1996)
36. H.B. Sun, Y. Xu, S. JuodKazis, K. Sun, M. Watanabe, S. Matsuo, H. Misawa, J. Nishii, Opt. Lett. **26**, 325 (2001)
37. J. Reif, F. Costache, M. Henyk, S.V. Pandelov, Appl. Surf. Sci. **197–198**, 891 (2002)
38. Y. Shomotsuma, P.G. Kazansky, J. Qiu, K. Hirao, Phys. Rev. Lett. **91**, 247405 (2003)

13

Free Electron Laser Synthesis of Functional Coatings

Peter Schaaf and Daniel Höche

Summary. Functional and smart surfaces and coatings play an increasingly decisive role in the applicability and performance of modern materials. From an industrial point of view, there is a great interest with respect to friction, wear, corrosion, and further properties. Many methods have been developed for the improvement of the respective surface and material properties. Traditionally, these treatments range from simple PVD and CVD processes to complicated plasma and hybrid methods. Recently, it has been established that short laser pulses of high energy can induce a direct laser synthesis of functional coatings if the material's surface is irradiated in a reactive atmosphere. The process is based on a complicated combination of laser plasma – gas – material surface interactions. Tests for steel, aluminum, magnesium, titanium, and silicon in nitrogen, methane, and hydrogen atmospheres have been carried out successfully; with these materials, interesting coatings can be produced by direct laser synthesis, for example, AlN and SiC. Various laser types can be used for this purpose: Excimer, Nd:YAG, CO_2 Laser, and even the free electron laser (FEL).

Despite the simplicity of the treatment itself, up to now, the process has neither been completely understood nor established as an industrial application, possibly due to the lack of high repetition rate pulsed high power lasers enabling fast and easy treatment of large areas and pieces. Here, the FEL with its unique properties is just the right tool to drive the earlier mentioned process into the direction of applicability. Its high power and flexibility in its temporal shaping motivated experiments with the FEL on the direct laser synthesis of functional coatings.

The produced coatings were investigated by means of a number of methods. The obtained results and properties are presented and discussed in connection with different laser specialties. For the FEL treatment, it was found that the ability to tune the pulse timing can be used to tailor the coating structure and properties (hardness, strain, grain-size, etc.). This will be discussed with the help of modeling the temperature and plasma evolution and the solidification behavior during the FEL irradiations.

13.1 Introduction

13.1.1 The Free Electron Laser

Surface irradiation with free electron laser (FEL) light is a new topic in material processing research. The Jefferson Lab FEL is a unique tool to modify materials and study basic processes during treatments. It is a light source, based on an energy recovered linac [1, 2]. Figure 13.1 shows the scheme of the facility at the Jefferson lab (from [3]). Electron bunches were created by photoemission of GaAs and injected into the superconducting linac with up to 10 MeV. Then they were accelerated up to 150 MeV and aligned with several beam optics. Before the electrons emit light, bunches were compressed in a magnetic chicane. Afterwards, a broadband THz beam was extracted, which could be used to investigate the functionality of the FEL. The short bunches itself were directed through the optical cavity and as a result of the electron acceleration in the wiggler, they emitted a tunable narrow-band light known as the laser beam. Additionally, the light source was coherent and had good polarization properties with a ratio about 6,000:1. All these flexibilities

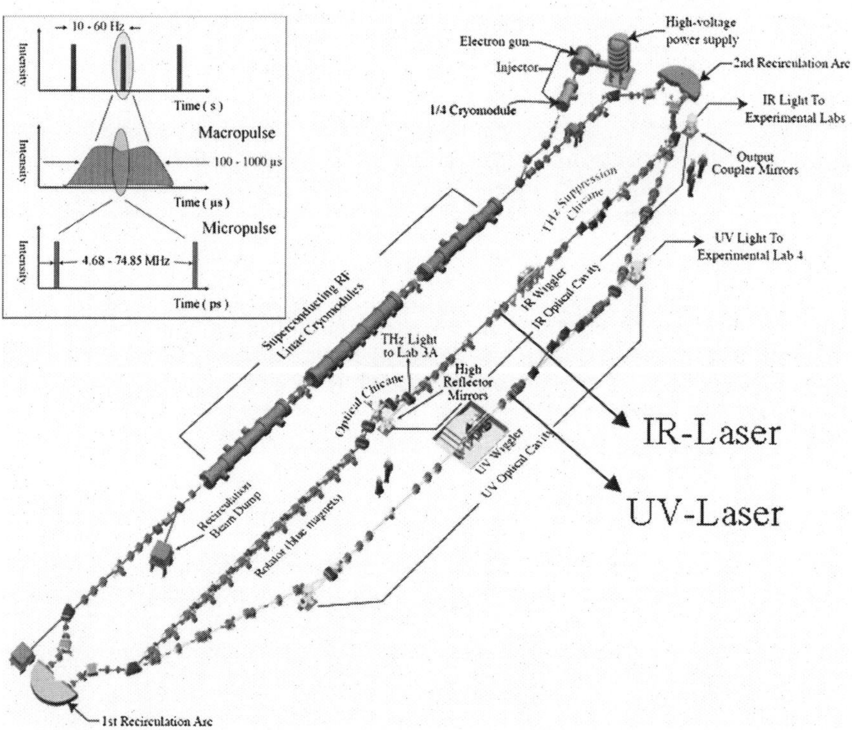

Fig. 13.1. Scheme of the Jefferson Lab free electron laser facility (from [3]) and the time structure at pulsed mode

Table 13.1. FEL beam parameters

	IR	UV
Wavelength range (μm)	1–14	0.25–1
Bunch length (FWHM ps)	0.2–2	0.2–2
Laser power/pulse (μJ)	100–300	25
Laser power (kW)	>10	>1
Repetition rate (cw operation, MHz)	4.68–74.85	4.68–74.85

initiate new application fields and research possibilities, especially for material processing.

Currently the 4th generation of the FEL is in use. The laser is capable of emitting cw – mode like trains of subpicosecond pulses (micro pulses of 0.2–2 ps in FWHM) up to an average power of 10 kW in a range of 1–14 μm in wavelength. The frequency of the micro pulses is tunable between 4.68 MHz and 74.85 MHz in steps of 2n where n is an integer from one to eight [4]. Alternatively the setup can be switch to pulse mode. In this mode, packages of micro pulses (macro pulse) were emitted with frequencies up to 60 Hz (see Fig. 13.1). The pulse duration therefore was some hundred microseconds. Moreover, it was possible to run the FEL in the UV – branch, but with decreasing power output. Table 13.1 shows the actual beam parameters at Jefferson Lab.

13.1.2 Direct Laser Synthesis

The direct laser synthesis is a simple and innovative process in relation to many surface cladding techniques. The worksheets were placed in a chamber fulfilled with a reactive gas, typically nitrogen or methane and then irradiated with the focused laser beam. As a result of the local heating, induced melting, and plasma formation, gas diffusion occurred into the sheets and a coating formation could be observed, whose thickness mainly would be determined by the diffusion coefficient and the melting depth. Many experiments and investigations have been performed successful for the nitriding of titanium with CO_2 lasers [5,6] or Nd:YAG lasers [7–11]. In the last case, the coatings have a thickness of about 2 μm. Other experiments show the successful synthesis of Fe_3N [12], AlN [13], TiC [14] and other compounds [15]. First synthesis of functional coatings by means of the FEL has been shown in [16–23]. The results show interesting coating properties and dependencies on the scan parameter. That indicates the possibility of tailoring the treatments for industrial applications.

Contrary to the simple technical process, the governing physical mechanisms are very complex. Laser material interactions like absorption, induce melting and phase transformations takes place. At high energy densities, the process will be assisted by plasma formation and expansion into the ambient gas. Due to the shock wave and the laser light absorption in dense gases or plasmas, the gas molecules or atoms can be dissociated or ionized [24].

This rapidly increases the activity of the gas which results in an amplified adsorption at the liquid surface and lastly the coating formation.

At a timescale long enough (milliseconds), convectional flow in the melt pool occurs. This is determined by the Marangoni and the recoil pressure induced force. As a result, the surface quality decreases but contrarily the gas atom flux into deeper regions gets higher. As regards the tribological properties such as wear resistance and hardness, the solidification process is the most important. The nucleation and phase formation in the modified tracks mainly determine the solid properties like grain size and stress. The following enumeration shows the involved processes during laser nitriding:

- Laser light absorption and heating
- Melting and evaporation (ablation, surface recession)
- Plasma expansion into the ambient gas (recoil pressure)
- Dissociation and ionization → increase of gas activity – plasma shielding
- Gas absorption (Sievert's law)
- Mass transport: melt (convective/diffusive) or in the solid (diffusive)
- Nucleation, solidification and phase formation → coating properties

Because of complicated interactions between the different physical processes, control is the main problem. Therefore, it is necessary to get quantified data about the interactions by means of experiments and in the case of experimental inaccessibility, by means of simulations.

13.1.3 Protective Coatings and TiN

Thin films, coating technologies, and cladding techniques are of great interest in current research. Thus, several methods like PVD, CVD or sputtering have been optimized and combined with other techniques (hybrid). New high-tech alloys have been developed with increasing mechanical, optical or electrical properties. They had extensive corrosion resistance, hardness or were used as thermal barriers. For industrial applications, it is necessary to protect strained components and assemblies in an effective way. In the case of commercial metallic components mostly based on iron, titanium or aluminum alloys, the simplest technique is nitriding or carburizing.

A typical example is the synthesis of titanium nitride. It can be used as a direct protection of titanium or as an additional coating on several alloys. The compound and its phase diagram have been studied as well. The nitrogen solubility in pure α-Ti was determined to be 23 at.%. In the case of stoichiometric δ-TiN, it has a cubic lattice (Fm3m, 225) with a lattice constant of 4.24 Å. Pure TiN has a golden like color and has hardness up to 25 GPa. Additionally, it has strong chemical resistivity and high melting point of $3,220°$K.

In order to carry out basic studies on the direct laser synthesis, the titanium–nitrogen system was chosen. It is a popular compound in material science and industries and allows comparisons with other coating methods.

13.2 Experiments

13.2.1 Sample Preparation and setup

Commercial titanium sheets (1 mm thickness, purity >99.98%) were cut into pieces of $15 \times 15\,\text{mm}^2$ size. For the laser treatments, the samples were placed in a chamber, first evacuated and then filled with nitrogen (purity 99.999%) to a pressure of 1 to 5×10^5 Pa. The focused beam reached the sample surface through a fused silica window. In order to treat the whole surface of the samples, the chamber was mounted onto a computer-controlled x-y table. A relative velocity v_{scan}, the lateral shift δ of the tracks, the spot size D, and the pulse frequency f are the main scan parameters. Figure 13.2 shows the experiments at lab 2 at the FEL facility.

The treatments were performed in cw – and pulse mode. Therefore some blind tests have been executed in order to get some information about the process parameters and their influences. Scan velocity, spot size, lateral shift, laser power or macro pulse duration have been varied in a parametric study. In pw-mode the sheets were treated in a meandering scheme. Table 13.2 presents

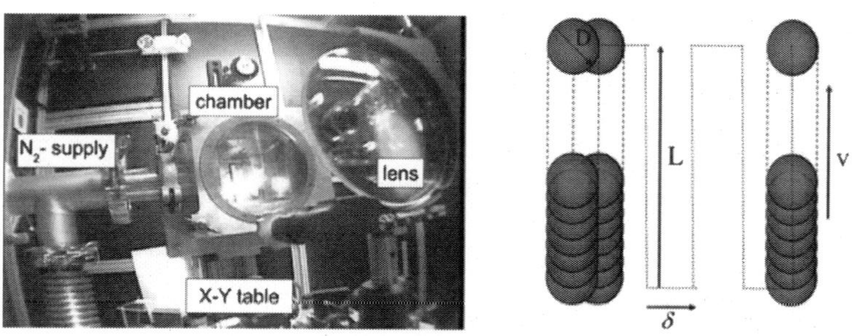

Fig. 13.2. Experimental setup at the FEL facility and meandering scan scheme

Table 13.2. Scan- and process parameters used during the treatments at the different modes

Parameter	CW	pulsed
Wavelength range (μm)	1.6	3.1
Bunch length (FWHM ps)	0.2	0.5
Laser power/pulse (μJ)	125	20
Laser power (W)	650	160–750
Repetition rate (cw operation, MHz)	4.68	37.4
Scan velocity v_{scan} $(\text{mm}\,\text{s}^{-1})$	24	0.5
Spot size D (μm)	600	440
Lateral shift δ (μm)	400–2,000	100–200
Macropulse duration (μs)	–	250–1,000
Gas pressure (atm.)	1.15	1

the scan parameters at the different laser modes, where coating formation was successfully observed.

13.2.2 Analysis Methods

The microstructure of the synthesized coating was analyzed by X-ray diffraction in grazing incidence (GIXRD), Bragg – Brentano and Rocking curve geometry using a Bruker AXS diffractometer equipped with a $Cu - K_\alpha$ tube and a thin film attachment. Peak analyses yielded lattice constants, average nitrogen contents, stresses, textures, and crystallite/grain size.

Nitrogen depth profiling was carried out by means of the resonant nuclear reaction analysis (RNRA) employing the reaction $^{15}N\,(p, \alpha\gamma)^{12}\,C$. The measurements were performed at the Göttingen IONAS accelerator. Details are given in [25, 26]. The nitrogen depth profiles were limited to a depth of approximately 500 nm due to the limited proton energy of maximal 500 keV.

The microhardness depth profiles were measured with a Nanoindenter (Fischerscope HV100). It operates with a Vickers diamond tip and a maximum load of 1N. Scanning electron microscopy (SEM) was performed for surface analyses attached with EDX measurements and imaging cross-section micrographs (FEI Nova 600).

13.3 Results

13.3.1 FEL Irradiation at CW-Mode

As demonstrated in Fig. 13.3 a melting track occurs during the irradiation. The nitrogen reacts with the melt and the synthesis of titanium nitride takes place. They get a gold like color and are quite inhomogeneous.

As expected, the track properties are mainly determined by the melt flow. Marangoni convection and pressure induced melt modifications resulted in a strong roughness. Humps and melt ejection were not being observed. Further a periodical structure is visible as a result of the equilibrium of the surface acting forces. Short wavelength structures can be observed too. Due to the oscillations on the liquid titanium such modifications are developed. They are formally known as Rayleigh–Taylor instabilities.

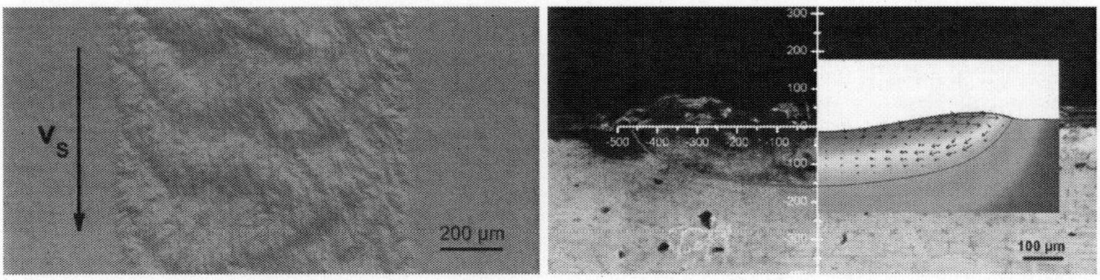

Fig. 13.3. Nitrated tracks in top view and as cross section

Numerical studies have shown a strong influence of the convection on the track or respectively coating properties [27]. The Marangoni force induced flow velocities up to $1\,\mathrm{m\ s^{-1}}$. Convective heat transfer becomes the main determining process. The describing number in fluid mechanics is the Peclet number (Pe). It reaches values of about 60. That is the reason for the low aspect ratio of the melted tracks. Further the diffusive nitrogen transport can be assisted by convection. Due to the mixing in the liquid pool the coating thickness will be determined by the melting depth. In the shown examples this depth was about 200 μm.

XRD measurement in grazing incidence geometry at 5° shows that only δ-TiN has been developed. The information depth is about 600 nm. Figure 13.4 presents the diffraction pattern of a selected sample with multiple tracks at a distance of 600 μm. The virgin α-Ti was observed too, but at a low content.

The Bragg–Brentano scan on the right-hand side at Fig. 13.4 shows a strong [200] peak which indicates the development of a weak [200] fiber texture. As a result the titanium nitride lattice is directed perpendicular to the surface.

Cross-section micrographs have been performed too (Fig. 13.5) in order to take into account the phase formation during solidification. They show the redirected TiN dendrites near the surface. Their distribution is quite inhomogeneous due to the varying temperature conditions during the treatments.

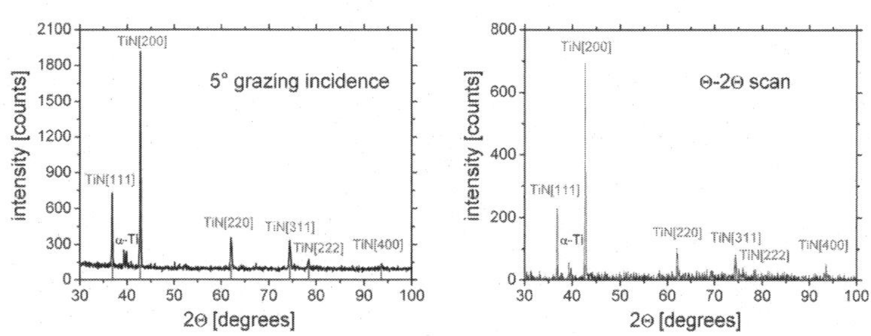

Fig. 13.4. Grazing incidence (*left*) and Bragg-Brentano (*right*) diffraction pattern of the selected sample

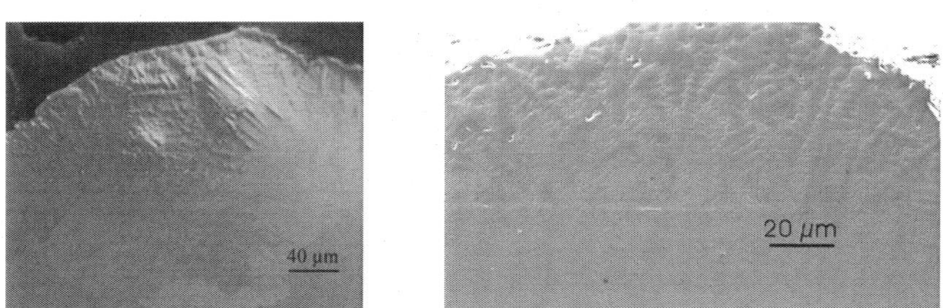

Fig. 13.5. Cross-section micrographs of titanium nitride tracks at different scales

Dendritic solidification seems to be the main mechanism during cw irradiation titanium with the FEL. The top-down direction is a result of the difference in melting temperature of Ti and TiN of about 1,250°K. The directed lattice and the stoichiometric TiN phase are the reasons for the improved tribological properties.

13.3.2 FEL Irradiation at Pulsed Mode

At these treatments coatings with very varying properties and thicknesses up to 20 μm have been generated. They have interesting and correlating properties to their scan and beam parameters. The most representable sheets have been investigated and are studied next. Their scan parameters are shown in Table 13.3.

The surface properties are very different and show a strong dependence on the scan parameters. Figure 13.6 presents SEM micrographs of the coatings. Samples 1 and 2 had solidified melting droplets and some cracks as a result of the remelted titanium and the resulting induced intrinsic stress. For longer macropulse durations melting droplets could be avoided as a result of exceeding the evaporation point. The energy entry for sample 3 was four times higher than for sample 4. As a consequence, sample 3 had many cracks and was still fragile. From a technical point of view, sample 4 has the best properties. The coating is relatively smooth and without any fractures.

In order to understand the melting behavior, numerical simulations have been performed by means of the finite element method (FEM). Heat transfer and phase transitions were studied. Detailed information is available in [18]. The results show that the surface temperature during the treatment

Table 13.3. Macropulse duration τ_{ma}, macropulse repetition rate f_{ma} and lateral shift δ (in y-direction) used for the FEL treatments

Sample	τ_{ma} [μs]	f_{ma} [Hz]	δ [μm]
1	250	60	100
2	750	30	100
3	1,000	30	100
4	1,000	10	200

The scan velocity in x-direction is 0.5 mm s^{-1}

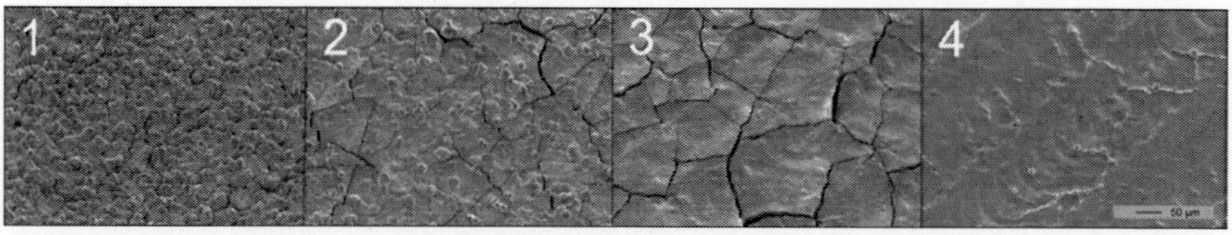

Fig. 13.6. Surface properties of selected samples

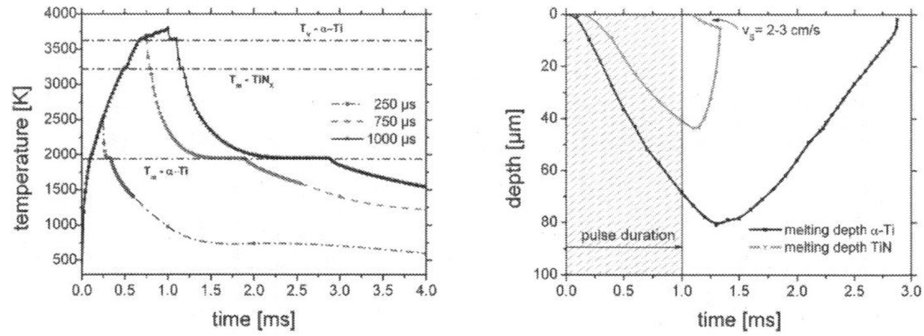

Fig. 13.7. Surface temperature development during a macropulse

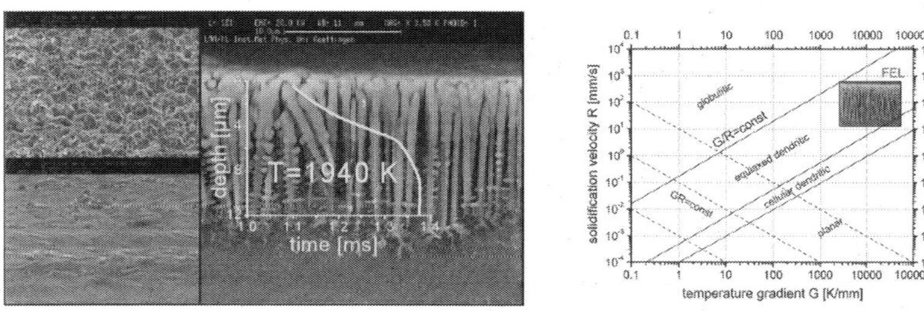

Fig. 13.8. Solidification behavior, melting temperature isotherm (TiN) of sample 4 (*left*) and cooling rate for FEL processing (*right*)

determines the properties. Figure 13.7 shows the temperature distribution during one macropulse. For longer pulse durations the surface temperature is high enough to evaporate titanium and remove the droplets. The increase in coating quality is a main result of the observations. Profilometrie of the samples shows decreasing roughness R_a from 3 to 1.2 µm.

The melting depth for the pure titanium and the generated TiN is shown on the right hand side of Fig. 13.7. The main information was obtained by the solidification behavior of the TiN melt front. Because of the big difference in thermal properties of titanium and its nitride, the solidification direction was achieved top-down. By means of the solidification velocity $R\sim\left(2\text{–}3\,\mathrm{cm\,s^{-1}}\right)$, the temperature gradient $\mathrm{G}\sim\left(10^4\,\mathrm{K\,mm^{-1}}\right)$, and the cooling rate $\mathrm{GR}\sim\left(20\times10^4\,\mathrm{K\,s^{-1}}\right)$ it is possible to determine the solidification mechanism.

Dendritic growth was observed close to the surface. Figure 13.8 shows the solidification behavior and its reasons. The dendrites are equiaxed and have side branching andarm spacing respectively in deeper regions. Because of the knowledge of G and R, it is possible to control the solid structure and at least the tribological properties.

With regard to mechanical loading, the coating hardness is the most important parameter. Wear resistance and friction coefficient can be improved in order to optimize several components or assemblies in technical applications. Figure 13.9 presents the results obtained for the selected samples.Becuase of bad surface quality (rifts, droplets), samples 1 to 3 are very inhomogeneous.

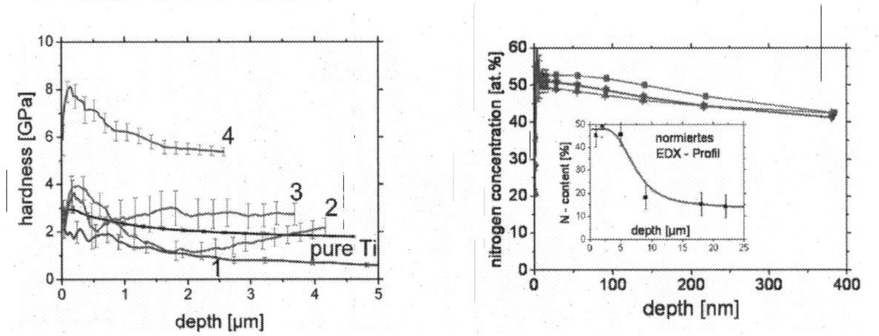

Fig. 13.9. Hardness depth profiles as measured by nanoindentation technique and nitrogen depth profiles

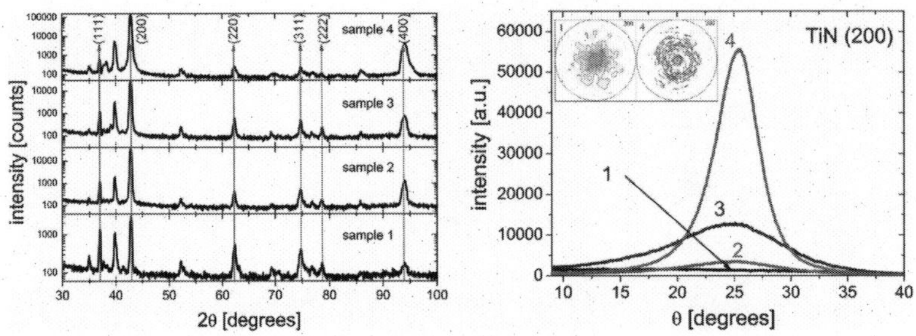

Fig. 13.10. Θ–2Θ scan, rocking curve and pole figures of the four selected samples. All indicated peaks are cubic TiN, the others belong to pure titanium

Sample 4 shows a strong improvement in the overall hardness to 8 GPa (film hardness 12 GPa). These properties are mainly determined by the phase transitions which are strong related to the nitrogen content. For maximal optimized coatings, stoichiometric TiN (50 at.%) has to be synthesized. Therefore, the RNRA measurements show results in agreement to that. At Fig. 13.9 the stable titanium nitride was observed over the whole measured range. Additional EDX investigations show the diffusion like profiles at deeper regions.

X-ray diffraction measurements resulted in correlating lattice properties. The development of a strong 200 fiber texture was observed by means of Bragg-Brentano and Rocking curve scans. Figure 13.10 shows that orientation behavior is assisted by pole figures.

These diffraction patterns verify the development of dendrites perpendicular to the surface. The directed lattice seems to be the reason for the improved mechanical properties and its strength.

13.4 Conclusions

The chapter represents the synthesis of functional coatings by FEL irradiation in reactive atmospheres. Therefore several investigations and experiments have been performed by means of the model-system titanium and nitrogen.

The running physical processes have been identified and were correlated to the synthesized titanium nitride. Coatings of 5 to 200 μm thickness have been produced. Their quality is mainly determined by the scan parameters. In the case of cw irradiation the coatings or tracks have high thickness but as a result of the strong acting forces their roughness and homogeneity is still improvable. The processing time is in the range of up to 20 s. Their tribological properties have been studied as well.

For the pulsed treatments, very satisfying tests have been performed. It was possible to synthesize coatings of 20 μm in thickness without fractures and melting droplets. They have low roughness of about one micron and have micro hardness up to 12 GPa [28]. Their properties are mainly determined by the dendritic solidification behavior and by the moderate heat entry, respectively. Nitrogen depth profiling confirms stoichiometric TiN at the near surface range. In deeper regions under stoichiometric titanium nitride in a stable phase was observed. The lattice has a strong 200 fiber texture and with an average induced strain at values up to 0.004.

Regarding cladding techniques, the investigations offer conclusions which are universally valid. They are necessary for effective synthesis of functional coatings in gas atmospheres by means of laser irradiation:

- Melting depth mainly determine the film thickness due to efficient diffusion in the liquid phase
- Exceeding the boiling point leads to plasma expansion and to activation of the reactive gas which ends in higher absorption rate and reduced roughness
- Convection can amplify the gas transport in the tracks and improve the mixing effect gradients
- Concentration is able to determine the solidification direction (top-down)

The synthesis process is an interaction of complicated physical and chemical mechanism. For optimized processing it is necessary to quantify them and correlate them to the scan parameters. Heat entries, that is, energy densities, have to be tailored in order to achieve the optimal conditions and to avoid fractures, stress, inhomogeneity, and roughness. From the industrial point of view, this new method is an alternative to other techniques due to the reduced processing duration of some seconds per square centimeter.

References

1. G. Williams, Rev. Sci. Instrum. **73**, 1461 (2002)
2. A. Thomas, G. Williams, Proc. IEEE **95**, 1679 (2007)
3. S. Benson, G. Biallas, J. Boyce, D. Bullard, J. Coleman, D. Douglas, F. Dylla, R. Evans, P. Evtushenko, A. Grippo, Nucl. Instum. Methods Phys. Res. A **582**, 14 (2007)
4. P. Evtushenko, J. Coleman, K. Jordan, J. Klopf, G. Neil, G. Williams, AIP Conf. Proc. **868**, 193 (2006)

5. I. Ursu, I.N. Mihailescu, A.M. Prokhorov, V.I. Konov, V.N. Tokarev, S.A. Uglov, J. Phys. D Appl. Phys. **18**, 2547 (1985)
6. M. Raaif, F. El-Hossary, N. Negm, S. Khalil, A. Kolitsch, D. Hoche, J. Kaspar, S. Mandl, P. Schaaf, J. Phys. D Appl. Phys. **41**, 085208 (2008)
7. D. Höche, H. Schikora, H. Zutz, A. Emmel, R. Queitsch, P. Schaaf, J. Coating. Tech. Res. **6**, (2008)
8. D. Höche, H. Schikora, H. Zutz, R. Queitsch, A. Emmel, P. Schaaf, Appl. Phys. A **91**, 305 (2008)
9. E. György, A. Perez del Pino, P. Serra, J.L. Morenza, Surf. Coating. Tech. **173**, 265 (2003)
10. H.C. Man, Z.D. Cui, T.M. Yue, F.T. Cheng, Mater. Sci. Eng. A **355**, 167 (2003)
11. J. Mori, P. Serra, E. Martianez, G. Sardin, J. Esteve, J. Morenza, Appl. Phys. A **69**, S699 (1999)
12. P. Schaaf, M. Han, K.-P. Lieb, E. Carpene, Appl. Phys. Lett. **80**, 1091 (2002)
13. E. Carpene, P. Schaaf, M. Han, K. Lieb, M. Shinn, Appl. Surf. Sci. **186**, 195 (2002)
14. B. Courant, J. Hantzpergue, L. Avril, S. Benayoun, J. Mater. Proc. Tech. **160**, 374 (2005)
15. P. Schaaf, Progr. Mater. Sci. **47**, 1 (2002)
16. E. Carpene, M. Shinn, P. Schaaf, Appl. Phys. A **80**, 1707 (2005)
17. D. Höche, G. Rapin, J. Kaspar, M. Shinn, P. Schaaf, Appl. Surf. Sci. **253**, 8041 (2007)
18. D. Höche, M. Shinn, J. Kaspar, G. Rapin, P. Schaaf, J. Phys. D Appl. Phys. **40**, 818 (2007)
19. E. Carpene, J. Kaspar, M. Shinn, P. Schaaf, J. Laser Micro/Nanoengin. **1**, 129 (2006)
20. E. Carpene, M. Shinn, P. Schaaf, Appl. Phys. A **80**, 1707 (2005)
21. M. Han, K. Lieb, E. Carpene, P. Schaaf, M. Shinn, Appl. Surf. Sci. **186**, 195 (2002)
22. E. Carpene, J. Kaspar, P. Schaaf, M. Shinn, in *International Conference on Laser Precision Microfabrication, LPM 2005, April 3–8, 2005, Williamsburg, VA, USA* (Japan Laser Processing Society, Osaka, 2005)
23. E. Carpene, J. Kaspar, P. Schaaf, M. Shinn, *3rd International WLT-Conference Lasers in Manufacturing, LIM 2005, June 2005, München* (AT-Fachverlag, Stuttgart, 2005)
24. D. Höche, G. Rapin, P. Schaaf, Appl. Surf. Sci. **254**, 888 (2007)
25. F. Landry, P. Schaaf, Nucl. Instrum. Methods Phys. Res. B **179**, 262 (2001)
26. M. Uhrmacher, K. Pampus, F.J. Bergmeister, D. Purschke, K.P. Lieb, Nucl. Instrum. Methods Phys. Res. B **9**, 234 (1985)
27. D. Höche, G. Rapin, S. Müller, M. Shinn, P. Schaaf, Metall. Mater. Trans. B **40**(4), pp. 497–507 (2009), DOI: 10.1007/s11663-009-9243-1
28. E. Carpene, M. Shinn, P. Schaaf, Appl. Surf. Sci. **247**, 307 (2005)

14

PLD of Piezoelectric and Ferroelectric Materials

Maria Dinescu

Summary. The direct piezoelectric effect consists of the generation of a macroscopic polarization in certain dielectric materials when subjected to stress. A class of the polar piezoelectrics is ferroelectrics. Ferroelectrics are characterized by a spontaneous polarization which can be switched by applying an electric field; its expression is a typical hysteresis loop polarization P–electric field E. They undergo a structural phase transition from a high temperature paraelectric into a low-temperature ferroelectric phase at Curie temperature. The preparation of these materials in thin film form is a challenging problem, due to their complex composition, the appearance of metastable phases, crystallinity, interface problems, defects, oxygen vacancies, etc. The possibility of transferring complicated stoichiometries directly from the target to a collector pushed pulsed laser deposition to a top position to be applied for piezoelectric and ferroelectric thin films growth. Results obtained in the field of pulsed laser deposition of piezoelectric and ferroelectric thin films are presented.

14.1 Introduction

Discovered by Pierre and Jacques Curie in 1880, the direct piezoelectric effect consists of the generation of a macroscopic polarization in certain dielectric materials when subjected to stress. The inverse effect appears when an external electric field applied on these materials produces a strain through the converse effect. For crystals the piezoelectricity is strongly determined by the symmetry [1]. Piezoelectrics can be divided into polar (which possess a net dipole moment) and non polar piezoelectric materials (whose dipolar moments summed in different directions give a null total moment). A material is said to be *ferroelectric* when it exhibits an electric dipole moment in the absence of an external electric field and spontaneous polarization can swap between the two states under the action of an electric field.

Classification of materials from the point of view of piezoelectricity and ferroelectricity is given in Fig. 14.1.

Piezoelectricity is characterized by several important coefficients. One of the most important is the square of the coupling factor k, which measures

Fig. 14.1. Materials classification

the fraction of stored energy in the conversion process and is defined by the following relationships [2]:

$$k_{m\lambda}^{2'} = \frac{d_{m\lambda}^2}{\varepsilon_{mm}^{TE'} S_{\lambda\lambda}} \tag{14.1}$$

$$D_i = di_\mu X_\mu + \varepsilon_{im}^T E_m \tag{14.2}$$

$$x_\lambda = s_{\lambda\mu}^E X_\mu + d_{m\lambda} E_m \tag{14.3}$$

where

1. d is the piezoelectric tensor which describes the direct piezoelectric effect (in (14.2)) and the inverse one (in (14.3)) (summation over the subscripts μ and λ extends from 1 to 6, and over the subscript m and l from 1 to 3)
2. s is the elastic compliance tensor
3. ε is the dielectric permittivity tensor

Another important coefficient is d_{33}, which represents the induced charge as a result of applied stress: d_{33} is measured in Coulomb/Newton.

The interest to obtain thin piezoelectric films was generated by the wide range of applications, like microsensors, micromotors, integrated electro-optic devices etc. [1,3–5]. Their development has also been stimulated by the need for piezoelectric transducers of very high frequency, which cannot normally employ bulk materials, since frequencies of about 1 GHz would normally require layers of few microns which cannot be obtained otherwise than by thin film deposition techniques.

Ferroelectric materials are characterized by a hysteresis loop which represents the polarization evolution with respect to the electric field applied (polarization $P\ (\mathrm{C\,m^{-2}})$–electric field $E\ (\mathrm{Vm^{-1}})$) [2,6]. The main part of ferroelectrics exhibits a phase transition between a ferroelectric phase at low temperature, and a non-ferroelectric phase (paraelectric) at higher temperature: the transition temperature is the Curie point of the material, T_C. At this value the dielectric permittivity has a maximum and then it decreases for higher temperature according to Curie–Weiss law:

$$\varepsilon = \varepsilon_0 + \frac{C}{T - T_0} \approx \frac{C}{T - T_0}$$

where T_0 ($T_0 < T_C$) is the Curie temperature and C the Curie constant. The phase transition can be of the first or the second order: a special class of materials is ferroelectric relaxors, which possess a diffuse phase transition. Applications are mainly related not only to the interest for non-volatile random access memories, but also for tunable microwaves applications such as oscillators, filters, resonators, phase shifter, actuators,and transducers, in medical diagnosis equipment (electrocardiography, ultrasounds) [7].

14.2 RF-Assisted Pulsed Laser Deposition

Pulsed laser deposition (PLD) is an excellent and versatile technique applicable to almost any material, in particular to compounds that are difficult or impossible to produce in thin-film form by other techniques.

A PLD typical setup is shown schematically in Fig. 14.2. The main components are the laser system, the reaction chamber with a vacuum system, and the target and the substrate holders - the latter being provided with a heating unit. This deposition technique consists of a sequence of processes: the removal (ablation) of material from the surface of a solid (or liquid) target irradiated with short and high energy laser pulses of specific wavelength followed by the condensation of the particles ejected on the surface of a substrate. These processes take place in a vacuum or in a gas atmosphere (inert or reactive gas).

Due to the high power density of the beam, a plasma plume perpendicular to the target surface is generated at the incidence point; the plume contains ions of the target and interacts with the surrounding atmosphere. The thin film is deposited on the surface of a substrate that is placed few centimeters apart from the target (usually 3–8 cm) and facing the top of the plasma plume [8,9]. The target is continuously rotated in order to ensure a uniform ablation and to avoid craters or cone formation [8]. This setup may be enriched by other different techniques for deposition assisting or for in situ characterization of plasma and/or thin films. In our particular case a radiofrequency source is added to the pulsed laser deposition setup. It generates (the RF plasma source) a beam containing excited and ionized species in the gas: this beam

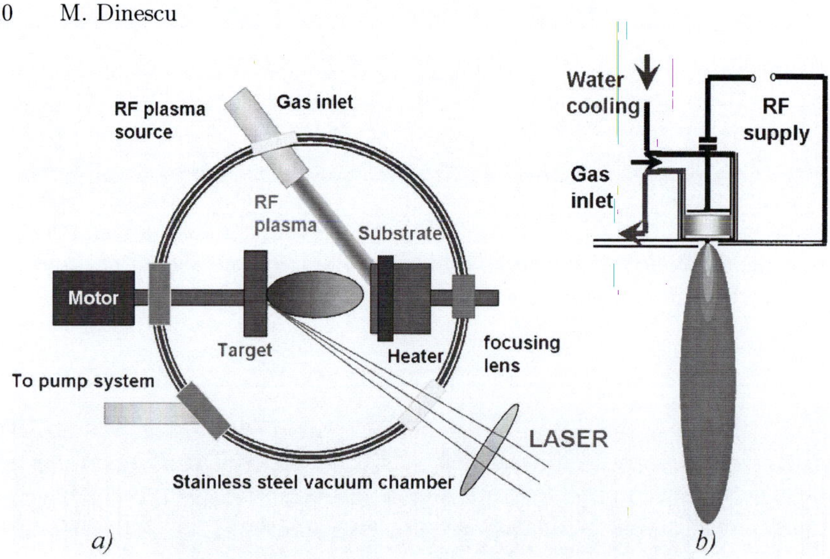

Fig. 14.2. (**a**) PLD typical setup; (**b**) RF plasma source

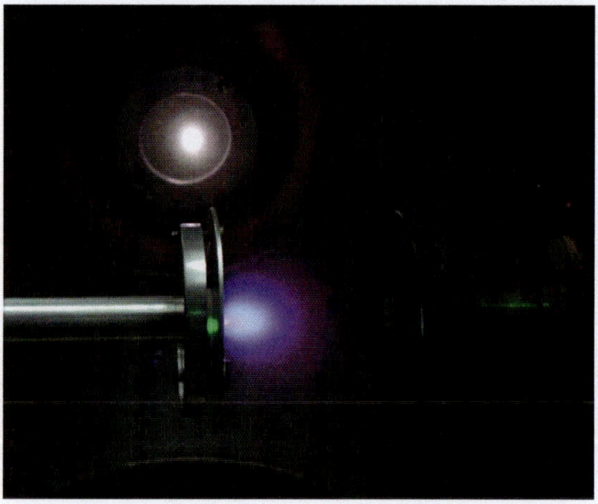

Fig. 14.3. Plasma plume during a deposition together with RF plasma source

can be directed to the substrate, to increase the reactivity during the growth process [8,9]. The RF system consists of a radio-frequency discharge generated at pressures in the range of 10–1,000 Pa between two parallel electrodes. The RF system is separated from the classic PLD system through an ejection nozzle. The plasma generated (in oxygen or other gas) expands insight into the deposition chamber, with the possibility to be oriented towards the substrate or the target, or the space in between. The RF system works simultaneously with the PLD system. Figure 14.3 presents a picture taken during an RF-PLD experiment.

The experimental parameter variation results in a strong influence on the film properties [10]. The temperature of the substrate during the deposition can be controlled by a heating unit in the range RT–1,273 K; temperature of the substrate was found to have high influence on the quality and orientation of the grown ZnO thin films [11–13].

Gas pressure strongly influences the crystalline structure and film surface morphology. It was shown [14, 15] that the growth of AlN thin films by PLD requires a background atmosphere free of oxygen, and/or the presence of nitrogen or ammonia reactive gas.

Results concerning pulsed laser deposition of thin films of several well known piezoelectric and ferroelectric compounds are presented in this paper. Materials like ZnO, AlN, PZT, PMN, and lead free (NKN, SBN, NBT–BT) are mainly emphasized.

14.3 Non-Ferroelectric Piezoelectrics

The best known piezoelectric materials which do not exhibit ferroelectric properties are ZnO and AlN.

14.3.1 ZnO

Zinc oxide presents high piezoelectric properties and high electromechanical coupling factor. In bulk, ZnO has the piezoelectric constant $d_{33} = 12.5\,\mathrm{pC/N}$. The crystalline structure is hexagonal (wurtzite type) (Fig. 14.4). In the wurtzite type cell, the layers 0, 2, 4 ... are exactly one on top of the other and the packing sequence is 01010101... The bonding with the closest neighbors is tetrahedral. The ZnO lattice constants are: $a = 3.2499 \pm 0.0001\,\mathrm{nm}$ and

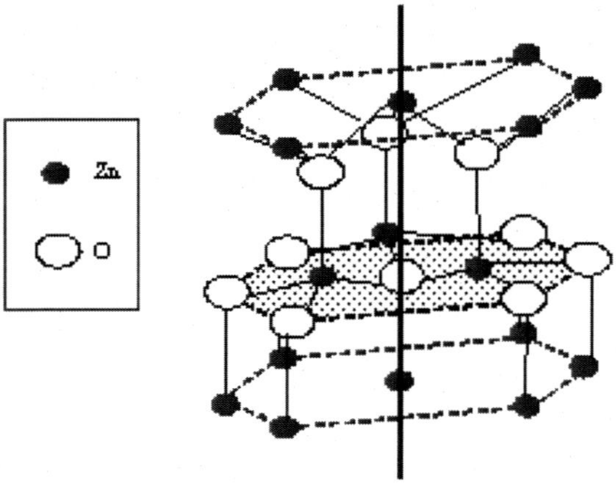

Fig. 14.4. Crystalline structure of ZnO

$c = 5.206 \pm 0.001$ nm. The elementary cell has four atoms, two zinc atoms at the sites (000) and $\left(\frac{1}{3}\frac{2}{3}\frac{1}{2}\right)$, and two of oxygen atoms at $\left(00\frac{3}{8}\right)$ and $\left(\frac{1}{3}\frac{2}{3}\frac{7}{8}\right)$.

Different methods were used to grow ZnO thin films: sputter techniques [16–18], chemical vapor deposition [19], molecular beam epitaxy [20, 21], and pulsed laser deposition [11,22]. Out of the typical applications based on piezoelectric properties (SAW and BAW devices), ZnO can also be used in several devices such as gas sensors, p–n nano-junctions, solar cells, UV laser diodes, nano-lasers, and transparent electronics.

PLD has become one of the most versatile methods for ZnO thin films growth. For the first time ZnO was grown as thin film at the beginning of the 1980s, by Shankur and Cheung [22]. A CO_2 laser was used working at $\lambda = 10.6\,\mu$m. The deposition took place in oxygen reactive pressure and the temperature was varied in the range 323–723 K. The films were crystalline, with c-axis perpendicular to the substrate surface.

Many other experiments were reported when nanosecond excimer lasers (ArF $\lambda = 193$ nm [23], KrF $\lambda = 248$ nm [24–27], and Nd-YAG [15, 16] [11, 12]) were used. Using UV radiation, the laser fluence was in the range 1–5 J cm^{-2}, whilst in IR, the fluence was much higher (20–30 J cm^{-2}). Femtosecond lasers were also used to deposit ZnO thin films, as reported by Millon et al. [28] and Okoshi [29].

As a target, ceramic ZnO pellets [24, 25] or metallic Zn [11, 12] in reactive oxygen atmosphere can be used for experiments. The distance between target to substrate was about 3–5 cm. The substrate temperatures are in a very wide range 473–773 K [24, 25], or from RT to 1,073 K [30]. The substrates used are a very important factor, due to the initial growth stages resulting from the mismatch film–substrate lattice [8]. Different substrates have been used: Si (100) and (111) [11,22,24], GaAs (100)(111) [22,25], Corning glass [22–24], sapphire [22, 31, 32], quartz [32] etc. Usually, the geometry is the standard geometry for PLD (the target parallel to the substrate), but different other settings can be used (Fig. 14.5).

Another important parameter for the morphology and structure of the ZnO films is the deposition gas pressure. In most of the reports, the gas was oxygen with pressure in a large range: 5×10^{-4}–10 Pa. In [31], Choopun et al. varied the oxygen pressure during the experiment, for a better control of the early stages of film formation. The pressure was 10^{-2} Pa for the first 10–50 nm after which it increased to 10 Pa.

The physical properties of zinc oxide films are highly dependent on their structure. Some results coming from reported studies on ZnO films deposited by PLD and RF-PLD in our laboratory will be discussed in detail.

Crystalline textured ZnO films with c-axis oriented perpendicular to the substrate are obtained usually, irrespective of the substrate nature and temperature. Different experimental conditions such as laser wavelength and fluence, oxygen pressure during deposition, and presence of the radiofrequency oxygen beam can influence the layers properties. An important effect of the

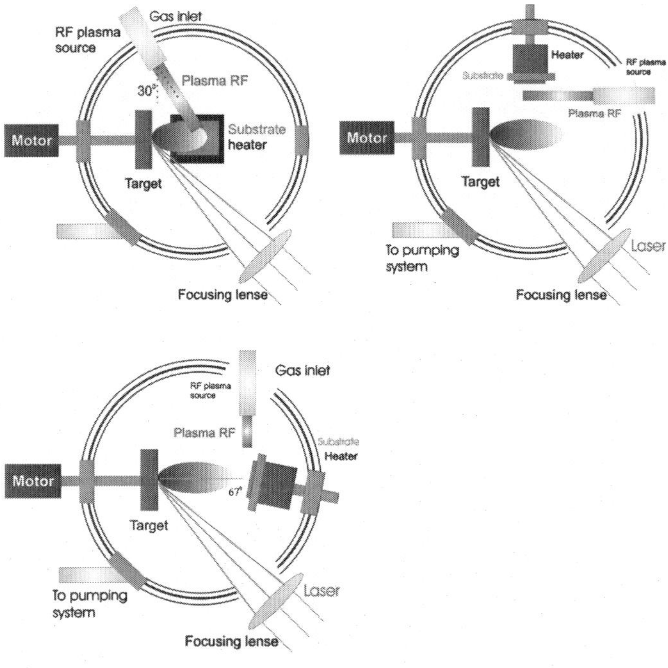

Fig. 14.5. Different geometries used for ZnO thin film deposition

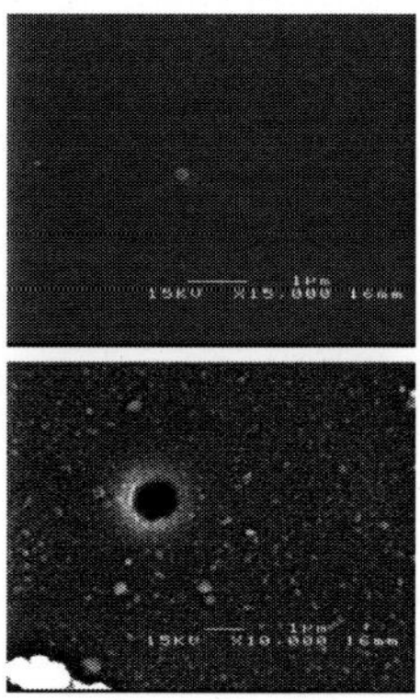

Fig. 14.6. SEM images on two samples prepared in the same experimental conditions $\lambda = 355\,\text{nm}$, 5 Pa oxygen, on Pt/SiO_2/Si substrate, laser fluence $4\,\text{J}\,\text{cm}^{-2}$, substrate temperature 573 K, with RF discharge ($P_{\text{RF}} = 200\,\text{W}$) (**a**) and without RF discharge (**b**) [33]

radiofrequency beam addition is the decrease in the number of droplets on the film surface. This was proved by SEM and AFM investigations, for different sets of experimental conditions and substrates. In Fig. 14.6a,b are presented the SEM images of the surfaces of two films prepared in the same experimental conditions except for the RF beam addition.

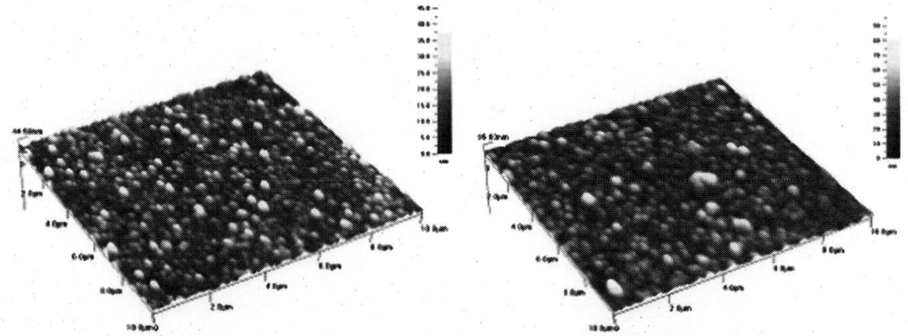

Fig. 14.7. AFM images $(10 \times 10)\,\mu m^2$ on ZnO layers deposited in the same experimental conditions (λ = 355 nm, oxygen pressure 60 Pa, laser fluence $5\,J\,cm^{-2}$, substrate temperature 773 K, with P_{RF} = 100 W on: (**a**) MgO (rms = 4.4 nm); (**b**) Pt/Si (rms = 3.9 nm)

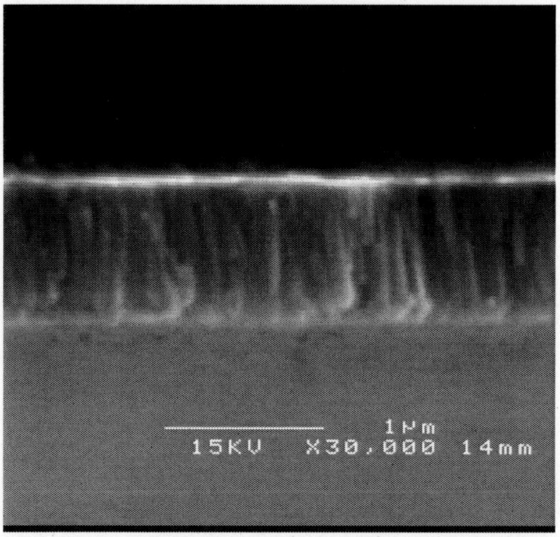

Fig. 14.8. Cross-section SEM on $ZnO/Ti/SiO_2//Si(100)$ [33]

Figure 14.7 presents the surface morphology of two layers deposited in the same experimental conditions on two different substrates. The smooth surface of the deposited layers can be pointed out.

Figure 14.8 is a cross-section SEM image of a ZnO thin film deposited at 10 Pa oxygen, 573 K. The columnar growth of the ZnO film can be easily observed, the columns being perpendicular to the Si substrate.

The transmission spectrum of a ZnO film grown on a Corning glass is presented in Fig. 14.9. A mean transmission of 85% in the visible range, with a maximum of 95% can be seen. The calculated optical band gap is E_g = 2.8–2.9 eV for films deposited at temperatures lower than 373 K and up to 3.23 eV for films deposited at 523 K.

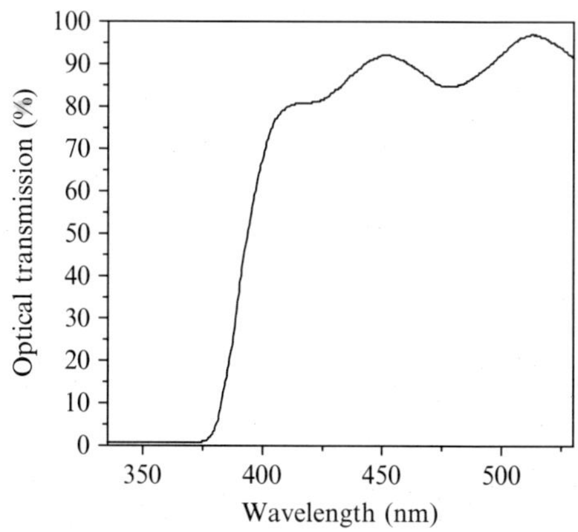

Fig. 14.9. Transmission spectrum for a film deposited on Corning glass

In common piezoelectric materials used to produce shear horizontal SAWs (SH-SAWs), a part of the energy is lost to a bulk acoustic wave which propagates normal to the surface, instead of along it. Propagation along the surface is best accomplished by using an acoustic wave guide made from a material for which the waves are trapped near the surface. ZnO can be used for this kind of applications, but for the generation of detectable SH-SAWs, it is mandatory to grow ZnO thin films with the c-axis tilted away from the normal to the film surface.

The ZnO cell orientation can be changed using different deposition geometries (Fig. 14.5). ZnO layers, a-axis oriented have been obtained on different substrates for a narrow window of experimental conditions specific to each type of substrate. The easiest way to achieve this goal is the use of r-cut sapphire, due to the lattices fit of the two crystals. For a substrate temperature of 573 K during deposition, RF power of 200 W with a beam tilted at 60° with respect to the substrate normal results in an a-axis grown layer, with a completely suppressed (002) orientation (Fig. 14.10a). A similar result was obtained using MgO substrate at room temperature: this time the a-axis had two perpendicular directions (Fig. 14.10b). In both cases a very sharp interface was noticed and an epitaxial growth was evidenced by HRTEM cross-section analysis [33].

A detailed characterization was performed on an MgO deposited ZnO layer. The specimen was cut perpendicular to a $\langle 100 \rangle$ crystallographic direction of the MgO substrate and parallel to the plasma jet during the film deposition. The aim of this cut was to reveal a possible film growth with columns inclined from the substrate normal. A low magnification image of the deposited film is presented in Fig. 14.11a. One can notice the vertical growth of the columns (around 50 nm wide). The columns are marked by a strong strain contrast. The corresponding selected area electron diffraction pattern (SAED) is presented in Fig. 14.11b, where the selected aperture included both

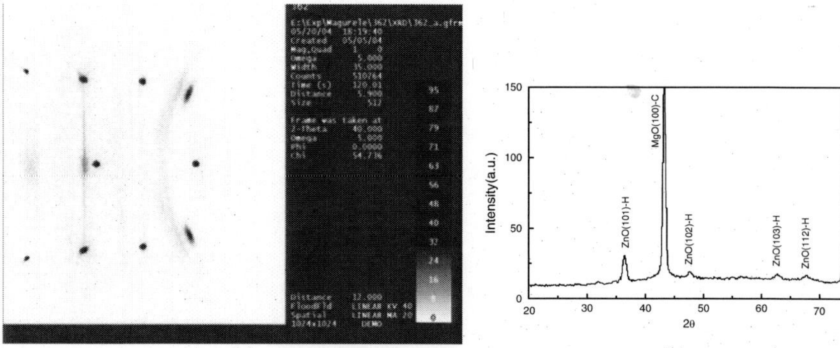

(a) (b)

Fig. 14.10. X-ray diffraction spectrum of ZnO film deposited on: (**a**) r-cut Sapphire at ($\lambda = 265\,\text{nm}, p_{O2} = 5\,\text{Pa}\ P_{RF} = 100\,\text{W}$, at 473 K), (**b**) MgO substrate, at 773 K, 60 Pa oxygen, $5\,\text{J cm}^{-2}$ laser fluence, $\lambda = 355\,\text{nm}$, $P_{RF} = 100\,\text{W}$

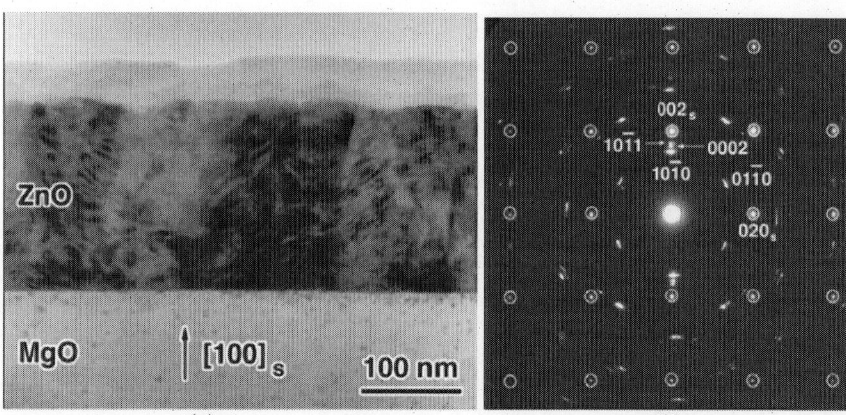

(a) (b)

Fig. 14.11. TEM image (**a**) and SAED pattern of a ZnO film (**b**) grown on a (001) MgO substrate

the ZnO film and the MgO substrate. The diffraction spots corresponding to the substrate (subscript "s" on the figure) are encircled. The SAED pattern reveals an accentuated film texture around the [0001] axis of the hexagonal ZnO, which means that the film has grown mainly with the crystallographic *c*-axis parallel to the substrate surface. Moreover, as one can observe on the diffraction pattern, the $(1\bar{0}10)$ planes of ZnO hexagonal are oriented parallel to the substrate: (1010) (001)s. The appearance of the $(1\bar{0}11)$ and (0002) diffraction spots, as marked on the SAED pattern, indicates the presence of a reduced number of ZnO grains in other crystallographic orientation, namely having the {1011} and {0001} families of planes parallel to the substrate surface.

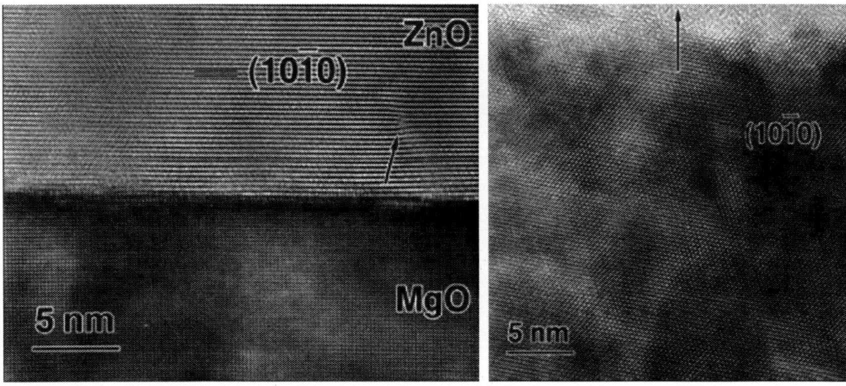

Fig. 14.12. (**a**) ZnO/MgO interface; (**b**) Top part of a ZnO grain

These results are confirmed by the high-resolution TEM (HRTEM) images. Figure 14.12a shows details of the film–substrate interface, with the MgO substrate in the [200] orientation. As indicated by the SAED pattern, the (1010) planes of ZnO are oriented parallel to the (002)s MgO planes. As the specimen at the interface was rather thick, the contrast of the HRTEM image is affected. On the right side of the image one can identify an edge-on dislocation in the ZnO film revealed by the end of an extra (1010) half-plane and marked. The HRTEM image in Fig. 14.12b representing the top part of the ZnO grain was recorded in the same specimen tilt conditions, i.e. along the [200] s zone axis of MgO. The ZnO grain from this image is oriented along the hexagonal [0002] zone axis, which means that the c-axis is parallel to the substrate, as it was in the region close to the interface (Fig. 14.12a). The arrow in Fig. 14.12b indicates the direction of the film growth which is parallel to the normal of the film–substrate interface. As one can see there is an 11° disorientation of the ZnO grain and the substrate. Because the specimen is much thinner in this region, the high resolution contrast is better.

Figure 14.13a reveals a HRTEM image of a region from the ZnO–MgO interface where the ZnO grain was oriented along the [2110] zone axis. The c-axis is marked on the figure and it is obviously parallel to the film–substrate interface. The top part of this grain near the film surface is presented in Fig. 14.13b.

Ultrasonic tests have been performed by implementing and testing SAW delay lines on the c-axis ZnO surface.

The delay line consists of two transducers composed of 15 finger pairs, with a spatial periodicity λ of 30 μm and a finger overlap of 40λ, placed at a center to center distance of 4.25 mm. The propagation direction has been chosen lying on the plane generated by the plate normal and ZnO c-axes. For a ZnO layer of 1 μm ($h/\lambda = 0.033$), theoretical calculations predict a phase and group velocity for the first Rayleigh mode of 4,345 and 4,060 m s^{-1}, respectively. The corresponding electromechanical coupling coefficient is $k^2 = 0.46\%$ [33]. The

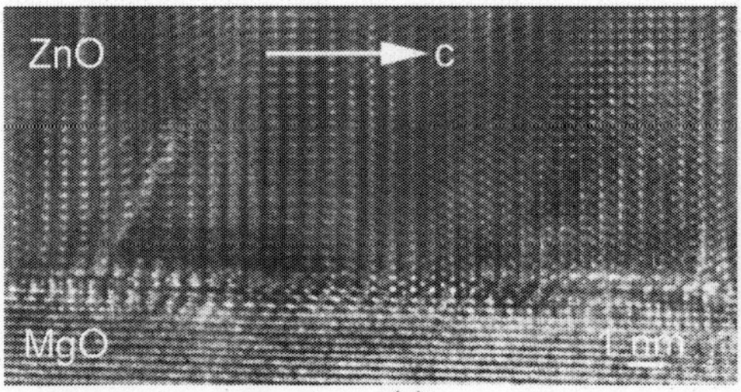

(a)

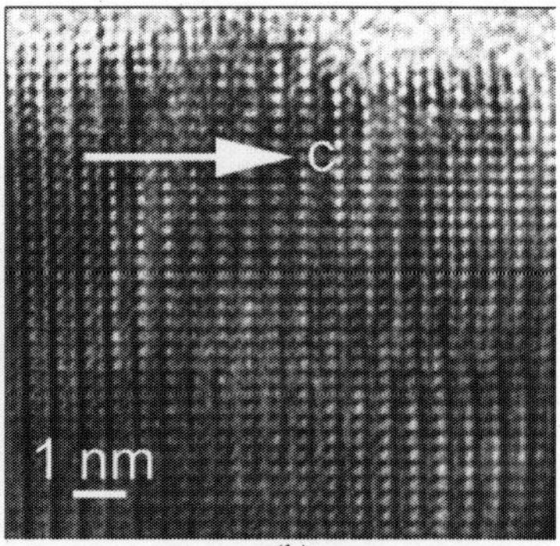

(b)

Fig. 14.13. HRTEM image on ZnO/MgO interface (**a**) and the top part of the ZnO grain (**b**)

experimental phase velocity v_{ph}, evaluated from the maximum in the electric transfer function of the delay line at 144.6 MHz (as shown in [33]), gives a value of $4,338\,\mathrm{m\,s^{-1}}$ ($\mathrm{v_{ph}} = f \cdot \lambda$) for the Rayleigh wave, in accordance (0.2%) with theoretical predictions, while calculated $\left(4,060\,\mathrm{m\,s^{-1}}\right)$ and experimental $\left(3,860\,\mathrm{m\,s^{-1}}\right)$ group velocities are in accordance within an error of 4%. Electric measurements reveal the absence of SH modes, thus confirming the perpendicular orientation of ZnO c-axes with respect to the surface of the substrate and the propagation direction. The piezoelectric properties were tested with a dedicated device, as described in [34]. The device measures the piezoelectric voltage corresponding to the propagation of the acoustic wave parallel to the normal to the film. For the ZnO films tested in the range 1–3 GHz, a d_{33} constant of 8–10 pC/N was obtained, which is close to the value for the ZnO crystal.

AlN

Aluminum nitride is a nonferroelectric piezoelectric material [35]. Aluminum nitride is an attractive semiconductor used for optoelectronic applications. The crystal structure of aluminum nitride AlN is hexagonal (wurtzite type). It has a good piezoelectric constant of 5.4 pC/N, high electrical resistivity of $10^{14}\,\Omega\mathrm{cm}$, measured band gap 6.1 eV, and the breakdown field $5 \times 10^{6}\ \mathrm{V\,cm^{-1}}$ [8,36].

The growth of AlN thin films by pulsed laser deposition involves good control on the growth parameters like gas pressure, substrate temperature, and laser fluence which allows the growth of high quality AlN thin films.

AlN thin films have been deposited by laser ablation of a metallic Al target in nitrogen atmosphere [37,38]. The advantage of growing AlN thin films by PLD is that it does not require a high substrate temperature but the drawback is that a background atmosphere almost free of oxygen is needed. Piezoelectric behavior comparable with those of layers obtained by traditional (magnetron sputtering) techniques were reported [39].

Lead Zirconate Titanate

The most studied and used piezoelectric ferroelectric material in research and industry is lead zirconate titanate, $Pb(Zr_xTi_{1-x})O_3$, commonly named PZT. The attention and interest devoted to this material are due to PZT's extraordinary properties: large polarization values $(30–100\,\mu C\,cm^{-2})$ [40], combined with relatively low crystallization temperature (around 923 K), high dielectric constant, low losses and high values of piezoelectric coefficients, in the range $50 \times 10^{-12} - 350 \times 10^{-12} m\,V^{-1}$ [41]; PZT exhibits a perovskite cubic structure [2] corresponding to the paraelectric state and a rhombohedral or tetragonal structure in the ferroelectric state. PZT ferroelectric and piezoelectric properties are strongly related to chemical composition and to crystallographic properties. The highest values of piezoelectric and ferroelectric coefficients were obtained for Zr/Ti ratio of about 53/47, close to the morphotropic phase boundary (MPB) (between rhombohedral and tetragonal phase).

PZT thin films have been deposited by many techniques: pulsed laser deposition, sputtering, and chemical routes. The most important issue in obtaining good PZT thin film is the preservation of the correct stoichiometry, more precisely the preservation of the lead content, which is easy to realize when PLD is used.

PZT thin films have been deposited by laser ablation starting from targets with different stoichiometries and on different combinations of substrate/bottom electrode-buffer layer. The main drawback, especially for nonvolatile ferroelectric memory application is poor fatigue endurance and high coercive field ($E_c = 120\,kV\,cm^{-1}$) when conventional platinum coated silicon substrates are used. This has been a reason for using oxide electrodes ($SrTiO_3$), because apparently the diffusion of oxygen coming from the oxide substrates fills the oxygen vacancies in the film [42]. For high-end devices the fatigue is set to a number of 10^{15} switching cycles and for today's devices a number of 10^{12} switching cycles, but the best value for PZT is reported to be around 10^{13} switching cycles by Schorn et al. [42], with the amendment that has been measured for PZT deposited on $SrRuO_3$. The industrial use of such combinations is difficult to be considered. The clamping effect present at the interface between the layer and the metallic substrates is also to be considered [43]. The fatigue can be partially removed by annealing the sample above the Curie temperature or by applying high electric fields.

Summaries of the most important results concerning pulsed laser deposition of PZT thin films can be found in [8, 36].

PMN and PMN-PT

The ferroelectric relaxors are characterized by a broad peak in the dielectric constant values with the temperature, explained by the existence in the material of nanopolar regions or clusters [44].

Lead magnesium niobate, PMN ($Pb(Mg_{1/3}Nb_{2/3})O_3$), is one of the most known ferroelectric relaxors, with high dielectric constant in bulk form ($\varepsilon_r = 20,000$), high electrostrictive coefficients, diffuse phase transition (around 270 K), and a strong frequency dispersion around and below this temperature. Solid solution with lead titanate (PT), PMN-PT, has been synthesized to increase the Curie temperature of PMN. The end members of the family PMN-PT are on one side of PT which is a highly anisotropic ferroelectric, with large tetragonal distortion and a typical ferro-paraelectric transition at $T_c = 763$ K, and with long range spontaneous ferroelectric order occurring below T_c; on the other side is PMN.

For PMN-PT thin films obtained with PLD, some interesting results have been obtained: a value of around 2,000 for dielectric constant has been obtained by Tantingate et al. [45]. The electrical properties of PMN/PT compounds with different ratios ($x = 0$, 0.07, 0.1, 0.2 and 0.3) have been reported by Donnelly et al. [46]. The highest electromechanical strain value (0.3%) was observed for PMN-PT 90/10 sample. Piezoelectric coefficient d_{33} was about 100 pmV^{-1} in PMN-PT 93/7 and 160 pmV^{-1} in PMN-PT 70/30 samples. Films with $x = 0.3$ show good remanent polarization (about 25 μC cm^{-2}).

Lead-Free Ferroelectrics

Lead-free ferroelectric materials are gaining much attention due both to their environmental-friendly characteristics and electric properties [9, 47–50]. The European Union has tried to limit the use of harmful elements in the industry (lead, cadmium, mercury, hexavalent chromium etc.), by a directive which specifies the year 2012 as a deadline for the use of these elements. Suitable materials, with similar properties have to be developed in order to replace lead-based ferroelectrics like $Pb(Zr_xTi_{1-x})O_3$ – lead and zirconium titanate – the most used ferroelectric in the industry. Studies regarding lead-free ferroelectric materials, e.g. $Na_xK_{1-x}NbO_6$ – sodium and potassium niobate, $Sr_xBa_{1-x}NbO_6$ – strontium and barium niobate, $Na_xBi_{1-x}TiO_3$ – sodium and bismuth titanate etc. are more and more systematically.

NBT–BT

Solid solutions based on sodium and bismuth titanate (NBT) are important substitutes for lead and zirconium titanate (PZT). Sodium and bismuth titanate are part of the bismuth based perovskites in which the A-site atom is replaced. The crystalline structure, phase transitions, and physical properties have been intensely studied since their discovery in 1960 by Smolensky

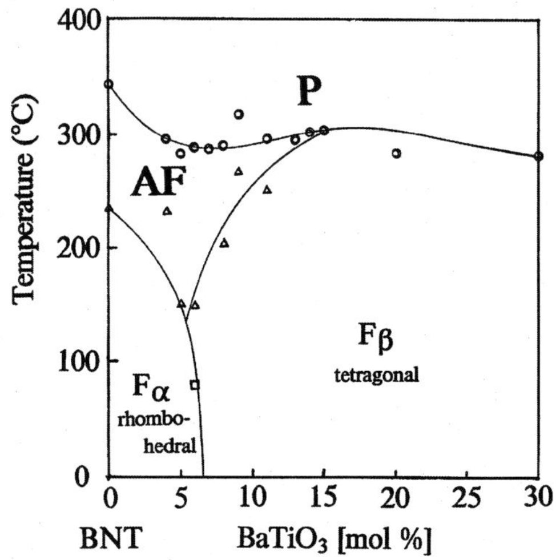

Fig. 14.14. Phase diagram for NBT–BT system (from [48])

[51, 52]. It has complicated phase transitions on which are also diverging opinions regarding the electronic states. For example, Sakata et al. [53, 54] observed a double hysteresis at high temperatures and concluded that this is due to the existence of an antiferroelectric phase during the diffuse phase-transition. Park and Hong [55], as well as Suchanicz et al. [56], suggest the creation of micropolar regions, and that the changes in the dynamics and size of these regions can cause the double hysteresis. For $(1-x)$ NBT–xBT solid solution an MPB between a rhombohedral ferroelectric relaxor structure and a tetragonal ferroelectric structure has been found at $x = 0.06$–0.07, as can be seen in Fig. 14.14 [9, 50–52]. Ceramics with this composition show structural coexistence of the two phases in a large temperature range and improved electromechanical properties.

Owing to the high value of the coercitive field and the high dielectric conductivity, sodium and bismuth titanate cannot be easily polarized, so the dielectric properties are not very good. However, in solid solution with $BaTiO_3$ (NBT–BT) the piezoelectric response was shown to be comparable to the response of lead and zirconium titanate (PZT), for a lower density ($6\,g\,cm^{-3}$ vs. 7.5–$8.5\,g\,cm^{-3}$). Sodium and bismuth titanate have a relatively high Curie temperature, $T_c = 503\,K$, high remanent polarization, $38\,\mu C\,cm^{-2}$ and piezoelectric coefficient $d_{33} = 125\,pC/N$ [57,58]. At MPB the properties of the material improve considerably: $d_{33} = 450\,pC/N$, $T_c = 593\,K$ and relative deformation up to 85%. This proves the dependence of the properties of the $(1-x)$ NBT–xBT system on the $BaTiO_3$ content, as shown by Chiang et al. and Craciun et al. [57–63].

Another point that must be considered for the NBT–BT thin films is the dependence of the dielectric properties on crystalline structure. Yilmaz et al. [62] demonstrated that the electromechanical properties (electro-mechanic coupling coefficients) for the $(Na_{0.5}Bi_{0.5})\,TiO_3$–5.5 mol% $BaTiO_3$ crystal are anisotropic and higher on (001) axis, which is similar to most rhombohedral perovskites, which means that the ceramic materials with oriented crystallites exhibit electro-mechanic properties superior to the one with random orientation.

Few studies are devoted to deposition of NBT–BT thin films by laser ablation [9, 49, 50, 59] mainly due to the complicated stoichiometry that makes difficult the transfer from the bulk form to thin films. To overcome these difficulties, NBT–BT enriched with Na and Bi was used, but using non stoichiometric targets which are not stable, have led to non-reproducible results. $(Na_{0.5}Bi_{0.5})\,TiO_3$–$BaTiO_3$ thin films deposited by PLD on (100) MgO exhibit a slight (100) preferential orientation [64]; single phase perovskite and (001)-oriented $Na_{0.5}Bi_{0.5}TiO_3$–$BaTiO_3$ thin films obtained by RF-PLD and PLD have been deposited on Pt(111)/Si substrates [59]. The dielectric constant values reported in these studies are slightly smaller than in bulk, $\varepsilon_r = 1,300$ for films on MgO substrates and $\varepsilon_r = 1,100$ for films on Pt/Si substrates. From the dielectric and loss tangent behavior at 1, 10, and 100 kHz respectively with temperature (Fig. 14.15), we can speculate that increase of dielectric permittivity above 433 K could be attributed to the transformation from relaxor ferroelectric (rhombohedral) to antiferroelectric (tetragonal ferroelastic) phase. The broad maximum of the dielectric constant at about 533 K decreases but does not shift with frequency, thus evidencing a phase transition behavior. This anomaly corresponds to the antiferroelectric–paraelectric (cubic) transition. The anomalous increase of the dielectric constant and loss of tangent at high temperature and low frequency is due to space charge polarization and ionic conductivity. The remanent polarization values reported are considerable smaller than for bulk, mainly due to high leakage currents in the films.

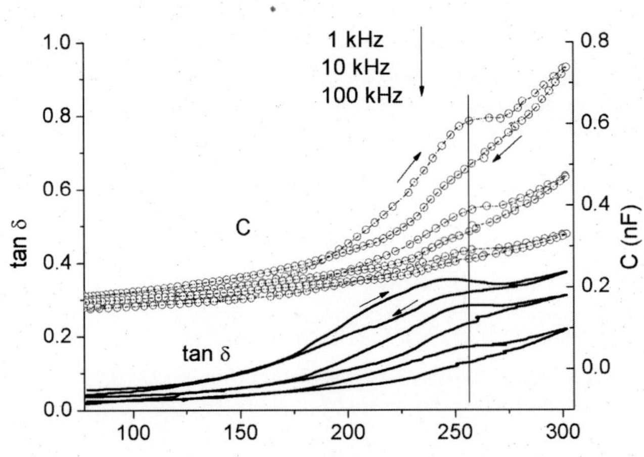

Fig. 14.15. Dielectric permittivity vs. temperature curves for films deposited by PLD for different frequencies (from [59])

NKN

$Na_xK_{1-x}NbO_3$ (NKN) is an attractive material for microwave applications and FRAM memories, both as bulk [65] and thin films [66–70]. In the vicinity of MPB (morphotropic phase boundaries)-$x = 0.5$, Fig. 14.16, $Na_{0.5}K_{0.5}NbO_3$ has a rhombohedral structure between 143 and 473 K, with values reported in literature for the coupling electro-mechanic coefficient K^2 up to 45%, relatively moderate dielectric permittivity $\varepsilon_r = 290$–420, and $d_{33} = 160$ pC/N, smaller than for the PZT ($d_{33} = 280$ and 380 pC/N). Among the first results on NKN thin films properties can be mentioned [71] where Wang et al. reported $\varepsilon_r = 580$ (higher than for AlN or ZnO), lower dielectric losses compared with PZT tan $\partial = 0.007$ and a tunability of 35% measured at 1 MHz for thin films deposited by magnetron pulverization on $LaAlO_3(001)$ starting from a non-stoichiometric target enriched by Na and K. With regard to pulsed laser deposition use for NKN thin films growth, we can mention studies of Cho et al. in 2002 [69] and Kharstev et al. in 2003 [72]. Deposition conditions are similar in both papers: KrF excimer laser, $\lambda = 248$ nm, 7 Hz, fluence: 4–6 Jcm^{-2}, 46–60 Pa oxygen pressure, $Pt_{80}Ir_{20}$ or SiO_2/Si substrate at a temperature of 923 K. Both cases demonstrated the possibility of obtaining $Na_{0.5}K_{0.5}NbO_3$ thin films with dielectric constant of $\varepsilon_r = 480$, low dielectric

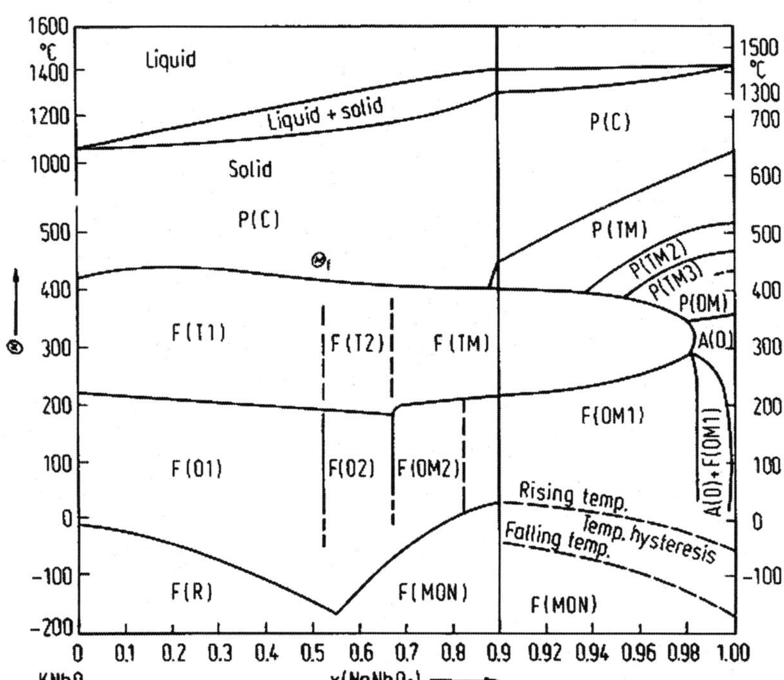

Fig. 14.16. Sodium potassium niobate phase diagram [73]

losses $\tan \partial = 0.025$–0.016, remanent polarization $P_r = 10\,\mu C\,cm^{-2}$, coercitive field $E_c = 20\,kV\,cm^{-2}$ and tunability between 39% and 47%.

SBN

$Sr_xBa_{1-x}Nb_2O_6$ (SBN: x) is a ferroelectric material with tetragonal tungsten-bronze structure (Fig. 14.17) and has the largest linear electro-optic coefficients of all known materials to date ($r_{33} \sim 1,300\,pm\,V^{-1}$), more than 50 times larger than $LiNbO_3$ which is the primary electro-optic material in industry. The pyroelectric coefficients of bulk tetragonal SBN are also higher than those of other well known ferroelectric materials [74]. The phase transition takes place in the temperature interval 333–623 K, depending on Sr content; SBN is the only known uniaxial ferroelectric with a displacive type phase transition. Because of these excellent properties, the applicability of high-quality SBN: x thin films can be quite high, mainly in electro-optic devices [75]. For many applications as pyroelectric detectors, electro-optic modulators or surface acoustic waves devices SBN is used in the single crystal form; this makes them expensive, as growth is quite difficult. For both miniaturization and cost reasons SBN heteroepitaxial thin films are very important and represent an alternative to single-crystal SBN. PLD [75–79] again demonstrates its capability to produce good layers on different substrates such as: SBN(001) (homoepitaxy), $LaNiO_3$ (100) /CeO_2/YSZ/Si (100) multilayer structure, $SrTiO_3$(001). However (001) oriented MgO substrate is the most common used [75, 76, 78, 79] mainly due to the much lower refractive index of MgO, the system SBN/MgO being studied for optical applications of SBN thin films. (001) oriented MgO single crystals are the appropriate substrates for the growth of SBN: x also because of the match between the good fit of

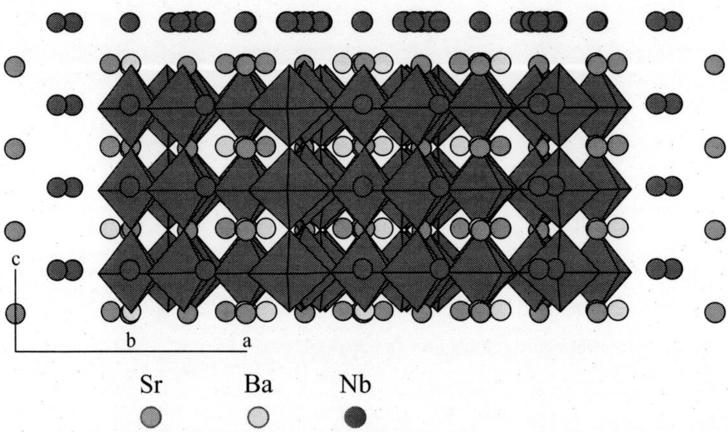

Fig. 14.17. Structural model of the tetragonal tungsten-bronze SBN (joined offered by Dr. Leona C. Nistor, INCDFM Bucharest)

the in-plane lattice parameter of the SBN unit cell ($a = 12.465$ Å, $x = 0.5$) and of the cubic MgO unit cell ($3a = 12.639$ Å). For $0.35 < x < 0.75$, the lattice mismatch δ between one c plane of SBN grown on three unit cells of MgO, increases from 1.4% to 1.8%. Highly c-axis oriented SBN thin films have reported to be deposited by PLD on MgO substrates [71,76,78]. The existence of three in-plane orientations of crystallites relative to the MgO (100) azimuth was noticed: 0°, arctan (1/3) = 18.43°, and arctan (3/5) = 30.96° [75,76,78,79]. Schwyn Thöny et al. [76] showed that for c-axis oriented SBN:61 thin films, the domains are twinned in-plane, with the SBN (100) axes parallel to the MgO (310) but the (130) equivalent axes at an angle of ± arctan $1/3d = \pm 18.43°$ to the MgO (100) direction.

Willmott et al. [75] demonstrated that the twinning can be promoted by the electrostatic interactions between MgO substrate and SBN, which will conduct to the apparition of $SrNb_2O_6$ (SNO) phase as buffer layer at the substrate-film interface (Fig. 14.18).

Another set of in-plane twinned domains in SBN:67 has been reported by Cuniot-Ponsard et al. [78], with the SBN (100) axes oriented at an angle of ±31° to the MgO (100) axis. Rouleau et al. [79] reported both twinned sets of domains in SBN:61: arctan (1/3) = 18.43°, and arctan (3/5) = 30.96° relative to the MgO (100) azimuth. Using a RF-PLD technique, Scarisoreanu et al. [80], reported in SBN:50 also both sets of twinned domains (Fig. 14.19), together with the presence of a thin SNO intermediate layer. Applying a slow cooling rate procedure ($5\,\mathrm{K\,min^{-1}}$) and in the presence of the oxygen radiofrequency plasma, SNO buffer layer was eliminated and a dominant 31.6° in-plane orientation of the grains relative to the MgO (100) azimuth was obtained (Fig. 14.8).

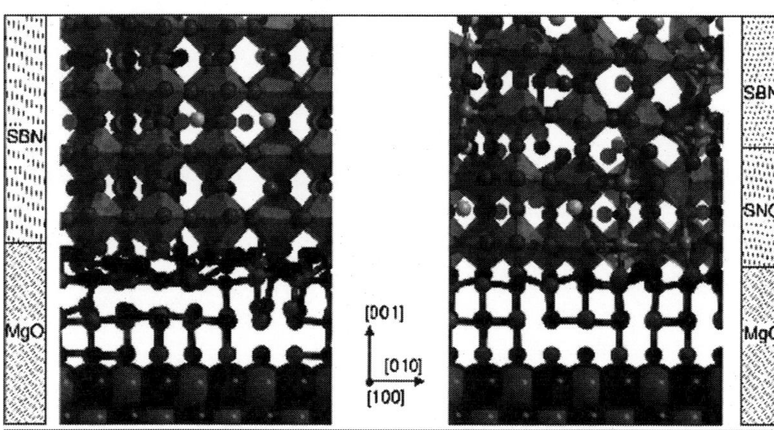

Fig. 14.18. SBN thin films on MgO with $SrNb_2O_6$ buffer layer at the substrate-film interface (from [74])

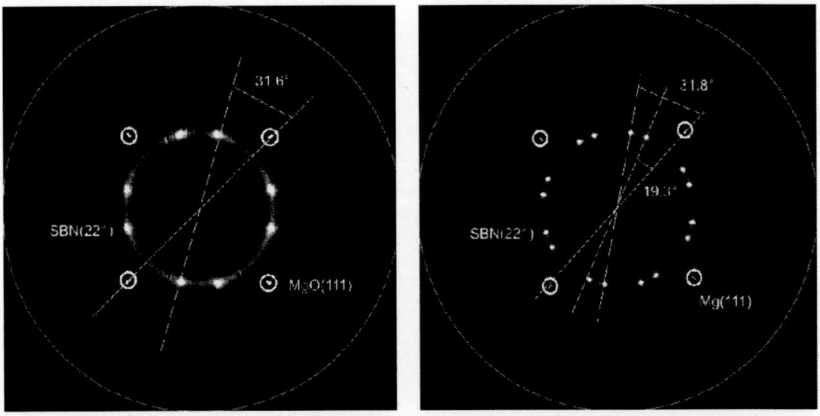

Fig. 14.19. X-ray pole figure analysis for two SBN thin films with different in-plane orientation (from [80])

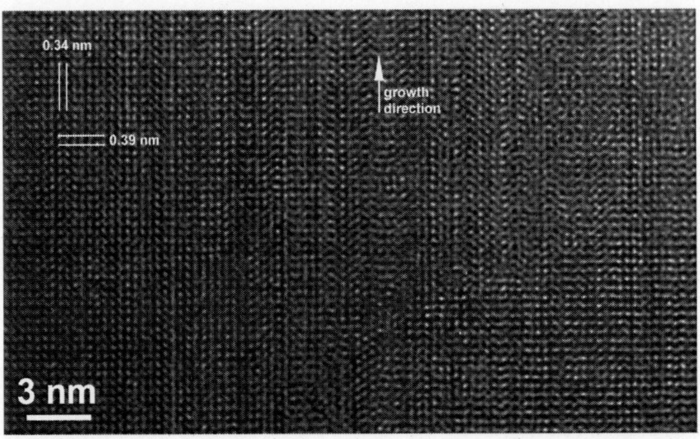

Fig. 14.20. HRTEM image of a columnar SBN crystal grain oriented close to the $[\overline{2}30]$ zone axis

In Fig. 14.20 the HRTEM image of an SBN crystal grain in the film is shown. As mentioned earlier, the columnar grains can show several orientations with respect to the substrate, with the *c*-axis always perpendicular to the substrate (growth direction parallel to *c*-axis). The grain in Fig. 14.20 is oriented along the $[\overline{2}30]$ zone axis. The $(001)_{\mathrm{SBN}}$ and $(320)_{\mathrm{SBN}}$ planes are indicated in the image, with $d_{001} = 0.395\,\mathrm{nm}$ and $d_{320} = 0.345\,\mathrm{nm}$. The presence of the planar defects running parallel to the *c*-axis may also be noticed. The traces of these planar defects delimitate vertical stripes where the HRTEM pattern varies from one stripe to the other. The large number of defects parallel to the *c*-axis is at the origin of the diffuse streaks showing in the diffraction patterns.

Depending on the Sr concentration, the electrical properties of the SBN material can be tailored from conventional ferroelectric, where the maximum of the susceptibility at a given temperature does not depend on frequency (for SBN40) to the extreme relaxor where the broad, characteristic type of susceptibility maximum strongly depend on the frequency, as Lukasiewicz et al. demonstrated for bulk single-crystal type [81].

In thin film form, due to the complicated crystalline structure it is difficult to overcome the problems arising from the high leakage currents – problems displayed in different studies. The ferroelectric properties reported are much smaller than for bulk form, with values for remanent polarization lower than $P_r = 32\,\mu C\,cm^{-2}$ in the case of SBN: 50 for example.

14.4 Conclusions

Pulsed laser deposition was shown to be effective for the growth of thin films with piezoelectric and ferroelectric properties. Its efficiency was demonstrated mainly for compounds with complicated stoichiometry, where the transfer of the composition from target to the films can be easily achieved even at low substrate temperature. The association of pulsed laser deposition with other techniques enriched its capabilities. The incorporation of oxygen in main cases (or nitrogen) can be improved by a radiofrequency discharge addition, which increases reactivity by creating excited and ionized species.

Acknowledgments

The author gratefully acknowledges N.D. Scarisoreanu, N.G. Epurescu and A. Matei. Part of results has been obtained in the frame of 3D DEMO STRP 033297 Project.

References

1. J.F. Nye, *Physical Properties of Crystals* (Oxford University Press, Oxford, 1979)
2. M.E. Lines, A.M. Glass, *Principles and Applications of Ferroelectrics and Related Materials* (Clarendon Press, Oxford, 1977)
3. S.K. Krishnaswamy, B.R. McAvoy, M.H. Francombe, in *Physics of Thin Films*, vol. 17, ed. by M.H. Francombe, J.L. Vossen (Academic, Boston, 1993), p. 145
4. M.H. Francombe, in *Physics of Thin Films*, vol. 17, ed. by M.H. Francombe, J.L. Vossen (Academic, Boston, 1993), p. 225
5. M. Sayer, in Proceedings of Ultrasonics Symposium (1991), p. 595
6. D. Damjanovic, Phys. Rev. B **55**, R649 (1997)
7. Materials Research Society, Symposium Proceedings, "Materials Issues for Tunable RF and Microwave Devices III," vol. 720, ed. by S.C. Tidrow, J.S. Horwitz, X.X.I.J. Levy (2002)

8. F. Craciun, M. Dinescu, in *Pulsed Laser Deposition of Thin Films: Applications in Electronics, Sensors, and Biomaterials*, ed. by R.W. Eason (John Wiley, Hoboken, 2007), p. 487
9. N. Scarisoreanu, F. Craciun, G. Dinescu, P. Verardi, M. Dinescu, Thin Solid Films **453–454**, 399 (2004)
10. D.B. Chrisey, G.K. Hubler (eds), *Pulsed Laser Deposition of Thin Films* (John Wiley, New York, 1994)
11. M. Dinescu, P. Verardi, Appl. Surf. Sci. **106**, 149 (1996)
12. P. Verardi, M. Dinescu, A. Andrei, Appl. Surf. Sci. **96–98**, 827 (1996)
13. P. Verardi, N. Nastase, C. Gherasim, C. Ghica, M. Dinescu, R. Dinu, C. Flueraru, J. Cryst. Growth **197**, 523 (1999)
14. M. Osiac, N. Scarisoreanu, M. Dinescu, J. Optoel. Adv. Mat. **10**, 2068 (2008)
15. S. Bakalova, A. Szekeres, A. Cziraki, C.P. Lungu, S. Grigorescu, G. Socol, E. Axente, I.N. Mihailescu, Appl. Surf. Sci. **253**, 8215 (2007)
16. S. Wanuga, J. Midford, J.P. Dietz, in Proceedings of IEEE Ultrasonics Symposium, Boston, 1965
17. G.L. Dybward, J. Appl. Phys. **42**, 5192 (1971)
18. T. Yamamoto, T. Shiosaki, A. Kawabata, J. Appl. Phys. **51**, 3113 (1980)
19. A.A. Scalisi, R.G. Toro, G. Malandrino, M.E. Fragala, G. Pezzotti, Chem. Vap. Dep. **14**, 115 (2008)
20. S.K. Lee, J.K. Kim, B.J. Kwon, Y.H. Cho, H.J. Ko, T. Yao, Solid State Comm. **147**, 498 (2008)
21. E. Cagin, J. Yang, W. Wang, J.D. Phillips, S.K. Hong, J.W. Lee, J.Y. Lee, Appl. Phys. Lett. **92**, 233505 (2008)
22. H. Shankur, J.T. Cheung, J. Vac. Sci. Technol. A **1**, 1806 (1983)
23. S. Hayamizu, H. Tabata, H. Tanaka, T. Kawai, J. Appl. Phys. **80**, 787 (1996)
24. V. Craciun, J. Elders, J.G.E. Gardeniers, I.W. Boyd, Appl. Phys. Lett. **65**, 2963 (1994)
25. V. Craciun, J. Elders, J.G.E. Gardeniers, J. Geretovsky, I.W. Boyd, Thin Solid Films **259**, 1 (1995)
26. R.D. Vispute, V. Talyansky, Z. Trajanovic, S. Choopun, M. Downes, R.P. Sharma, T. Venkatesan, M.C. Woods, R.T. Lareau, K.A. Jones, A.A. Iliadis, Appl. Phys. Lett. **70**, 2735 (1997)
27. R.D. Vispute, V. Talyansky, S. Choopun, R.P. Sharma, T. Venkatesan, M. He, X. Tang, J.B. Halpern, M.G. Spencer, Y.X. Li, L.G. Salamalanca-Riba, A.A. Iliadis, K.A. Jones, Appl. Phys. Lett. **73**, 348 (1998)
28. E. Millon, O. Albert, J.C. Loulergue, J. Etchepare, D. Hulin, W. Seiler, J. Perriere, J. Appl. Phys. **88**, 6937 (2000)
29. M. Okoshi, K. Higashikava, M. Hanabusa, Appl. Surf. Sci. **154–155**, 424 (2000)
30. Washington, H.C. Ong, J.Y. Dai, R.P.H. Chang, Appl. Phys. Lett. **72**, 3261 (1998)
31. S. Choopun, R.D. Vispute, W. Noch, A. Balsamo, R.P. Sharma, T. Venkatesan, A. Iliadis, D.C. Look, Appl. Phys. Lett. **75**, 3497 (1999)
32. V. Srikant, D.R. Clarke, J. Appl. Phys. **81**, 6357 (1997)
33. M. Benetti, D. Cannatá, F. Di Pietrantonio, E. Verona, P. Verardi, N. Scarisoreanu, D. Matei, G. Dinescu, A. Moldovan, M. Dinescu, Superlattice Microst. **39**, 366 (2006)
34. P. Verardi, F. Craciun, Rev. Sci. Instrum. **74**, 4453 (2003)
35. J. Bjurstrom, D. Rosen, I. Katardjiev, V.M. Yantchev, I. Petrov, IEEE Trans. Ultrason. Ferroelectr. Freq. Control **51**, 1347 (2004)

36. F. Craciun, P. Verardi, M. Dinescu, in *Handbook of Thin Film Materials*, vol. 3. *Ferroelectric and Dielectric Thin Films*, ed. by H.S. Nalwa (Academic, New York, 2002), p. 231
37. P. Verardi, M. Dinescu, C. Gerardi, L. Mirenghi, V. Sandu, Appl. Surf. Sci. **109–110**, 371 (1997)
38. P. Verardi, M. Dinescu, C. Stanciu, C. Gerardi, L. Marenghi, V. Sandu, Mater. Sci. Eng. B **50**, 223 (1997)
39. J. Meinschien, G. Behme, F. Falk, H. Stafast, Appl. Phys. A **69**, S683 (1999)
40. H. Morioka, S. Yokoyama, T. Oikawa, H. Funakubo, K. Saito, Appl. Phys. Lett. **85**, 3516 (2004)
41. H. Du, D.W. Johnson Jr., W. Zhu, J.E. Graebner, G.W. Kammlott, S. Jin, J. Rogers, R. Willett, R.M. Fleming, J. Appl. Phys. **86**, 2220 (1999)
42. P.J. Schorn, D. Bräuhaus, U. Böttger, R. Waser, G. Beitel, N. Nagel, R. Bruchhaus, J. Appl. Phys. **99**, 114104 (2006)
43. V. Nagarajan, Appl. Phys. Lett. **87**, 242905 (2005)
44. G.A. Samara, J. Phys. Condens. Matter **15**, R367 (2003)
45. C. Tantigate, J. Lee, A. Safari, Appl. Phys. Lett. **66**, 1611 (1995)
46. N.J. Donnelly, G. Catalan, C. Morres, R.M. Bowman, J.M. Gregg, J. Appl. Phys. **93**, 9924 (2003)
47. J. Kreisel, P. Bouvier, B. Dkhil, P.A. Thomas, A.M. Glazer, T.R. Welberry, B. Chaabane, M. Mezouar, Phys. Rev. B **68**, 014113 (2003)
48. T. Takenaka, Jpn. J. Appl. Phys. **30**, 2236 (1991)
49. Y.M. Chiang, G.W. Farrey, A.N. Soukhojak, Appl. Phys. Lett. **73**, 3683 (1998)
50. Z.H. Zhou, J.M. Xue, W.Z. Li, J. Wang, H. Zhu, J.M. Miao, Appl. Phys. Lett. **85**, 804 (2004)
51. G.A. Smolensky, V.A. Isupov, A.I. Agranovskaya, N.N. Krainik, Fiz. Tverd. Tela **2**, 2982 (1960)
52. G.A. Smolensky, V.A. Isupov, A.I. Agranovskaya, N.N. Krainik, Trans. Sov. Phys. Solid State **2**, 2651 (1961)
53. K. Sakata, Y. Masuda, Ferroelectrics **7**, 347 (1974)
54. K. Sakata, T. Takenaka, Y. Naitou, Ferroelectrics **131**, 219 (1992)
55. S.E. Park, K.S. Hong, J. Appl. Phys. **79**, 383 (1996)
56. J. Suchanicz, K. Rolender, A. Kania, J. Handerek, Ferroelectrics **77**, 107 (1988)
57. Y.M. Chiang, G.W. Farrey, A.N. Soukhojak, Appl. Phys. Lett. **73**, 3683 (1998)
58. T. Takenaka, H. Nagata, Key Eng. Mater. **157–158**, 57 (1999)
59. N. Scarisoreanu, F. Craciun, V. Ion, S. Birjega, M. Dinescu, Appl. Surf. Sci. **254**, 1292 (2007)
60. A. Purice, G. Dinescu, N. Scarisoreanu, P. Verardi, F. Craciun, C. Galassi, M. Dinescu, J. Eur. Ceram. Soc. **26**, 2937 (2006)
61. F. Craciun, M. Dinescu, P. Verardi, N. Scarisoreanu, C. Galassi, D. Piazza, Ferroelectrics **302**, 313 (2004)
62. H. Yilmaz, G.L. Messing, S. Trolier-McKinstry, J. Electroceramics **11**, 217 (2003)
63. F. Craciun, M. Dinescu, P. Verardi, N. Scarisoreanu, A. Moldovan, A. Purice, C. Galassi, J. Eur. Ceram. Soc. **25**, 2299 (2005)
64. N.D. Scarisoreanu, M. Dinescu, F. Craciun, P. Verardi, A. Moldovan, A. Purice, C. Galassi, Appl. Surf. Sci. **252**, 4553 (2006)
65. A.M. Margolin, Z.S. Surovyak, I.N. Zakharchenko, V.A. Aleshin, L.K. Chernysheva, M.G. Radchenko, V.P. Dudkevich, Sov. Phys. Tech. Phys. **33**, 1435 (1988)

66. C.-R. Cho, A. Grishin, Appl. Phys. Lett. **75**, 268 (1999); J. Appl. Phys. **87**, 4439 (2000)
67. C.-R. Cho, J.-H. Koh, A. Grishin, S. Abadei, S. Gevorgian, Appl. Phys. Lett. **76**, 1761 (2000)
68. S. Abadei, S. Gevorgian, C.-R. Cho, A. Grishin, J. Andreasson, T. Lindback, Appl. Phys. Lett. **78**, 1900 (2001)
69. C.-R. Cho, I. Katardijev, M. Grishin, A. Grishin, Appl. Phys. Lett. **80**, 3171 (2002)
70. C.-R. Cho, A. Grishin, Appl. Phys. Lett. **75**, 268 (1999)
71. X. Wang, U. Helmersson, S. Olafsson, S. Rudner, L.D. Wernlund, S. Gevorgian, Appl. Phys. Lett. **73**, 927 (1998)
72. S. Khartsev, A. Grishin, J. Andreasson, J.-H. Koh, J.-S. Song, Integr. Ferroel. **55**, 769 (2003)
73. H.H. Landolt, R. Börnstein, *Crystals and Solid State Physics*, vol. III/16a, Ferroelectrics (2002)
74. L. Hesselink, M.C. Bashaw, Opt. Quant. Electron. **25**, S 611 (1993)
75. P.R. Willmott, R. Herger, B.D. Patterson, R. Windiks, Phys. Rev. B **71**, 144114 (2005)
76. S. Schwyn Thöny, K.E. Youden, J.S. Harris, L. Hesselink, Appl. Phys. Lett. **65**, 2018 (1994)
77. K. Tanaka, O. Nakagawara, M. Nakano, T. Shimuta, H. Tabata, T. Kawai, Jpn. J. Appl. Phys. Part 1 **37**, 6142 (1998)
78. M. Cuniot-Ponsard, J.M. Desvignes, B. Ea-Kim, E. Leroy, J. Appl. Phys. **93**, 1718 (2003)
79. C.M. Rouleau, G.E. Jellison, D.B. Beach, Appl. Phys. Lett. **82**, 2990 (2003)
80. N.D. Scarisoreanu, G. Dinescu, R. Birjega, M. Dinescu, D. Pantelica, G. Velisa, N. Scintee, A.C. Galca, Appl. Phys. A **93**, 3 (2008)
81. T. Lukasiewicz, M.A. Swirkowicz, J. Dec, W. Hofman, W. Szyrski, J. Cryst. Growth **310**, 1464 (2008)

15

Lasers in Cultural Heritage: The Non-Contact Intervention

Wolfgang Kautek

Summary. Conservation and protection of works of art as well as of rare remnants of natural history has turned more and more into a race against time. Mechanical and chemical methods are involved in traditional conservation treatments. Contactless cleaning by lasers, on the other hand, is a new and prospering field of laser material processing. It allows avoiding mechanical disruption and the disadvantage of cleaning fluids which could cause potentially long-term degradation of the substrate or health hazards. The high-precision deliverance of laser radiation to morphologically and chemically inhomogeneous artefact surfaces – even down into the nanometre range – allows an unprecedented treatment quality by providing the base for repeatability of this treatment during the upcoming centuries of civilization.

Laser cleaning basically is a phase separation process driven by laser radiation. There exists experience in paint removal, and the restoration of building facades, stone and metal artefacts. More recently, the contactless laser cleaning of biogenetic fibre substrates such as paper, parchment and textiles has been approached. These biogenetic surfaces constitute the most sensitive and chemically fragile substrates studied up to date. They require ultimate thermal energy localization such as, e.g., minimized heat-affected zones. Therefore, the application of femtosecond laser pulses may play a future role in this context.

15.1 Introduction

Pulsed lasers are becoming new tools in the hands of restorers [1–8]. Most activities were concerned with the laser cleaning of stone artefacts, wall paintings and facades. There also exist fundamental knowledge and experience on the cleaning of technical surfaces relevant, e.g., in the electronic industry [9, 10], or the high-precision removal of inorganic and organic films, e.g., in biosensoric technology [11]. The contactless laser cleaning of biogenetic surfaces such as parchment and paper, on the other hand, has been approached only in recent years [1–8, 12–30]. Traditional cleaning methods, e.g., stone facades are water jet or microsand blasting, whereas dry as well as aqueous cleaning technologies for paper include mechanical scalpel scratching and the use of a brush, eraser or draft clean powder, i.e., a granulated eraser. Paper may also be cleaned with water, organic solvents, enzymes etc. [31,32]. Satisfactory results are often obtained by a treatment using cellulose ethers such as carboxymethyl cellulose (CMC), methyl cellulose (MC) or hydroxypropyl cellulose (HPC). Nevertheless, the current aqueous cleaning methods are not always sufficient, in particular if the paper is coated with sensitive printing as well as writing media.

15.2 Architectonic Structures and Sculptures

From the Mona Lisa and Parthenon of Athens to the entombed warriors of the Qin dynasty, the mediaeval treasures of Venice and Florence, the cathedrals of France and Austria and the early churches of Egypt and Romania, laser cleaning has been proven as an unreached approach of restoration since the pioneering activities in the 1970s [33]. A dependable variety of laser equipment has been developed which can sustain crude and lofty working conditions (Fig. 15.1) [8,34]. Laser cleaning of architectural facades and sculptures is a widely accepted restoration procedure and has attained an established place on the high-quality restoration market [1–8]. A perfect etch-stop can be achieved when laser parameters (e.g., fluence and number of pulses) are optimized (Figs. 15.1b and 15.2) [35].

Its application to polychromed surfaces, however, is a subject of concern due to the sensitivity to light of painting materials, including pigments and binders (Fig. 15.3) [35, 36]. Discolouration and degradation caused by laser irradiation constitute the object of ongoing research. Application of lasers for the elimination of dirt and contamination layers from polychromes requires always a preliminary and systematic study of the adequate conditions to guarantee the preservation of the artwork (Fig. 15.4) [35]. Discolouration and loss of adhesion between layers upon irradiation with a free running Nd:YAG laser (1,064 nm) was, e.g., observed in contrast to good cleaning results with a Q-switched system (6 ns pulses, fundamental and up to the third harmonic) [36].

From the economical point of view, the comparison between traditional and high-tech cleaning systems is decisive. Often the best cleaning result can be achieved by combining three or more cleaning techniques including also laser treatment for special purposes (Table 15.1). It is of major importance for the decision for laser cleaning in competition with conventional techniques, that the productivity of the laser procedures is practically the same as conventional approaches (e.g., microsand blasting, ca. 20 dm^2 min^{-1}). However, the laser allows an unprecedented treatment quality providing the base for repeatability during the upcoming centuries which is certainly not the case with e.g., microsand blasting where an etch stop almost never can be reached.

(a)

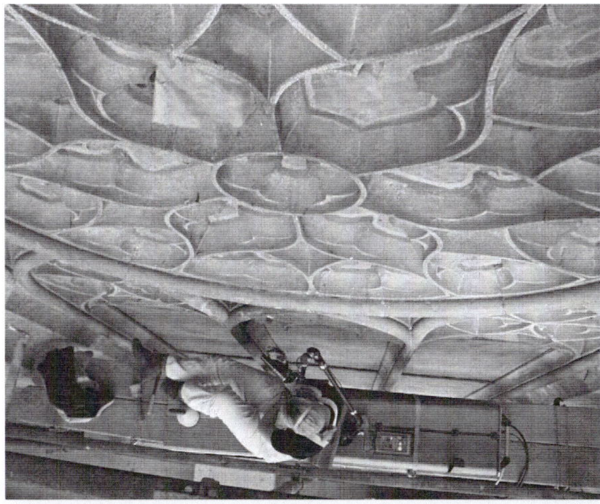

(b)

Fig. 15.1. (**a**) Laser cleaning and conservation of the gothic soutgate, "Primtor" of the St. Stephan's cathedral in Vienna, Austria (Atelier Erich Pummer, Stone Conservation. Rossatz, Austria [34]). (**b**) Laser cleaning in operation at portal of Cathedrale de Nantes (W. Kautek)

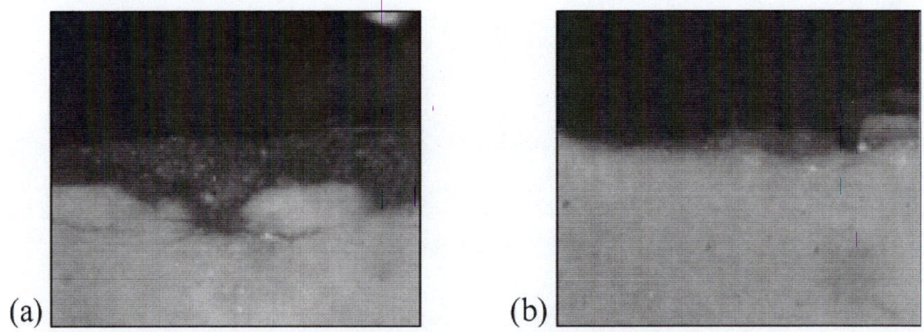

Fig. 15.2. Microscopical cross-sections of stone before (**a**) and after (**b**) laser cleaning [35]

Fig. 15.3. Mural paintings on mortar (Church of Cervera de la Cañada, Zaragoza) [36]

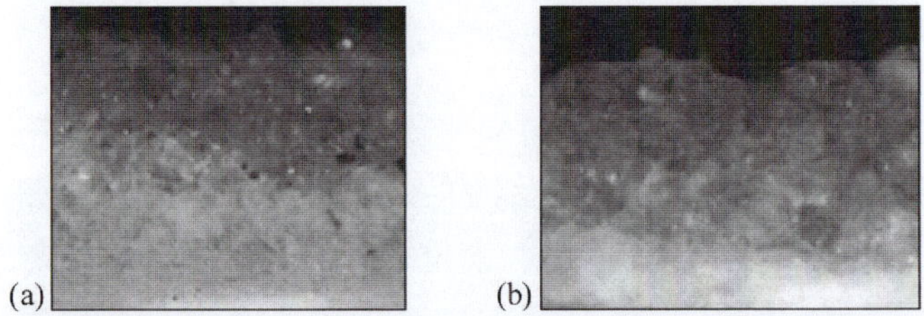

Fig. 15.4. Microscopical cross-sections of stone with azurite pigment layer before (**a**) and after (**b**) laser cleaning. Only the black crust has been removed without alteration of the azurite zone [35]

Table 15.1. Comparison of cleaning technologies [35]

Tool or material	Location	Application	Time [min dm^{-2}]
Abrasive gums, wish-up and fine brushes	Ornamented vaults, ornaments indoors	Pre-cleaning	30
Vacuum cleaner	All surfaces	Before pre-cleaning	1
Compresses (9×) and fine brushes	Columns, ornamented vault, flat surface indoors	Pre-cleaning (only of salty areas)	45
Precise whirl-blasting (low pressure with dolomite powder)	Flat surface outdoors	Pre-cleaning	10
Microsand-blasting with various powders	Flat surface indoors, cornices	Pre-cleaning	15
	Flat surface outdoors	Final cleaning	20
Laser cleaning with high energy	Sculptures and ornaments outdoors, flat surface indoors	Final cleaning	20
Laser cleaning with low energy	Ornamented vault, columns, sculptures tympana (all details indoors)	Final cleaning	15
Compresses with ammoniumcar-bonate	Figurative consoles outdoors	Additional cleaning	20

15.3 Metallic Artefacts

Innumerable metallic cultural objects as well as archaeological finds exist, which survival is threatened by processes of corrosion caused by the environment. Classical mechanical cleaning methods such as water jet and microsand blasting cleaning or chemical cleaning agents often result in incomplete ablation of deposits or corrosion products including irreversible changes or damage to the material. Nd:YAG laser treatments can cope with patina, black and brown incrustations, impurities, old conservation materials etc. Metals can be iron, copper, pewter, silver, brass (Fig. 15.5), various bronze alloys, gold leaf and firegilding, plating etc. [1–7, 37].

Fig. 15.5. Laser cleaning procedure of metallic objects with nitrogen gas treatment. (Brass cherub, Neptune fountain, Herrenhäuser Gardens, Hannover, seventeenth/beginning of eighteenth century) [37]

15.4 Biogenetic Substrates

Biogenetic materials such as parchment, paper, and textiles, constitute one of the most important substrates for long-term cultural heritage data storage. Laser cleaning of parchment (which is processed skin) to some extent parallels dermatological laser applications such as tattoo removal at a first superficial glance (Fig. 15.6) [38, 39]. There, selective photothermolysis of pigmented subsurface structures, such as melanin particles, enlarged blood vessels, and tattoo ink particles, or char-free vaporization of skin takes place. Then the skin's natural physiological mechanisms break down and remove the laser-altered remnants. In parchment, i.e., dead skin, cleaning, however, natural resorption of laser-altered remnants is absent, and contaminants have to be removed completely without any irreversible morphological and chemical conversions of the substrate (Fig. 15.7).

The cleaning objects, paper and parchment, belong to the chemically most fragile substrates exposed to high-power laser radiation. The main constituent of paper is cellulose. The overall structure is of aggregated fibrils with extensive pores capable of holding relatively large amounts of water by capillarity. The main constituent of parchment collagen forms long ropes and tough sheets. The collagen molecules are associate side-by-side, like fibres in a rope, to form tough fibrils.

15.5 Technology

Laser beam delivery has been realized commonly either via an optical fibre or an articulated optical arm to a hand-held output optics common in facade laser cleaning [7, 8]. An evaluation of this near-UV radiation versus visible

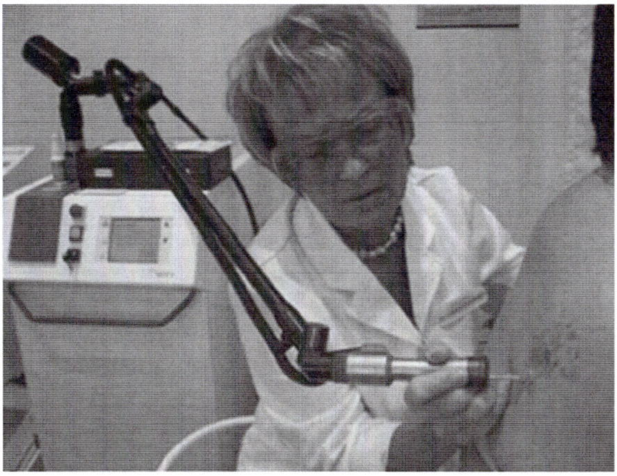

Fig. 15.6. Laser tattoo removal (Dr. Gabriele Grabner, Vienna, Austria)

Fig. 15.7. Laser cleaning of manuscript parchment. Preliminary testing on mediaeval document, AD 1428 (E. König and W. Kautek)

radiation of a solid state Nd:YAG laser at 532 nm indicated that cellulose and paper degraded after UV-laser treatment as well as after a period of accelerated humid oven ageing due to depolymerization [12, 18]. No detrimental effects of a Nd:YAG laser treatment at 532 nm was observed. Cellulose is more or less transparent in the visible region, however strongly absorbs UV radiation, e.g., the C-C-bond at 347 nm, the C-O-bond at 333 nm, and the C-H-bond at 289 nm. Photolysis may occur as a result of photon absorption

Fig. 15.8. Scheme of paper and parchment laser cleaning system. Two wavelengths, scanner and hand-held fibre optics [28]

which in turn leads to severe immediate depolymerization. However, it is more likely that the absorption of a near UV photon results in excitation of electrons in a chemical bond, facilitating photo-oxidation reactions.

With this knowledge, a computerized laser cleaning system suitable for high-precision cleaning of flat large area substrates was developed (Fig. 15.8) [28]. It allowed the restoration of artefacts of organic materials, such as paper, parchment, leather, textiles, wood, but also inorganic materials, such as metals, alloys, and ceramics. It was based on a compact high pulse energy diode pumped Q-switched Nd:YAG laser operating at 1,064 nm and 532 nm with a pulse duration of 8 ns and a repetition rate of up to 1,000 Hz. The laser-processing compartment with an integrated exhaust system provided Laser Class I conditions, so that the operator did not require safety goggles. The workstation featured on-line visible, ultraviolet and fluorescence imaging for the identification and documentation of visible and chemical changes of the irradiated substrate areas. The laser spot was scanned over the objects through a remote computer control system. The operator followed the process on the computer screen through a camera system and controlled it manually through keyboard, mouse and/or foot pedal operation or by automatic laser beam scanning operation. As an alternative, an optical fibre with an ergonomic hand piece could be used for manual cleaning of 3D objects.

The workstation featured on-line diagnostic tools such as visible, ultraviolet and fluorescence imaging for the identification and documentation of visible and chemical changes of the irradiated substrate areas. A multi-spectral imaging system operated in a spectral range from 320 nm up to 1,550 nm. Several imaging modes were possible: visible reflection, infrared reflection, visible fluorescence, and ultraviolet reflection.

Colorimetry allows relative, not absolute, comparisons in respect to both lightness changes ΔL and saturation and hue changes given by the chromaticity coordinates Δa^* and Δb^* according to CIE-$L^*a^*b^*$ colour coordinates. The vector between two data sets in the colour sphere is the colour difference ΔE. It is very useful for semiquantitative assessments of colour changes:

$$\Delta E = \sqrt{L^{*2} + a^{*2} + b^{*2}}.$$

Lightness changes ΔL quantified by this relative technique could be correlated with the cleaning status. The overall colour difference ΔE included also the colour changes for which the human eye is particularly sensitive.

15.6 Case Studies and Diagnostics

Modern analytical chemistry offers valuable tools to investigate laser-induced degradation, i.e., morphological and chemical changes, of biogenetic fibre materials. On-line laser-induced plasma spectroscopy (LIPS) and on-line laser-induced fluorescence (LIF) during laser cleaning of model systems consisting of multilayer composites of carbon containing black material, pigments and parchment have been reported before [12–16]. Since LIPS relies on the destruction of a substrate, it provides limited applicability in the in situ monitoring of non-destructive laser cleaning of parchment and paper. However, a substantial list of mainly inorganic pigments shows strong LIF emission intensity, and provide good spectroscopic separation from the LIF emission of the parchment substrate. Further, Diffuse Reflectance Infrared Fourier Transform Spectroscopy (DRIFT) is a powerful technique to verify any laser-induced chemical alterations both of the pigment-binder coatings and the parchment [15,16].

The case studies presented in the following provide a good insight into the potentials of laser cleaning of biogenetic substrates but also into the urgent need for diagnostic tools in addition to the inspection by naked eye and optical and electronic microscopy. There are threshold fluences to be considered which deviate from the mere morphological destruction threshold which are, e.g., melting and chemical conversion (such as depolymerization and oxidation). Figure 15.9 schematically shows the thus diminished "cleaning window" remaining for practical conservation.

A parchment sheet of music of a psaltery from fifteenth century, which was heavily soiled during a bombing attack at very end of Second World War, could be precisely cleaned (Fig. 15.10) [12,14]. There, hand-written ink letters, notes, and lines of the staff often have to be preserved from the converting action of the laser beam. Again, the cleaning result demonstrates the feasibility of this ultra-precise "contactless rubber."

Multi-spectral imaging has proven to be a powerful tool to monitor laser cleaning of biogenetic materials [20]. Cotton paper, e.g., was treated with increasing fluence on the patches numbered with 1, 2, and 3 (Fig. 15.11). The visible image showed no alteration below the ablation threshold fluence above 1.4 J cm^{-2}. At a drastically higher value, at 5.0 J cm^{-2}, the visible image only

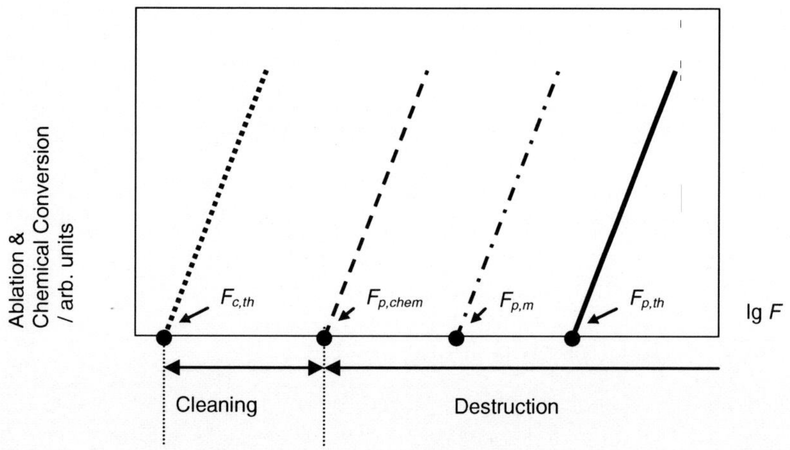

Fig. 15.9. Schematics of laser cleaning action leading to a cleaning threshold fluence ($F_{c,th}$) and destructive processes, i.e., chemical conversion ($F_{p,chem}$), morphological conversion ($F_{p,m}$), and ablation ($F_{p,th}$) vs. the logarithm of fluence

Fig. 15.10. Precision-cleaned ancient parchment (fifteenth century, Southern Germany; private W. Kautek)

indicated a faint darkening. Invisible diagnostic tools, however, allow a more sensitive judgement on the degradation status. The reduction of the IR and UV reflectivity indicate an increased absorption in IR and UV due to chemical changes. This UV absorption increase is accompanied by a drastic fluorescence increase.

The ablation threshold fluences F_{th} for fresh and pre-aged papers with all laser wavelengths have been determined [25]. The pre-aged papers are more sensitive against laser radiation and therefore, their ablation threshold fluences are lower. Especially for the infrared laser, all papers show in the aged stadium

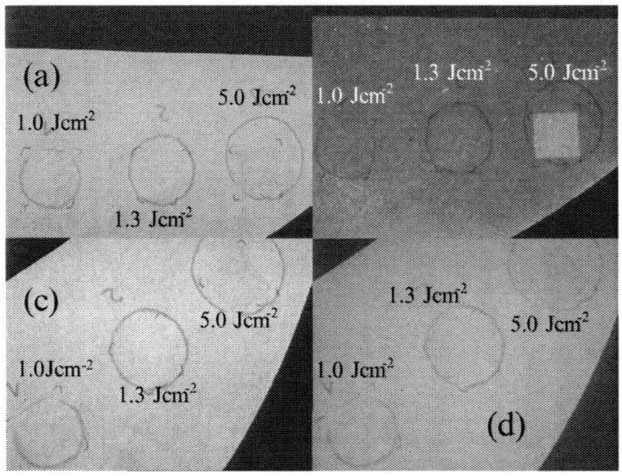

Fig. 15.11. Multispectral imaging diagnostics of laser-treated cotton paper (Whatman filter paper). (**a**) VIS reflectivity, (**b**) fluorescence, (**c**) IR reflectivity, (**d**) UV reflectivity

a decrease of resistance against laser radiation. Size exclusion chromatography (SEC) was used to characterize the molecular mass distribution of cellulose in laser treated areas and untreated background. From the chromatogram average molecular masses and the corresponding average depolymerization (DP) values could be reported [25]. A comparison between the background and the lasered matrix areas showed that the ultraviolet laser radiation resulted in a significant decrease in the M_w value for higher fluences and higher number of pulses. Whereas the green and the infrared laser light interaction caused no DP effect.

Chemiluminescence experiments (CL) were carried out at paper samples was recorded during dynamic experiments performed in nitrogen [25]. Paper discs were conditioned for 10 min at 40°C, and then heated up to 200°C. A comparison of fresh bleached sulphite softwood cellulose paper and additive-free cotton linters cellulose paper to post-aged samples for the laser treatment with the green wavelength showed no significant differences between the CL responses of treated and untreated areas. The bleached sulphite softwood cellulose paper, in contrast, showed the typical CL peroxide decomposition peak centred at 125°C. Its position and intensity is not altered by laser treatment and/or artificial ageing. Contrary to the results obtained with the green laser, significant differences between the CL responses of all papers were observed with UV radiation. Laser treatment causes the CL intensity of the fresh paper to increase. Artificial ageing considerably reduced the differences between the CL response of lasered and non-lasered areas. These results suggest the generation of emitting species as a result of UV laser treatment, which react further into non-emitting products in the course of artificial ageing.

More and more, destructionless processing based on off-line and on-line diagnostics based on systematic multi-method approaches become important. Scanning electron microscopy (SEM) and Transmission Electron Microscopy

(TEM) play a role in observing the threshold of morphological effects on originally intact fibres [30]. Moreover, DRIFT, Pyrolysis Capillary Gaschromatography (PY-CGC), Electron Spin Resonance (ESR), Electron Spectroscopy for Chemical Analysis (ESCA), Energy-dispersive X-ray Fluorescence (EDX), the determination of the hydrothermal stability by the micro-hot-table technique as well as the analysis of hydrolysis products allow to establish an in-depth chemical diagnosis of laser-induced chemical changes and degradation processes in dependence on laser fluence and wavelength [30]. At the visible (532 nm) and the infrared wavelengths (1,064 nm), the photochemical phenomena were absent, and the only result was a decrease of the solubility due to thermally induced crosslinking of the collagen fibres. Both photochemical (308 nm) and photothermal (532 nm, 1,064 nm) alterations occurred already at fluences below the respective ablation thresholds. SEM detected massive phase changes like melting and boiling that accompanied the ablation (vaporization) of the material above F_{th} (Fig. 15.12). However, no changes were observed at $F < F_{\mathrm{th}}$, where the water-soluble degradation products of collagen already were affected by the laser treatment and exhibited changes on the molecular level.

The hydrothermal stability by the micro-hot-table technique relies on heating collagen above the helix-coil transition temperature, which causes a collapse of the rod-like three-stranded collagen unit into random coils, which then constitute gelatine [30]. When collagen hide fibres are heated in water they will deform and this deformation was seen as shrinkage of the fibres (shrinkage temperature, T_{s}). The hydrothermal stability of the fibres decreased in proportion to the state of deterioration.

The quantity of water-soluble degradation products of collagen is affected by laser radiation already below the ablation threshold fluence, F_{th}, at which material removal is detected under microscopic inspection (Table 15.2). The solubility of the modern parchment increased with 308 nm at low fluences (0.3 J cm^{-2}) possibly due to photochemical degradation. The ancient

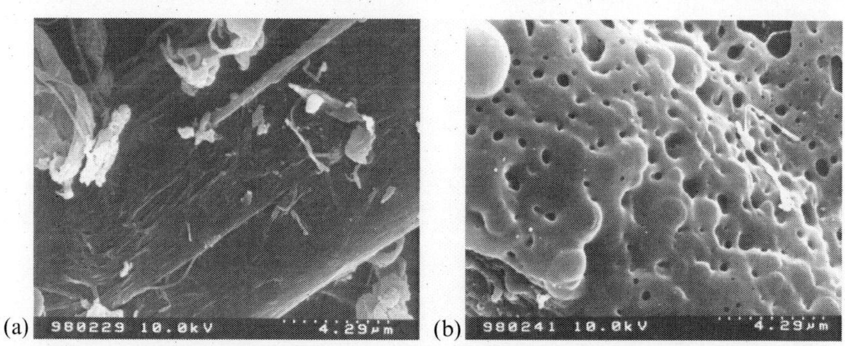

Fig. 15.12. Scanning electron micrographs of laser treated calf parchment. 308 nm, 17 ns, ablation threshold $F_{\mathrm{th}} \sim 0.7\,\mathrm{J\,cm^{-2}}$. (**a**) $F = 0.0\,\mathrm{J\,cm^{-2}}$, (**b**) $F = 3.6\,\mathrm{J\,cm^{-2}}$ [30]

Table 15.2. Ablation threshold fluences of parchment determined by electron microscopy [30]

	Ablation threshold fluence/J cm^{-2}		
Parchment	308 nm	532 nm	1,064 nm
Modern	0.7	3.0	3.0
Ancient	0.5	0.8	0.8

specimen already showed increased solubility at even lower fluence (0.1 Jcm^{-2}), but exhibited a decreased value at 0.3 J cm^{-2}, where the modern sample still showed molecular degradation. Obviously, the decrease of soluble products indicates crosslinking or other analogous phase changes.

The photochemical deterioration at 308 nm observed by the increase of soluble products and the decrease of T_s may be related the oxidation of collagen. Both oxidative and acidic deterioration is reflected in the measurements of the increase of soluble products and the shrinkage temperature. These methods are therefore valuable indicators for the total degree of deterioration of a parchment sample. Recently, shrinkage temperatures determined by micro-thermomechanical analysis was correlated with the amino acid residue composition of parchment, in particular proline and hydroxyproline [40].

Oxidative breakdown processes of parchment base on heat and light [41]. Parchment starts degrading at tripeptides in clusters of charged amino acids following the pattern: (1) loss of mainly the basic amino acids Lysine, Arginine, Hydroxylysine, and the imino acids Proline and Hydroxyproline, (2) gain of acidic amino acids, (3) formation of small amounts of breakdown products. Autoxidation of parchment occurs only in the presence of light [42]. Accordingly, one can conclude from our findings that hydroxyl radicals are generated by UV laser irradiation. These may attack carbon atoms in peptide side chains. Polar groups, particularly carboxylic acid functions (besides some conjugated double bonds) are then formed.

Biodeterioration of organic cultural heritage materials is a common problem [43, 44]. There are several types of fungal damage on paper, such as (1) surface damage caused by obstruction of any image by the growth of colonies, and embedded fruit bodies. (2) Discoloration may be caused by pigments either produced in fruiting bodies, or located in mycelium, or secreted into the paper substrate. (3) Structural deterioration of the paper may be due to the enzymatic digestion of cellulose. Many species of moulds are involved in cellulose decomposition, but none are more widespread than species of the ascomycete genus *Chaetomium*.

While the conventional sanitizing techniques for removal of fungal material from paper using chemical or physical means have proved sufficient in many cases [44], the removal of discoloration caused by fungal pigments is yet a problem in paper conservation. There are very rare reports on the removal of fungi from paper by lasers [45, 46]. A far-UV krypton fluoride (KrF) excimer laser was able to remove *Aspergillus niger* mould from filter paper while viable spores and mould fragments were released into the atmosphere [47]. Actually UVB radiation of a 308 nm XeCl excimer laser can be used for the treatment of skin mycosis fungoides [48]. That means that UV laser radiation can deteriorate fungi. However, irradiation of cellulose with even a near-UV excimer laser at 308 nm resulted in photo-oxidative degradation of the paper substrate, accompanied by an increase in oxidized groups content (carbonyl or carboxyl) and a severe decrease in degree of polymerization [18, 19]. That means that successes of anti-fungal laser treatments with UV lasers are accompanied by photochemical paper deterioration.

Conventional anti-fungal treatment consists of (1) mechanical removal of fungal bodies (brush, suction devices, scalpel), of (2) disinfection by fungicides (ethanol 70%, fungicide, bleaching), and removing of discolorations of the object (solvents, bleaching) [45].

In this context, sanitizing by ethanol (70%) with 0.2% PHB-esters was done, though it does not act very sporicidal [49]. This process was followed by laser irradiation at a visible wavelength of 532 nm known to be the least aggressive to the paper matrix [18, 19]. Mechanical removal by rubber and scalpel of all fungi remnants after the alcohol treatment showed practically no success in almost all cases. The discoloration could not be reduced, only paper fibres were removed in an uncontrolled way. Chemical bleaching by $KMnO_4$ after the alcoholic treatment led to acceptable improvements of the sample appearance in the cases of *Trichophyton*, *Epicoccum* and *Penicillium.* Limited success was observed with *Cladosporium*, *Chaetomium*, *Alternaria*, and *Aspergillus.*

Laser irradiation after the ethanol leaching resulted in drastic improvements of the specimens' appearance to the naked eye in respect to discoloration with all fungi types except *Cladosporium* and *Aspergillus* (Fig. 15.13). *Alternaria* caused substantial problems with mechanical and chemical treatments but yielded good results by laser treatment. This grows in surface-near regions and therefore can be removed without paper ablation. In contrast, *Chaetomium* grows deeply in the paper matrix with its hair-like appendices. Therefore, conventional approaches were completely unsuccessful, whereas the

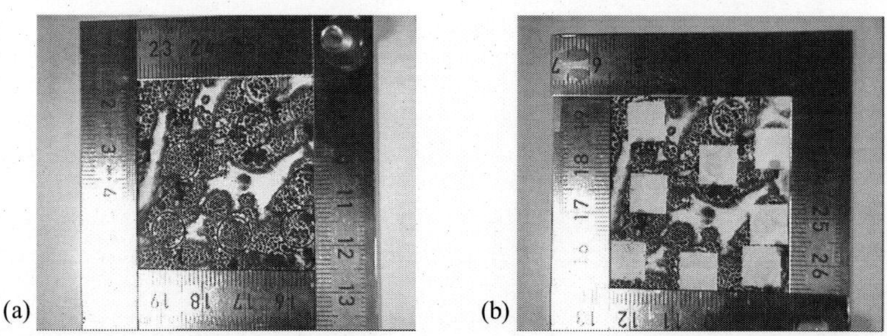

Fig. 15.13. Penicillium on paper (Whatman). (**a**) after ethanol treatment; (**b**) after laser treatment. 0.8–5.0 J cm^{-2}

laser could yield an acceptable cleaning affect. *Epicoccum* exhibits leaching remnants deeply distributed in the paper matrix with various colours. Bleaching resulted in relatively good in-depth discoloration. The laser could remove coloured material near the surface, and deep contaminant regions were left over causing a brownish colour. *Penicillium* material was converted into almost black material by the alcohol treatment. The laser was very successful at least in the surface-near regions. *Cladosporium* grows deeply in the fibre material forming dark leaching products that could not be affected by neither mechanical nor chemical treatment. The laser yielded a limited success near the surface.

Multispectral imaging allowed documenting laser-cleaning results in comparison to cut paper sections, which have undergone bleaching in permanganate instead. The applied laser fluences ranged below the ablation thresholds and above. In the latter case, the laser was used as a "contactless scalpel." The laser turned out to be more efficient than the aggressive bleaching process. On cotton, the best success was observed again with the dark types *Alternaria* and *Penicillium*, but not *Cladosporium*. Even *Chaetomium* and *Trichophyton* could be removed so that the cleaning result almost resembled that of the pure cotton paper. On rag, the darkest species *Penicillium* shows the best cleaning success followed by *Cladosporium* and *Alternaria*.

Soiled biogenetic fibrous substrates such as paper and parchment consist of a phase mixture of particulates, condensed foreign phases and cellulose or protein polymers, respectively. Particle decontamination on flat solids such as silicon wafers has already been studied systematically [10, 50–53]. The phase separation of these condensed films and particles from the matrix is the purpose of laser cleaning. Model experiments with microspheres on silk fibres allowed a first insight into the fundamental phase separation processes on fibrous substrates (Figs. 15.14 and 15.15) [54].

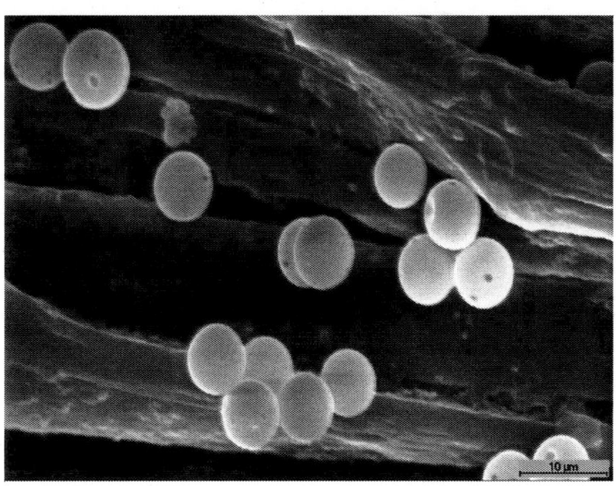

Fig. 15.14. Polystyrene microspheres on a silk fibre [54]

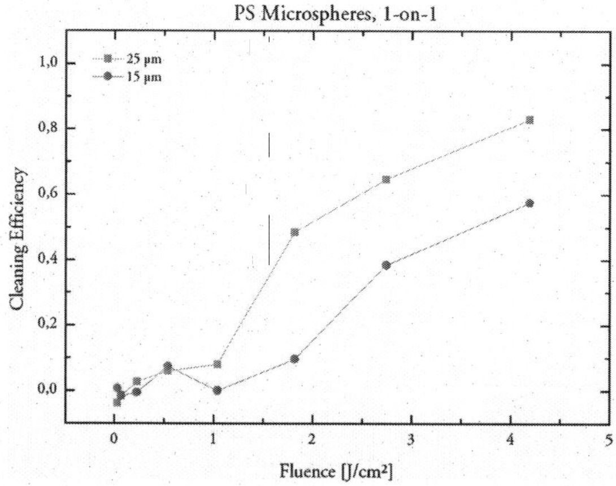

Fig. 15.15. Laser cleaning efficiency of polystyrene microspheres on a silk fibre. Dependence of sphere radius [54]

15.7 Conclusions

Laser cleaning of architectural facades and sculptures is a widely accepted restoration procedure and has attained an established place on the high-quality restoration market. An optimum etch-stop can be achieved when laser parameters are optimized. Its application to polychromed surfaces, however, requires a preliminary and systematic study of the adequate conditions to guarantee the preservation of the artwork. Often the best cleaning result can be achieved by combining three or more cleaning techniques including also laser treatment for special purposes. Laser treatment can also be applied successfully on metallic surfaces such as iron, copper, pewter, silver, brass, various bronze alloys, gold leaf and firegilding, plating etc. involving patina, black and brown incrustations, impurities, old conservation materials etc.

Know-how and knowledge about the contactless laser cleaning of biogenetic surfaces such as parchment and paper has been reviewed. Inspection by optical microscopy and even by SEM is only sensitive to massive phase changes like melting and boiling accompanying ablation (vaporization) above the threshold fluence. The weight-average molecular mass of cellulose significantly decreases by laser interaction with ultraviolet wavelengths. Treatment with visible (532 nm) is negligible and gave the most promising results, with no discolouration and no other visible alteration, nor detectable chemical changes. UV laser light causes photochemical degradation, i.e., molecular degradation also on parchment. Laser-induced alterations of parchment occur already below the ablation threshold fluence and are indicative for changes on the molecular level. Thermally induced crosslinking of the collagen fibres is increasingly observed with increasing fluence of IR radiation. The photochemical deterioration in the UV may be related to oxidative breakdown processes, which occur only in the presence of light.

Laser cleaning of fungi-overgrown paper sanitized by ethanol showed success in contrast to chemical bleaching by $KMnO_4$. Remnants after the removal of hyphae and mycelia and the deactivation of the remaining conidia by ethanol could be removed by the laser at least in surface-near regions. This is an excellent example that laser techniques may be successfully applied when intelligently combined with conventional techniques in order to achieve otherwise unreached conservation results.

References

1. W. Kautek, E. König (eds), *Restauratorenblätter* (Special Issue). *Lasers in the Conservation of Artworks I* (Verlag Mayer, Wien, 1997)
2. *Lasertechnik in der Restaurierung*, Restauro **104**, 6 (1998)
3. M. Cooper, *Laser Cleaning in Conservation* (Butterworth-Heinemann, Oxford, 1998)
4. C. Fotakis, W. Kautek, M. Castillejo (eds), *Laser Chemistry*, Special Issue, *Lasers in the Preservation of Cultural Heritage* (Hindawi, New York, 2006)
5. C. Fotakis, D. Anglos, V. Zafiropulos, S. Georgiou, V. Tornari, *Lasers in the Preservation of Cultural Heritage: Principles and Applications* (Taylor & Francis, New York, 2006)
6. D.M. Kane (ed.), *Laser Cleaning II* (World Scientific, NJ, London, Singapore, 2006)
7. J. Nimmrichter, W. Kautek, M. Schreiner (eds), *Springer Proceedings in Physics*, Vol. **116**, *Lasers in the Conservation of Artworks VI* (Springer, Heidelberg, 2007)
8. M. Schreiner, M. Strlic, R. Salimbeni (eds.), *Handbook on the Use of Lasers in Conservation and Conservation Science* (COST Office, Brussels, 2008), http://www.science4heritage.org/COSTG7/booklet/
9. D. Bäuerle, *Laser Processing and Chemistry*, 3rd edn. (Springer, Berlin, 2000)
10. R. Oltra, E. Arenholz, P. Leiderer, W. Kautek, C. Fotakis, M. Autric, C. Afonso, P. Wazen, Proc. SPIE **3885**, 499 (2000)
11. J. Bonse, S. Baudach, J. Krüger, W. Kautek, Proc. SPIE **4065**, 161 (2000)
12. W. Kautek, S. Pentzien, P. Rudolph, J. Krüger, E. König, Appl. Surf. Sci. **127–129**, 746 (1998)
13. P. Rudolph P., S. Pentzien, J. Krüger, W. Kautek, E. König, Restauro **104**(6), 396 (1998)
14. W. Kautek, S. Pentzien, J. Krüger, E. König, in *Lasers in the Conservation of Artworks I,* ed. by W. Kautek, E. König (Verlag Mayer, Wien, 1997), p. 69
15. W. Kautek, S. Pentzien, P. Rudolph, J. Krüger, C. Maywald-Pitellos, H. Bansa, H. Grösswang, E. König, in *Optics and Lasers in Biomedicine and Culture, Optics Within Life Science Series*, ed. by C. Fotakis, T. Papazoglou, C. Kalpouzos (Springer, Heidelberg, 2000), p. 100
16. W. Kautek, S. Pentzien, P. Rudolph, J. Krüger, C. Maywald-Pitellos, H. Bansa, H. Grösswang, E. König, J. Cult. Herit. **1**, S233 (2000)

17. W. Kautek, S. Pentzien, D. Müller-Hess, K. Troschke, R. Teule, Proc. SPIE **4402**, 130 (2001)
18. J. Kolar, M. Strlic, S. Pentzien, W. Kautek, Appl. Phys. A **71**, 87 (2000)
19. J. Kolar, M. Strlic, D. Müller-Hess, A. Gruber, K. Troschke, S. Pentzien, W. Kautek, J. Cult. Herit. **1**, S221 (2000)
20. W. Kautek, S. Pentzien, D. Müller-Hess, K. Troschke, R. Teule, Proc. SPIE **4402**, 130 (2001)
21. D. Müller-Hess, K.K. Troschke, J. Kolar, M. Strlic, S. Pentzien, W. Kautek, Restauro **8**, 604 (2001)
22. W. Kautek, J. Cult. Herit. **4**, 163s (2003)
23. W. Kautek, S. Pentzien, A. Conradi, D. Leichtfried, L. Puchinger, J. Cult. Herit. **4**, 179 (2003)
24. J. Kolar, M. Strlic, D. Müller-Hess, A. Gruber, K. Troschke, S. Pentzien, W. Kautek, J. Cult. Herit. **4**, 185 (2003)
25. P. Rudolph, F.J. Ligterink, J.L. Pedersoli Jr., M. van Bommel, J. Bos, H.A. Aziz, J.B.G.A. Havermans, H. Scholten, D. Schipper, W. Kautek, Appl. Phys. A **79**, 181 (2004)
26. P. Rudolph, F.J. Ligterink, J.L. Pedersoli Jr., H. Scholten, D. Schipper, J.B.G.A. Havermans, H.A. Aziz, V. Quillet, M. Kraan, B. van Beek, S. Corr, H.-Y. Hua-Ströfer, J. Stokmans, P. van Dalen, W. Kautek, Appl. Phys. A **79**, 941 (2004)
27. H. Scholten, D. Schipper, F.J. Ligterink, J.L. Pedersoli Jr., P. Rudolph, W. Kautek, J.B.G.A. Havermans, H.A. Aziz, B. van Beek, M. Kraan, P. van Dalen, V. Quillet, S. Corr, H.Y. Hua-Ströfer, Springer Proc. Phys. **100**, 11 (2005)
28. W. Kautek, S. Pentzien, Springer Proc. Phys. **100**, 403 (2005)
29. E. Pilch, S. Pentzien, H. Mädebach, W. Kautek, Springer Proc. Phys. **100**, 19 (2005)
30. L. Puchinger, S. Pentzien, A. Conradi, R. Koter, W. Kautek, Springer Proc. Phys. **100**, 51 (2005)
31. O. Wächter, in *Restaurierung und Erhaltung von Büchern, Archivalien und Graphiken* (Hermann Böhlaus NachfolgerWien, Graz–Wien–Köln, 1982)
32. D. van der Reyden, J. Am. Inst. Conserv. **31**, 117 (1992)
33. J.F. Asmus, J. Cult. Herit. **4**, 56 (2003)
34. E. Pummer, in *Lasers in the Conservation of Artworks VII*, ed. by M. Castillejo (Taylor & Francis, The Netherlands, Leiden, 2008), in press
35. J. Nimmrichter, R. Linke, in *Lasers in the Conservation of Artworks VI*, Springer Proceedings in Physics, vol. 116, ed. by J. Nimmrichter, W. Kautek, M. Schreiner (Springer, Heidelberg, 2007), p. 75
36. M. Castillejo, C. Domingo, F. Guerra-Librero, M. Jadraque, M. Martín, M. Oujja, E. Rebollar, R. Torres, in *Lasers in the Conservation of Artworks VI*, Springer Proceedings in Physics, vol. 116, ed. by J. Nimmrichter, W. Kautek, M. Schreiner (Springer, Heidelberg, 2007), p. 185
37. A. Gervais, M. Meier, P. Mottner, G. Wiedemann, W. Conrad, G. Haber, in *Lasers in the Conservation of Artworks VI*, Springer Proceedings in Physics, vol. 116, ed. by J. Nimmrichter, W. Kautek, M. Schreiner (Springer, Heidelberg, 2007), p. 37
38. J.G. Manni, Biophotonics International, May/June, 40 (1998)
39. H. Hogan, Biophotonics International, November, 62 (2000)

40. M. Odlyha, N.S. Cohen, G.M. Foster, A. Aliev, E. Verdonck, D. Grandy, J. Therm. Anal. Calorim. **71**, 939 (2003)
41. F. O'Flaherty, W.T. Roddy, R. Lollar, *The Chemistry and Technology of Leather*, vol. 4 (Reinhold, New York, 1965)
42. E. René de la Rie, Stud. Conserv. **33**, 53 (1988)
43. B. Zyska, Intern. Biodeter. Biodegrad. **40**, 43 (1997)
44. M. Nittérus, Restaurator **21**, 25 (2000)
45. H. Szczepanowska, C. Lovett Jr., J. Am. Inst. Conserv. **31**, 147 (1992)
46. H.M. Szczepanowska, W.R. Moomaw, J. Am. Inst. Conserv. **33**, 25 (1994)
47. T.R. Friberg, V. Zafiropulos, M. Kalaitzaki, R. Kowalski, J. Petrakis, C. Fotakis, Lasers Med. Sci. **12**, 55 (1997)
48. R. Soda, A.P. Vidolin, M. Esposito, M.S. Chimenti, A. Di Stefani, L. Bianchi, Exp. Dermatol. **11**, 279 (2002)
49. M. Nittérus, Restaurator **21**, 101 (2000)
50. M. Mosbacher, V. Dobler, J. Boneberg, P. Leiderer, Appl. Phys. A **70**, 669 (2001)
51. T. Fourrier, G. Schrems, T. Muehlberger, J. Heitz, N. Arnold, D. Baeuerle, M. Mosbacher, J. Boneberg, P. Leiderer, Appl. Phys. A **72**, 1 (2001)
52. Y.W. Zheng, B.S. Luk'yanchuk, Y.F. Lu, W.D. Song, Z.H. Mai, J. Appl. Phys. **90**, 2135 (2001)
53. N. Arnold, Appl. Surf. Sci. **197**, 904 (2002)
54. W. Kautek, Maedebach (to be published)

Index